STRUCTURAL ASPECTS OF BUILDING CONSERVATION

D1584479

STRUCTURAL ASPECTS OF BUILDING CONSERVATION

Second edition

Poul Beckmann and Robert Bowles

ELSEVIER
BUTTERWORTH
HEINEMANN

AMSTERDAM • BOSTON • HEIDELBERG • LONDON • NEW YORK • OXFORD
PARIS • SAN DIEGO • SAN FRANCISCO • SINGAPORE • SYDNEY • TOKYO

Elsevier Butterworth-Heinemann
Linacre House, Jordan Hill, Oxford OX2 8DP
200 Wheeler Road, Burlington, MA 01803

First published by McGraw-Hill International (UK) Limited 1995
Second edition 2004

British Library Cataloguing in Publication Data
A catalogue record for this book is available from the British Library

Library of Congress Cataloguing in Publication Data
A catalogue record for this book is available from the Library of Congress

ISBN 0 7506 5733 2

For information on all Elsevier Butterworth-Heinemann publications
visit our website at http://books.elsevier.com

Typeset by Integra Software Services Pvt. Ltd, Pondicherry, India.
Printed and bound in Great Britain

CONTENTS

FOREWORD

This second edition of Structural Aspects of Building Conservation has benefited from revisions and the additional contributions of Robert Bowles. Like Beckmann, he worked with Arup after taking an engineering degree at Cambridge. He is now a partner in Alan Baxter and Associates, and is working on St Paul's Cathedral where, as a boy he served as a chorister.

Poul Beckmann, starting in the office of Ove Arup & Partners, has been concerned with the conservation of historic and not-quite-so-historic buildings for over 35 years. His baptism was total immersion in the foundation problems of York Minster. This was a complicated project which, in the early days of computer analysis, took months to analyse. Unfortunately the observed deformations in the fabric did not agree with those calculated by the computer so, with great courage, Beckmann said 'the computer is wrong', and tried a different approach. This time there was almost complete correlation between the deformations, apart from a hiccup, which was explained by the events of 1406, when the former central tower had collapsed. Beckmann and his colleague David Dowrick were awarded the Institution of Civil Engineers' Telford Gold Medal for their report on this project. The message was vital 'Let the building speak to you'. This message is encapsulated in the important report of the Institution of Structural Engineers on the 'Appraisal of existing structures' [1980] to which he contributed. Following 7 years of difficult—if not desperate—work, the foundations of the Minster were stabilized, and since then the movements of the fabric have been monitored with completely satisfactory results.

Subsequently, Beckmann, who always had a valuable sense of proportion, has reviewed many buildings. Investigation had to be purposeful—research should never be indulged in for its own sake. His rather blunt manner might have discouraged some tender souls, but he liked architects who asked difficult questions. He worked from fundamentals, and was not bound by Codes of Practice, although he understood them very well. He is an engineer to his fingertips, who could illustrate his ideas with superb sketches. It is not surprising that the Royal Institute of British Architects made him an honorary member.

UNESCO used his experience when he was commissioned to report on the structural condition of the Taj Mahal. Two days were spent in careful consideration of this wonderful building. The cracks and deformations were assessed as being exactly what could be expected, and of no structural significance. The Archaeological Survey of India also employed American experts who, using a powerful computer for eight months, came to the same conclusion.

The above remarks are necessary for the reader to appreciate the wealth of experience that informed this important book. It has been refined by lecturing and teaching for the conservation courses at the Architectural Association, London, and the former Institute of Advanced Architectural Studies in York, and at the International Centre for Conservation, Rome [ICCROM].

Each chapter of this book deals clearly with its subject—starting with basic principles and the 'inevitability of deformations', which may surprise some readers. Next the authors give the aims, process and philosophy of structural appraisal. Structural appraisal is a process which encompasses the following: document research, inspection, measurements and recording and structural analysis; sometimes it includes testing of materials, and occasionally load testing of entire structures is involved. Structural appraisal is a vital step in the conservation of the building in question. When appraising an existing building, the engineer must, besides considering defects of general design, look for significant deterioration of structural members and for past damage to the structure from cutting away to insert service pipes or ducts. Incidentally, few people realize the amount of energy that is locked up in the buildings, so its continued use is a factor in reducing CO_2 emissions, together with the 'Greenhouse Effect'.

Having dealt with generalities, the authors give chapters dealing with the principal structural materials: masonry, timber, iron, steel and concrete. Repair and strengthening techniques are described and a contribution is made in describing fire problems and remedies. The chapters on steel and concrete review the contemporary problems of conserving framed buildings, so enabling the conservationist to face current problems, as in the future more and more architectural work will consist of refurbishing such buildings. Foundations are reviewed, and a final chapter contributed by Robert Bowles is devoted to the difficult but important subject of temporary works. The authors have decided that 1980 should be the 'cut off' date for their book. This is wise, as it generally takes at least 20 years for a building to show its faults.

The book is essentially practical, giving valuable advice on procedures and inspections, and reporting in subsequent chapters. The engineer will be part of a team, so communication of his or her ideas is an important skill. The engineer should offer the conservation team alternative solutions, which can be discussed and evaluated, so finding the 'minimum necessary interventions'. It takes a great deal of thought to find simple and elegant solutions to complex problems but Beckmann and Bowles have done so. This book is to be recommended to all young engineers, and to architects, whom it would help in their discussions with engineers, and give them the moral and technical confidence to question engineering proposals, so promoting a creative dialogue.

This is a new edition of a fundamental book.

Bernard M. Feilden

PREFACE

The first edition of this book grew out of the personal experiences of the author and his nearest colleagues in Ove Arup & Partners. This being so, it was perhaps inevitable that some points were taken for granted and one aspect not being included. When this present second edition was mooted, it was tempting to invite the critic, who had first pointed out the shortcomings, to fill the gaps and bring his, more recent, experience to bear on the subject, on the principle that two heads are usually better than one.

In this way the present co-authorship was born. We have each dealt with different buildings with different problems and this has coloured our respective experiences, but we share a common aim and a common approach to the structural problems met in building conservation.

The present edition is concerned with the same issues as those stated in the preface to the first edition, and is similarly addressed to architects and planners, as well as structural engineers, engaged in conservation of buildings. It incorporates all of the material of the first edition. The new material includes treatment of stability and robustness, the importance of engineer's drawings, the permits required to comply with the tangled mesh of planning, listing and scheduling regulations, and a complete new chapter on temporary works and phasing of operations. As buildings, constructed in the 1960s, are currently being listed in the UK, the scope of the book has been extended to include materials and methods of construction, used up to and during the 1970s. Essentially, however, the spirit of the book is the same, but the flesh has acquired a bit more muscle.

Several of the people who contributed to the first edition, and who were mentioned in the preface to that, have again assisted with advice. The authors are also indebted to members of the graphics group at Alan Baxter & Associates, who have produced the artwork for the new illustrations.

As with the first edition, our aim is not to present an academic treatise, but to provide background knowledge and technical advice to those who may be confronted with the structural problems that accompany most attempts at building conservation in the real world.

Poul Beckmann and Robert Bowles

PREFACE TO FIRST EDITION

The connection between building conservation and structural engineering is not obvious at first glance: What business have the designers of today's steel and concrete towers to meddle with the masterpieces of craftsmanship of the past, which make up such a large part of our cultural heritage?

A few moments' reflection should however suffice for the realization that unless the structure of the building is sound, its architectural and artistic glories are bound to perish. Outstanding specimens can sometimes be preserved as museum pieces, with very circumscribed uses, by the application of gentle maintenance techniques. There is however a far greater number of buildings, which can only be conserved if their existing use can continue, so as to earn revenue to pay for their maintenance, or if a new profitable use can be found for them. For such uses they have to satisfy today's user requirements in respect of, amongst other things, structural safety and serviceability.

For such buildings (and for great monuments, that appear threatened by collapse for whatever reason) a structural engineer, who is capable and willing to assess how the building really works, can be of great assistance in limiting the amount of 'invasive surgery' that may have to be carried out on the structure.

It has to be admitted that some structural engineers in the past have considerably damaged the character of some historic buildings, by introducing strengthening devices that counteracted the inherent mechanics of the original structure, in order to create a structural system conforming to their conventional textbook patterns and the letter of the codes of practice. The issue in 1980 of the Institution of Structural Engineers' report 'Appraisal of existing structures' did however constitute the first steps on the path towards the reconciliation between the modern mathematical/analytical methods and the 'common sense' approach to structural safety in existing buildings and helped engineers to establish a rational set of criteria, on which to base their engineering judgement.

Interestingly, modern computer methods now allow engineers to analyse complex three-dimensional structures that in the past were beyond calculation, and whilst such analysis is only rarely called for, it can in the right hands be of great help when assessing structures in marginal cases.

Older buildings do however have idiosyncrasies, which are not often found in modern structures and it is essential for the successful application of structural techniques, that the engineer becomes familiar with these peculiarities. Conversely, the architect in charge of maintenance, repair, restoration or refurbishment of an older building must realize that the laws of gravity and structural mechanics apply to all structures, regardless of their antiquity or historic significance, and he will find that understanding the basic structural behaviour will greatly help him to make the right decisions at an early stage and may well enable him to challenge the engineer effectively, if he feels that the latter is trying to blind him with science.

This book is therefore addressed to architects, planners and structural engineers engaged in conservation of buildings.

For the purposes of this book, conservation is assumed to encompass *preservation* of the building in its existing state, *repairs and restoration* necessary for continuation of its existing use and *adaptation or conversion* for new uses.

The book covers most forms of construction that have been in common use up to the 1960s. For the more recent innovations: cable-suspended fabric roofs, fibre composites, etc., the design assumptions are sophisticated, and realistic enough for subsequent appraisal and their materials' science is worthy of a book in its own right. (Some of it is covered in the *Construction Materials Reference Book*, published by Butterworth-Heinemann 1992.) These 'hi-tech' structures are therefore not included.

I have dealt with subjects of which I have had personal experience during a period of over 25 years of, albeit intermittent, involvement in conservation of buildings ranging from medieval cathedrals to twentieth-century office buildings.

The contents of this book do however reflect contributions from many of my collaborators over the years, to whom I owe a debt of gratitude: First and foremost Sir Bernard Feilden, who trusted me with the work on York Minster, who involved me in a number of investigations after that and then encouraged me to write this book and subsequently patiently read and commented on several drafts; David Dowrick, who did most of the detail design work on York Minster, including the computer analysis and who subsequently gave much practical advice on seismic matters; my colleagues in Ove Arup & Partners, particularly John Blanchard, who explained the basics of dynamic analysis of vibration response and the statistics of sampling and testing; Bob Cather, Alan Cockaday and Paul Craddock, who checked the materials' science in the chapters on concrete, iron and timber respectively; Peter Ross, who gave me deeper insight in the assessment of timber structures; Andrew Lord, who checked the chapter on foundations; Margaret Law, who guided me on aspects of fire resistance and Michael Bussell, who not only gave some very useful comments on the first chapters but also, being an avid book collector unlike myself, provided me with many ancient references in addition to those that were patiently provided by the Library staff of Arups. I am also grateful to Charmaine Isles, who typed the first draft of chapters one to four, prior to my biting the bullet and trying to learn word processing myself.

Much of the material in this book has been presented in lectures at conservation courses at the International Centre for Conservation in Rome (ICCROM) and at the Institute for Advanced Architectural Studies (IAAS) in York.

My aim is however not to present an academic treatise, but to provide background knowledge and technical advice to those who may be confronted with the structural problems that accompany most attempts at building conservation in the real world.

Putney, February 1994.   *Poul Beckmann*

ACKNOWLEDGEMENTS

Figures 1.14, 5.1 and 5.2 are reproduced from E. Suenson: Byggematerialer, Vols 1 (1920) and 2 (1922) respectively and Fig. 6.1 is adapted from Vol. 1, by permission of G.E.C. Gad, Copenhagen, successors to the original publishers.

Figures 2.1, 2.2 and 2.3 and Table 3.1 are reproduced, slightly adapted, by permission of The Institution of Structural Engineers from the Institution's report 'Appraisal of existing structures'(1996).

Figures 3.1, 3.9, 3.10, 3.11 (adapted), 3.12, 3.13, 4.8, 4.17 and 8.10 are reproduced from the Proceedings of the Institution of Civil Engineers, Paper 7415S (1971) by permission of Thomas Telford Publications. Figure 8.18 is reproduced from the same paper with permission from the Concrete Society.

The following figures are reproduced by permission of Ove Arup International: Figures 3.3, 3.5, 3.6 and 3.8 from P. Beckmann: 'Structural analysis and recording of ancient buildings', *The Arup Journal*, Vol. 7, No. 2, June 1972. Figures 4.7 and 4.17 from P. Beckmann: 'Strengthening techniques', *The Arup Journal*, Vol. 20, No. 3, Autumn 1985. Figure 4.16 from P. Beckmann and J.C. Blanchard: 'The Spire of Holy Trinity Church, Coventry', *The Arup Journal*, Vol. 15, No. 4, December 1980. Figures 6.8*a*, 6.9 and 6.11 from J.C. Blanchard, M.N. Bussell and A. Marsden: 'Appraisal of existing ferrous metal structures, Part 1', *The Arup Journal*, Vol. 17, No. 4, December 1982. The tables of *Twelvetrees Materials properties and Safety Factors*, the permissible working stresses from the 1909 London Building Act and from BS 449, together with Fig. 6.10 are reproduced from 'Appraisal of existing ferrous metal structures, Part 2', *The Arup Journal*, Vol. 18, No. 1, April 1983.

Figure 3.4 is reproduced from the 'Avongard' catalogue by permission of Avongard Tell-Tale, Ltd.

Figure 4.19 is reproduced from E.G. Warland: *Modern Practical Masonry*, Second edition, 1953, by permission of the Stone Federation of Great Britain.

Figures 5.3*a* and 5.4 are reproduced from Richard Harris: *Discovering Timber-Framed Buildings*, Shire Publications, Ltd, 1979, with the author's permission. Figure 5.3*b* is adapted from R. Brunskill: *Timber Building in Britain*, by permission of Victor Gollancz, Cassel Publishers.

Figure 5.5 is adapted from F.W.B. Charles and Mary Charles: *Conservation of Timber Buildings*, Stanley Thornes (Publishers) Ltd, 1985, by permission of Mr Martin Charles.

Figures 5.6, 5.7, and 5.9 are reproduced from James Newlands: *The Carpenter's and Joiner's Assistant*, Blackie and Son, 1857, by permission of the Random House Group.

Figures 5.15 and 5.16 are adapted from Ann E. Stocker and Leonard Bridge: 'Church of St Mary the Virgin, Sandwich, Kent: The Repair of the seventeenth century Timber Roof'; paper presented to the *Structural Repair and Maintenance of Historic Buildings* Conference, Florence, 1989 by permission from Wessex Institute of Technology.

The tables of 'green' and 'dry' stresses and the grading rules for timber in Appendix 5A have been reproduced by permission of the British Standards Institution.

Figures 6.3, 6.4, 6.5 and 6.6 are reproduced from T.K. Derry and T.I. Williams: *A Short History of Technology*, Clarendon Press, 1960, by permission of Oxford University Press.

Figure 7.1 is reproduced from *Cassell's Reinforced Concrete*, The Waverley Book Company, Ltd, 1920.

Figure 7.9 is reproduced from P.G. Fookes, C.D. Comberbach and J. Cann: 'Field Investigation of Concrete Structures in South West England', *Concrete Magazine*, April 1983, by permission of the Concrete Society.

Figure 8.6 has been reproduced from the CIRIA Report 111 *Structural Renovation of traditional Buildings*, 1986, by permission of the Construction Industry Research and Information Association.

Figure 8.7 has been adapted from Jack Stroud and Raymond Harrington: *Mitchell's Building Construction; Structure and fabric, Part 2*, by permission of Pearson Education.

The photograph on the frontispiece of Chapter 6 is reproduced by permission of Mr. Michael Bussell.

The photographs on the frontispiece of Chapters 7 and 8 are reproduced by permission of Arup.

STRUCTURAL BEHAVIOUR: BASIC PRINCIPLES

Approach Spans, Forth Railway Bridge: Overall, the steel lattice girders act as beams in bending, carrying the loads from the trains, and their own weight, to the stone piers, which act as columns. The bending of the girders is resisted by tension and compression in the horizontal chords; the shear forces by the diagonals of the lattice.

The basic principles of structural engineering are fairly simple. They are, however, often made to appear difficult by the jargon that structural engineers (like so many other professionals) are prone to lapse into when asked to explain their problems. This chapter aims at describing, in layman's terms, the factors that govern the behaviour of structures that are likely to be found in ordinary buildings. Readers who have studied (and can remember!) the theory of structures and strength of materials may find the following exposition rather simplistic. It may, however, help those for whom these subjects are new (or forgotten) and it should be adequate for practical purposes.

1.1 EQUILIBRIUM

This is the simplest and, at the same time, the most crucial principle for structures: If equilibrium is not present, objects that should stay in one place, take off into space.

1.1.1 The General Principle

For a building to be safe and useable, its structural parts such as the roof, floors and walls must remain stationary. This requires that the forces acting on them are equal and opposite. In structural engineering jargon this is known as equilibrium of forces and is just a re-statement of Newton's first law. This equilibrium must be resistant to disturbance by small extraneous influences; the overall structure and each of its parts must have stability (see Sec. 1.5).

The following simple examples of this general principle may help to explain some of the concepts and terms.

1.1.2 Equilibrium of External Forces

The lantern in Fig. 1.1 stays up only because the downward pull of Earth's gravity W is equalled by the upward pull of the (tensile) force in the hanging rod. From the point of view of the hanging rod or any part of it, it is being pulled downwards by the gravity force W on the lantern (or in everyday parlance, the weight of the lantern) but it stays where it is because it is being pulled upwards equally strongly by the bracket (Fig. 1.2).

The bracket in Fig. 1.1 is being pulled downwards at its tip by the hanging rod. To stay up, it needs an upward force equal to W, which is provided by the wall at its inner end. This is, however, not enough: as the two forces are not in the same line, they would, on their own, spin the bracket round, anticlockwise (Fig. 1.3). In order to stay still, the bracket must therefore receive not only an upward force W at its inner end, but also a 'moment', that is, a turning force, acting in this case clockwise and equal to W times the length of the bracket (Fig. 1.4).

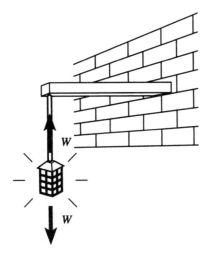

Figure 1.1 Equilibrium of forces acting on the lantern.

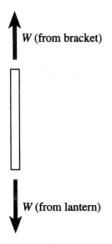

Figure 1.2 Equilibrium of forces acting on the hanging rod.

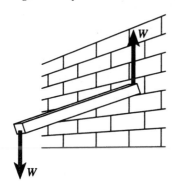

Figure 1.3 No equilibrium, because forces are not in line.

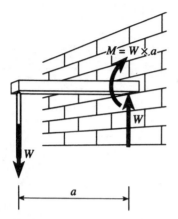

Figure 1.4 Equilibrium of forces and moments.

1.1.3 Equilibrium of Internal Forces

The example has so far dealt with the equilibrium of entire structural components under external forces. Each part of a component, however, requires equilibrium of the internal forces acting on it—otherwise it will not stay in place. For example, the hanging rod in Figs 1.1 and 1.2 is in equilibrium as a whole under the external forces W: the weight of the lantern (acting downwards) and the upholding force from the bracket (acting upwards). However, each part of the length of the hanging rod must also have equilibrium of the internal forces W acting at each end of it (Fig. 1.5).

1.1.4 Elementary General Conditions of Equilibrium

The example of the bracket in Fig. 1.4 shows that there are two conditions for equilibrium: Forces must be of equal size and act in opposite directions and, in addition, the moments of those forces must be of equal size and act in opposite directions.

1.1.5 Further Examples

The moment equilibrium enables us to calculate the forces in the posts under the bench in Fig. 1.6, on which two characters, weighing S and B kgf, are seated. There are two unknowns: the force in the left-hand post R_L and the force in the right-hand post R_R (R stands for 'reaction', the engineering term for forces resisting 'actions', such as the weight of people sitting on benches).

Consider the moments about a point on the left-hand post. R_L acts in line with the post and hence has no moment about that point. There then remain the following moments:

Clockwise: $\qquad S \times \frac{1}{4}l + B \left(\frac{1}{4}l + \frac{1}{2}l\right) = \left(\frac{1}{4}S + \frac{3}{4}B\right)l$

Anticlockwise: $\quad R_R \times l$

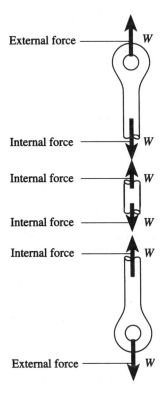

External force ——— *W*

Internal force ——— *W*

Internal force ——— *W*

Internal force ——— *W*

Internal force ——— *W*

External force ——— *W*

Figure 1.5 Equilibrium of internal and external forces.

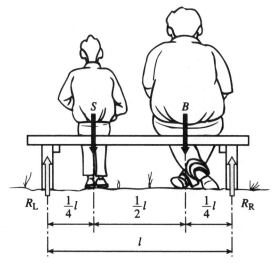

R_L $\frac{1}{4}l$ $\frac{1}{2}l$ $\frac{1}{4}l$ R_R

l

Figure 1.6 Actions and reactions.

Clockwise and anticlockwise moments must be numerically equal: $(\frac{1}{4}S + \frac{3}{4}B)l = R_R l$ or $R_R = \frac{1}{4}S + \frac{3}{4}B$.

For a point on the right-hand post, the anticlockwise moment is $B \times \frac{1}{4}l + S(\frac{1}{4}l + \frac{1}{2}l) = (\frac{1}{4}B + \frac{3}{4}S)l$ and the clockwise moment, $R_L \times l$. Equalling clockwise and anticlockwise moments, $(\frac{1}{4}B + \frac{3}{4}S)l = R_L l$, or $R_L = \frac{3}{4}S + \frac{1}{4}B$.

As a check, one can add up the reactions: $R_L + R_R = \frac{3}{4}S + \frac{1}{4}B + \frac{1}{4}S + \frac{3}{4}B = S + B$, so that the sum of the downward forces acting on the bench seat equals the sum of the upward forces; this is reassuring, as it is confirmation of the first condition of equilibrium.

Having calculated the reactions, one can now calculate the internal forces and moments in the bench seat: For any part of the seat between the left-hand post and Mr S to remain stationary, its right-hand end must receive a downward force equal to R_L and in addition (to prevent it spinning round) an anticlockwise moment equal to R_L multiplied by the distance from the centreline of the post to the point on the seat that is being considered. This moment is internal to the seat as a whole (although it is being treated as external to the left-hand part), and it is internal moments along the seat that cause it to bend. Internal moments, such as these, are referred to as bending moments by structural engineers.

Once our considerations move to a part of the seat to the right of Mr S, things get a little more complicated: To the left of the point under consideration (which is at the distance b from the left-hand post), the forces acting on that part of the seat are R_L upwards and S downwards. To achieve equilibrium, there must therefore be a downward force of $R_L - S$ acting on the end of the part of the seat to the left, and, in addition, an anticlockwise moment of $R_L \times b - S(b - \frac{1}{4}l)$ (Fig. 1.7). S and B are external forces that would usually be referred to as loads. The internal force $R_L - S$ is called a shear force.

Figure 1.8 shows how this type of calculation can be used to derive the formula for the midspan bending moment of a simply supported beam under a uniformly distributed load p per unit length: As far as the midspan point of the beam is concerned, the uniformly distributed load (Fig. 1.8a) can be replaced by two point loads at the quarter points of the span, each one equal to half the total load on the beam (the point loads have to act at the quarter points so that each will coincide with the resultant of the distributed load on its half span) (Fig. 1.8b).

The equilibrium of the beam as a whole demands that each reaction equals the load on half the span: $\frac{1}{2}pl$. With respect to the midspan point, the equilibrium of forces on the left-hand half of the beam requires no vertical force, as the reaction and the point load cancel out each other; but if one considers the moments about the midspan point of the external forces, one gets a clockwise moment of $R \times \frac{1}{2}l = \frac{1}{2}pl \times \frac{1}{2}l = \frac{1}{4}pl^2$ and an anticlockwise moment of $\frac{1}{2}pl \times \frac{1}{4}l = \frac{1}{8}pl^2$. These two moments add up to an external clockwise moment of $\frac{1}{8}pl^2$. For the half beam not to rotate under this external moment, there must act an internal anticlockwise moment at the midspan point equal to $\frac{1}{8}pl^2$. (Many

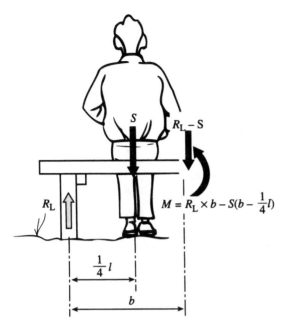

$$M = R_{\rm L} \times b - S(b - \tfrac{1}{4}l)$$

Figure 1.7 External and internal forces and moments.

structural engineers in the past earned most of their income from applying this formula.)

1.1.6 The Triangle of Forces

In the previous examples, all the forces under consideration were vertical. This is not always the case; in triangulated structures such as roof trusses, the forces follow the directions of the members and are influenced by the angles between them.

The roof truss in Fig. 1.9a receives its total load L through the purlins, as indicated, and will be in equilibrium as a whole as long as the reactions from the supporting walls together are equal and opposite to L; in this symmetrical case, they will each be equal to $\tfrac{1}{2}L$. If, however, the junction of the rafter and tie beam over the left-hand support (Fig. 1.9b) is considered, there is a downward force of $\tfrac{1}{8}L$ from the purlin and an upward reaction of $\tfrac{1}{2}L$; these cannot, on their own, ensure equilibrium. The loads from the higher parts of the truss must be transferred down the rafter by the force C. This force, being inclined, would push the junction outwards, were it not for the tie beam that holds it in with a force T.

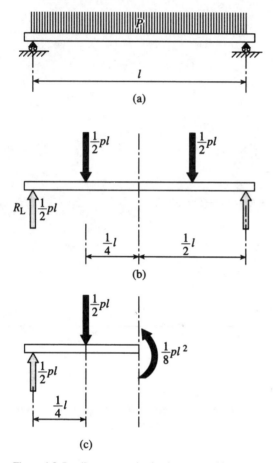

Figure 1.8 Bending moment in simply supported beam.

Figure 1.9c shows how the magnitudes of C and T can be determined: The net upward force on the junction (or 'node' in structural jargon) is obtained by subtracting the purlin load $\frac{1}{8}L$ from the reaction. The force C is assumed to be parallel to the centreline of the rafter and the force T parallel to the tie beam. (These assumptions are approximations, but perfectly adequate for most practical purposes.) One can now draw the triangle of forces on the left-hand part of Fig. 1.9b; this gives the magnitudes of C and T. The directions of C and T follow from the rule that in such a force diagram, the arrows must always describe a closed loop going in one direction only, either clockwise or, as in this case, anticlockwise.

From the force triangle in Fig. 1.9c it can be seen that C and T geometrically add up to a downward vertical force of size equal to the upward net force acting on the node. From this, and from the examples dealing with moments, one can re-state the conditions for equilibrium as follows in 1.1.7.

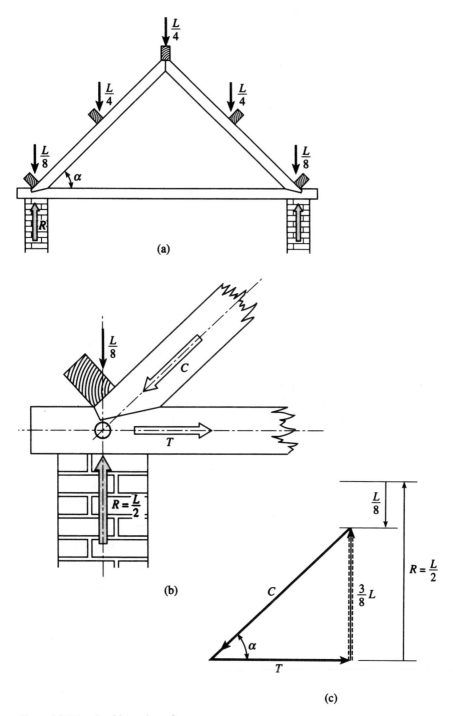

Figure 1.9 Triangle of forces in roof truss.

1.1.7 General (Expanded) Conditions of Equilibrium

1. The geometric sum of all forces acting on a structure, or part of a structure, must equal zero.
2. The geometric sum of the moments about any given point of all the forces acting on a structure or part of a structure must equal zero. ('Geometric' in relation to moments refers to direction, clockwise or anticlockwise.)

1.2 THE RESPONSE OF MATERIALS TO FORCE

When forces are applied to an object, it will respond in various ways. If the forces are in equilibrium, it will remain stationary but it will deform. The deformation may be invisible to the naked eye, but it will be there and will be accompanied by internal forces. The relation between the deformations and the internal forces depends on the nature of the applied forces, e.g. tension or compression, and on the properties of the material.

1.2.1 Behaviour under Tension

'Stress' and 'strain' are terms that, when used by engineers, often signify the 'black art' of engineering to the lay person. This is in spite of the fact that we all experience these phenomena whenever we stretch a rubber band to go round a parcel.

The concepts behind these terms may perhaps be most easily understood by considering the following imaginary experiment: A metal bar is fixed in a clamp at its top end and has a tray, on which weights can be put, at its bottom end (Fig. 1.10a). Before any weights are put on the tray, the distance between two punch-marks on the bar is measured and found to be l. When weights are put on the tray, the distance between the punch-marks will be found to increase, and to increase more when more weights are applied.

If the increase of the distance between the punch-marks is called Δl and the total of the weights on the tray is called P, and if one were to plot the corresponding values of Δl and P, one would obtain a graph like Fig. 1.10b. It will be seen that for moderate values of the weights P, the points fall on a straight line, indicating that the weight (or force) is proportional to elongation. This relationship was first postulated in 1678 by Robert Hooke who, as was customary then, put it in Latin in his notebook: 'Ut Tensio Sic Vis' ('As the stretch, such the force').

If one were to divide the elongation Δl under any force P with the original length l, one would obtain a ratio ε which is called the strain under that force. If one were to divide any value of the force P with the cross-sectional area of the bar A, one would obtain a quantity σ called the stress in the bar when subjected to the force P:

$$\text{Strain } \varepsilon = \frac{\Delta l}{l} \quad \text{and} \quad \text{Stress } \sigma = \frac{P}{A}$$

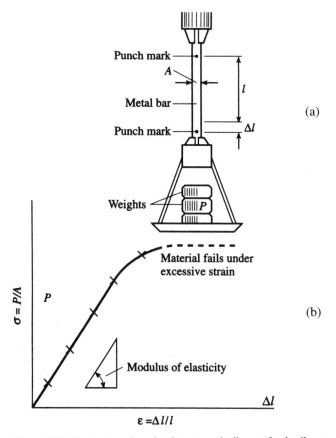

Figure 1.10 Tension 'experiment' and stress–strain diagram for ductile material under tension.

If one were to plot corresponding values of ε and σ, one would get a diagram of similar shape to that of Δl and P. (This is because l and A for practical purposes can be assumed to remain constant independent of the value of P.)

Up to a certain level of strain (and stress) Hooke's law can be stated as: stress is proportional to strain. The factor of proportionality is

$$\frac{\sigma}{\varepsilon} = E$$

where E is called the modulus of elasticity or Young's modulus.

If one goes on increasing the force P beyond a certain limit, the elongation Δl will start to increase at a greater rate than the force until a point at which the elongation will continue increasing *without any more force being applied*. This is called yielding and the stress at which it occurs is called the yield point.

The yielding may, for some materials such as mild steel, temporarily stop so that further load can be applied and the bar will carry a moderately increased

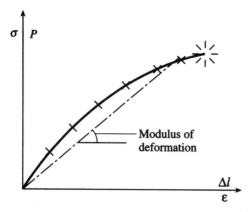

Figure 1.11 Stress–strain diagram for brittle material under tension.

load, but any further loading will quickly result in the cross-section starting to narrow at one point; this 'necking' is immediately followed by fracture. Alternatively, yielding may continue uninterrupted until fracture. This type of fracture, called *ductile*, is found in many metals such as copper and mild steel. These can be said to fail through excessive elongation or strain.

There is another class of materials, examples of which are stone, concrete and cast iron, with which if one were to try the previous experiment, the diagram of the relationship between force and elongation, and hence between stress and strain, would look like the heavy line in Fig. 1.11; unlike the stress–strain diagram for ductile metals, it is curved right from the beginning and stops suddenly.

It is important to note that this means that fracture is sudden; *there is no preceding yielding to give warning*. This type of failure is called *brittle*, a term that is also used to describe the materials. The load (or stress) at which brittle materials fail in tension is much lower than that at which they fail in compression (this is in contrast to the ductile metals, described above, for which the yield stress in tension is equal to, or slightly bigger than that in compression).

1.2.2 Behaviour under Compression

If a cube or a cylinder of a material is put in a hydraulic press and a force is applied to the end surfaces, the cube or cylinder will get shorter: the original height h will be reduced by Δh and as the force, P, increases, Δh will increase and h decrease progressively (Fig. 1.12).

At the same time as h decreases, the width of the cube or the diameter of the cylinder will increase by a very small amount which is proportional to Δh. (In fact, in the tensile experiment in Sec. 1.2.1, the diameter of the bar would have decreased, but the reduction would have been marginal and would not have affected the subsequent events in the tensile case.)

If the test specimen (cube or cylinder) is of a ductile material, the plot showing the corresponding values of stress and strain (the stress–strain diagram) will

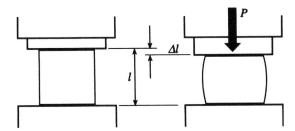

Figure 1.12 Behaviour of ductile material under compression.

be a mirror image of the one obtained from the tensile experiment in Sec. 1.2.1, except that no definite failure occurs—the material just squashes (Fig. 1.13).

If the test specimen is of a brittle material, what happens as the pressure increases will depend on the testing set-up as well as on the material. If the friction between the plates of the testing machine and the end faces of the specimen prevents sideways spreading (of the specimen ends), the sides of the specimen will spall away, leaving a waisted body, which subsequently crumbles if the pressure is maintained (Fig. 1.14*a*).

If the plates have been greased so as to allow the ends of the specimen to expand laterally, the specimen will split into slender prisms, or miniature pillars, which then fail (Fig. 1.14*b*).

In the second case the specimen may fail at a load that is barely more than half that of the first. This is important, because in real structures there are locations where there is no restraint against lateral expansion, e.g. at mid-height of a stone column, so that it will be the lower strength, such as would be obtained

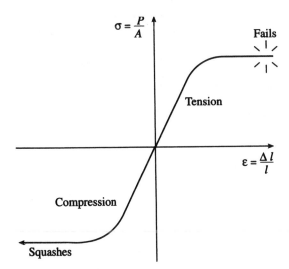

Figure 1.13 Complete stress–strain diagram for ductile material.

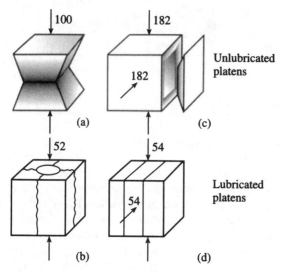

Figure 1.14 Failure modes and relative strengths of cubes of brittle material tested in compression under different test conditions (*after Suenson, 1920*).

from tests with greased plates, that will determine the load that the structural member can carry. Tests using greased plates tend however to give inconsistent results. Tests have therefore usually been carried out on prisms with a height about twice that of the width. This largely eliminates the end effect.

The importance of the tests with greased plates is that they demonstrate that what is called a *compressive*, or crushing, failure of a brittle material is in fact a transverse splitting, i.e. a *tensile*, failure.

A typical relationship between loads/stresses and deformations/strains for brittle materials is shown in Fig. 1.15. Both the tension 'branch' and the

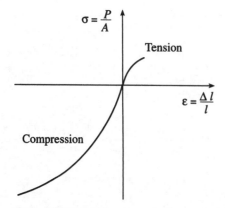

Figure 1.15 Complete stress–strain diagram for brittle materials.

compression 'branch' of the diagram are curved from the (unstressed) outset but the tensile failure strain and stress are both much smaller than the compressive values.

It is worth remembering that loads and elongations/shortenings can be directly measured. Stresses and strains are, however, calculated quantities, and the calculations of these quantities, particularly of stresses, rely on partly unverifiable assumptions about the distribution of stresses and strains across the section.

1.2.3 Behaviour under Bending

In Secs 1.2.1 and 1.2.2 it was demonstrated that a body subjected to a tensile force will elongate and one subjected to compression will shorten. In Secs 1.1.4 and 1.1.5 it was shown that a beam (in that case a bench seat) subjected to external moments has to develop internal moments in order to preserve internal equilibrium. What stresses and strains do these internal moments produce in a beam?

To avoid lengthy mathematical proofs, it may be easier to look at the problem 'back to front'. In Fig. 1.16*a*, the bench seat in Fig. 1.7 is imagined to be made of 'plywood' having eight layers (all with the grain in the same direction). The top four layers are imagined to be subjected to compressive forces, which increase in proportion to the distance of the layer from the centre plane of the beam. The bottom four layers are similarly attacked by tensile forces, which grow with the distance from the centre in the same proportion as the compressive forces.

It can be shown that the bottom four tensile forces are equivalent in their effect to a single tensile force *T* and that the four compressive forces equal a single compressive force *C* (Fig. 1.16*b*). *C* and *T* are numerically equal in size, opposite in direction and acting a distance apart, hence they equal a moment (Fig. 1.16*c*). Conversely, a bending moment in a beam causes internal forces, or stresses, that are parallel to the axis of the beam and which, at moderate stress levels, vary linearly with the distance from the centre plane of the beam or, more precisely, with the distance from the centroid of its cross-section.

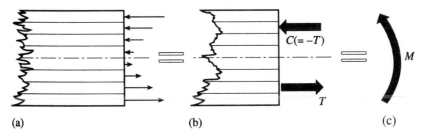

Figure 1.16 Stresses, varying linearly over a cross-section, equal a bending moment.

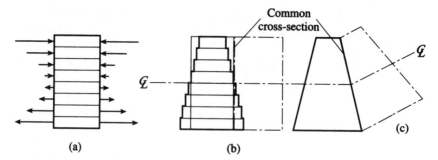

Figure 1.17 Internal forces, or stresses, which are equivalent to a bending moment cause bending deformation.

If one now considers a very short length of the beam, as in Fig. 1.17*a*, and assumes that the material follows Hooke's Law (the straight-line part of the stress–strain diagram, Fig. 1.13), then each layer will shorten or elongate in proportion to its distance from the centreline. What started as a rectangle bounded by two parallel (vertical) 'cuts' now forms a 'stack' of rectangular strips, which is approximate to the shape of a trapezium (Fig. 1.17*b*).

If instead of eight layers of finite thickness one assumes an infinite number of layers, each of infinitely small thickness, the deformed shape of the rectangular element would be a true trapezium. If one now considers two such rectangular elements next to each other and subjects both elements to the pattern of forces shown in Fig. 1.17*a* (which is equivalent to the moment in Fig. 1.16*c*), then they must both deform into trapeziums. They must, however, also stay in contact along their common cross-section, otherwise the elements would separate, which means that the beam would fall apart. By staying together, the two rectangles with an originally common, straight centreline (Fig. 1.17*b*) become two trapeziums with a 'kinked' common centreline (Fig. 1.17*c*). This rather crude illustration demonstrates that when a structural element is subjected to a moment, it will bend. For a beam this means that it will deflect.

1.2.4 Behaviour under Shear

In Sec. 1.1.5, a 'shear force' is mentioned which, being an internal force, acts vertically, at right angles to the bench seat. Such a force will cause stresses on the section on which it acts, and these shear stresses will appear to act at right angles to the beam, in contrast to the bending stresses which, as shown in Sec. 1.2.3, are parallel to the beam.

If, however, one considers a small square element, subjected to equal and opposite shear forces (or stresses) on opposite sides, it can be seen, by comparison with Fig. 1.3, that in the absence of other forces, the element will spin around (Fig. 1.18*a*). As there are no other forces to stop the element from rotating, a set of equal and opposite forces, acting on the other two opposing sides, are required for moment equilibrium to be satisfied (Fig. 1.18*b*). If one

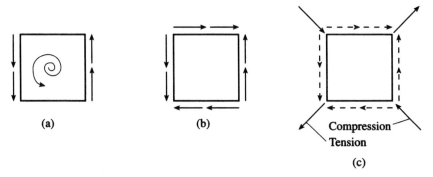

Figure 1.18 For equilibrium, vertical shear stresses (*a*) need to be accompanied by horizontal ones; (*b*) together they are equal to inclined compressive and tensile stresses (*c*).

now resolves the forces acting on the element by drawing triangles of forces similar to Fig. 1.9*c*, it can be seen that shear stresses, acting parallel to the faces of an element, are equivalent to simultaneous compressive and tensile stresses acting at 45°.

The behaviour of materials subjected to shear can therefore largely be deduced by reference to their behaviour under tension and compression. Ductile materials, being capable of withstanding substantial tensile stresses, tend just to deform, whilst brittle materials, having generally low tensile strengths, crack at right angles to the inclined tensile stresses (Fig. 1.19).

As 'shear failures' of brittle materials in reality are tensile failures, they are sudden without any preceding, visible, large deformations as warning.

1.2.5 Buckling

Any gardener trying to push a bamboo stick into hard soil will have experienced the following: if the stick is pushed from the top only, it will suddenly bow sideways rather than penetrate the ground, whereas if held at mid-height as well as at the top, it will remain straight, even under a larger 'push'.

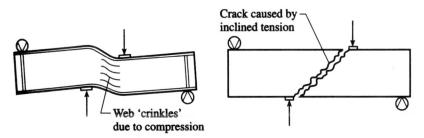

Figure 1.19 Failure modes of beams tested in shear: (*a*) ductile, e.g. steel I-beam; (*b*) brittle, e.g. stone or concrete.

What is happening to the bamboo stick, when pushed at the top only, is the structural phenomenon called 'buckling': at a load well below that which would cause squashing or splitting of the material, a slender compression member will become laterally unstable and fail to carry (further) load. The 'two-hand' push on the bamboo stick shows that the load that can be carried increases when the free length between lateral restraints is reduced. The substitution of a significantly thicker stick would show that for a given free length, a substantially greater load can be carried by a wider cross-section.

The buckling load of a slender compression member is in fact inversely proportional to the square of its unrestrained length and proportional to a section property called 'moment of inertia'. For a solid square with side a, the moment of inertia is $\frac{1}{12}a^4$; for a solid circle with diameter d, the moment of inertia is $\frac{1}{64}\pi d^4$, or approximately $\frac{1}{20}d^4$. The buckling load of slender compression members is also governed by the material properties, in this case not the compressive strength, but the elastic modulus (see Sec. 1.2.1).

In practice, a slender column can carry a load slightly in excess of its theoretical buckling load, but as it will assume a bent shape whilst so doing it is paradoxically the *tensile strength* of the material that determines the load-carrying capacity for many 'compression members', e.g. cast iron columns. However, a small eccentricity of the applied load will drastically reduce the carrying capacity below that predicted by the elastic theory, as will any slight manufacturing imperfection, such as a small curvature.

1.3 THE RESPONSE OF MATERIALS TO THEIR ENVIRONMENT

Forces or stresses are not alone in causing dimensional changes to materials; all materials expand and contract with changes in their temperature, and porous ones shrink and swell with varying moisture content. Temperature and moisture content of an object will depend on its environment.

1.3.1 Behaviour Due to Temperature Fluctuations

All materials tend to expand when their temperature rises and contract when it drops. The degree of the temperature movement varies: it is very high for some plastics, considerable for some metals like aluminium, quite low for most timbers and extremely low for Invar, an alloy used in precision measuring tapes and instruments.

The amount of movement is, for practical purposes, proportional to the change of temperature and to the original dimension, so that an unrestrained element of length l, subject to a temperature rise of t °C, will elongate by $\alpha \times t \times l$, where α is a factor characteristic for the material (the coefficient of thermal expansion).

The following are approximate values of α for some common building materials:

Lead	$29 \times 10^{-6}\,°C^{-1}$
Aluminium	$24 \times 10^{-6}\,°C^{-1}$
Copper	$17 \times 10^{-6}\,°C^{-1}$
Iron and steel	$12 \times 10^{-6}\,°C^{-1}$
Concrete	$11 \times 10^{-6}\,°C^{-1}$
Glass	$6–9 \times 10^{-6}\,°C^{-1}$
Timber	$3–9 \times 10^{-6}\,°C^{-1}$
Granite	$8 \times 10^{-6}\,°C^{-1}$
Limestone	$8 \times 10^{-6}\,°C^{-1}$
Clay brick	$5 \times 10^{-6}\,°C^{-1}$
PVC (hard)	$55 \times 10^{-6}\,°C^{-1}$

For example, an iron roof gutter 13 m long which is heated by the sun from 10 to 45 °C will expand by

$$12 \times 10^{-6}\,(45 - 10) \times 13 = 0.00546\ \text{m}$$

or nearly 6 mm; if the middle is held in position, each end will move outwards by 3 mm.

Temperature changes can have another effect: if a structure such as a tower is heated from one side only, for instance by the sun shining on it, its temperature will not rise uniformly; the sunny side will heat up more quickly and there will be a temperature 'drop' or 'gradient' across the width of the tower. This means that the side of the tower facing the sun will get longer than that facing away from the sun. Figures 1.17*b* and *c* show that when one side of a beam gets elongated relative to the other (in that case due to bending strains) the beam curves with the elongated side being convex. The same happens when strains or elongations are due to temperature changes: the tower will bend away from the sun, the heated side becoming convex, the cooler side concave.

This effect has been observed not only on slender minarets in the Middle East but also on 'Big Ben' in London. This 'thermal' bending depends on the *difference* in temperature between the two sides and will therefore be most pronounced for materials with low thermal conductivity (for which there will be a greater temperature difference between the heated and the unheated face).

For most materials, temperature movements are reversible; unrestrained elements expand when heated and contract by the same amount when cooled to their original temperature. Certain natural stones (mainly limestones and marbles) do, however, retain a small proportion of their thermal expansion when they cool down. The proportion is very small, but the effect is cumulative, so that structures subjected to largely one-sided heating from the sun, such as certain minarets, eventually display permanent curvature.

Certain stones contain minerals with significantly different coefficients of thermal expansion; this causes them to weather badly in climates with frequent severe temperature fluctuations.

It should be remembered that both reversible and irreversible thermal movements are governed by the temperature changes in the material—not by the fluctuations of the ambient temperature. Solar gain can raise the surface temperatures of some materials, particularly dark ones, by 30–40 °C above the ambient temperature and this temperature rise will gradually work its way inwards.

1.3.2 Behaviour Due to Fluctuations of Moisture Content

All porous materials hold moisture in their pores. The amount of that moisture will adjust itself, depending on the relative humidity (RH) of the atmosphere surrounding the material: if the RH goes up, the material will absorb moisture; if it goes down, the material will give off moisture.

Fluctuations in the moisture content of the material will be accompanied by dimensional changes: absorption of moisture will be accompanied by swelling; drying out will give rise to shrinkage.

The relationships between RH, moisture content and dimensional changes are far more complicated than those between ambient temperature, temperature of the material and thermal movement. For materials such as mortars, concrete and calcium-silicate bricks, there is a substantial initial irreversible shrinkage. For ceramic products, such as fired clay bricks, there is an initial swelling after removal from the kiln and a residual expansion that can go on for several years. This movement is most pronounced for lightly fired bricks and for some bricks made from clays and shales found between coal seams. The effect on the building is most serious if the bricks are laid too soon after removal from the kiln.

Timber may move twice as much tangentially to the growth rings as radially; the radial movement, in turn, may be 50 times as much as that parallel to the grain (see Sec. 5.1.1).

In a manner similar to that in which non-uniform temperature change through a material will cause originally straight components to curve, uneven drying out or moisture uptake will cause bending or warping.

1.3.3 The Effect of Restraints

It has so far been assumed that temperature and moisture movements could take place freely. As the results of these movements are often inconvenient, it is sometimes suggested that they should be restrained. This is rarely practical, as the following example will show.

Assume that the cast iron gutter in the example in Sec. 1.3.1 is 150 mm wide and 80 mm deep and that the metal is 3 mm thick. Its cross-sectional area will be $(2 \times 80 + 150) \times 3 = 930$ mm². Its thermal expansion over the 13 m length was 5.46 mm. If one wanted to restrain that expansion, it would be tantamount to pushing back that elongation. Referring to Sec. 1.2.1, it can be seen that the shortening under a compressive force C would be

$$\Delta l = \frac{C}{EA} l$$

where E is the elastic modulus of the material, A is the cross-sectional area and l is the length. E for cast iron is about 10 000 kgf/mm² (kilogram force per square millimetre), A is 930 mm² and l is 13 000 mm.

$$\Delta l = -5.46\,\text{mm} = \frac{-C \times 13\,000}{10\,000 \times 930}$$

from which one obtains $C = 3906$ kgf or nearly 4 tonnes. It is clearly not realistic to try to devise gutter fixings capable of providing restraining forces of this magnitude.

Other materials may have lower values of elastic modulus, but they will have correspondingly lower strengths and will hence be used in much more massive sections; restraint of thermal and/or moisture movements will therefore still require very large forces.

Materials with nearly equal tensile and compressive strengths will expand and contract, even when some restraint is imposed. However, materials or elements with a tensile strength substantially less than their compressive strength, e.g. stone or masonry, will expand when heated, but cannot contract by the same amount when cooled, and may crack because their tensile strength is not enough to resist the force from the restraint.

1.4 THE INEVITABILITY OF DEFORMATIONS

In Sec. 1.2 it was shown that tension leads to elongation, compression to shortening and that moment, which in Sec. 1.1.5 is seen to arise from beam action, causes curvature and hence deflection. Section 1.3 dealt with the dimensional response of materials to their environment.

Observed deformations in existing structures often give rise to anxiety. It is therefore appropriate to conclude these sections with the following tenets that should always be borne in mind when dealing with structures of any kind:

No beam, slab or plate carries load without deflecting.
No column, wall or strut carries load without shortening.
No footing or pile carries load without settling.
No component changes temperature without changing dimensions.
No component changes moisture content without changing dimensions.
No change of dimension can be restrained without using or creating large forces.

At the same time, it should be remembered that shear deformations are usually imperceptible to the naked eye, and that shear failures are often sudden.

1.5 STABILITY AND ROBUSTNESS

For a building to be fit for its purpose, two of the most important requirements are that the structural members must be safe and must be serviceable. This means that under the forces due to gravity, the loads arising from the use of the building and the environmental influences that can be expected:

1. the stresses in the members must not exceed certain values, and
2. the deformations must not be greater than what the user can reasonably tolerate.

These aspects of structural behaviour have been considered in Secs 1.1–1.4.

Before worrying about stresses and strains, however, one should remember that even if all the members of a structure satisfy the above conditions, it may develop a dangerous overall loss of shape and/or it may collapse if subjected to an unforeseen disturbance, if it does not possess adequate overall stability and robustness.

Stability can be achieved in two ways:

1. A structural member may have inherent stability, due to its geometry. For example, a castle wall 1 m thick and 5 m high on a good foundation is stable in itself, because of its low height-to-thickness ratio.
2. A structure may have interactive stability, due to its members being so arranged that they stabilize each other. This interaction will be familiar to anyone who has had to put together a 'flat-pack', 'home-assembly' book-case or wardrobe. With only the sides and the top connected, the whole arrangement sways from one side to the other in an uncontrollable way (Fig. 1.20a).

Once the back panel has been attached, the whole assembly becomes stable (Fig. 1.20b). Similarly, in a traditional brick-built house, the flank walls stabilize the facades and vice versa. In such a house, however, timber floors and partitions can also significantly help to stabilize walls.

This is a fact that must be borne in mind when dealing with older properties because, whilst lack of stability can be due to poor original design, it is more often caused by subsequent alterations, carried out to suit changed user requirements (Fig. 1.21).

Fitness for purpose also imposes another requirement, that of robustness. This means that in addition to being able to resist the forces imposed by gravity and the use of the building, the structure, as a whole, should be able to absorb accidental forces, such as may be created by vehicle impact or explosions, without suffering damage that is disproportionate to the cause. For example, if somebody forgets to

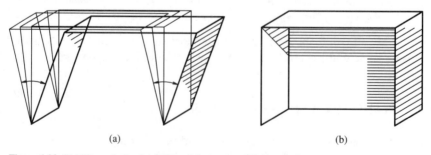

(a)   (b)

Figure 1.20 Stability, or lack of stability, of 'home assembly' wardrobe.

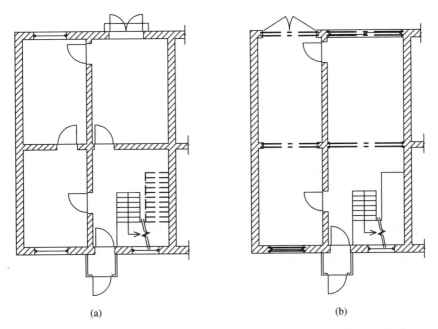

(a)   (b)

Figure 1.21 Plan of traditional dwelling house: (*a*) original, good cellular wall layout; (*b*) after conversion, stability compromised.

apply the brake when driving their car into the integral garage of their two-storey house, it may be acceptable that the end wall of the garage is 'demolished', but the rooms above and next to the garage must not collapse.

Robustness is most often achieved by ensuring that structural elements are tied together, in some way or other. Such tying together also imparts a substantial resistance to the effects of earthquakes.

More details of stability and robustness problems are given in the chapters dealing with the particular forms of construction.

TWO

AIMS, PROCESS AND PHILOSOPHY OF STRUCTURAL APPRAISAL

Houses in St Alban's: Past adaptations of the ground floors to shops appear to have succeeded in providing rental income to pay for their upkeep. Any further conversions or upgrading ought, however, to be preceded by a structural appraisal, but this need not necessarily involve sophisticated calculations.

Structural adequacy in a new building is nowadays achieved through the process by which the client's needs are first merged with the statutory requirements for structural performance, resulting in the design; this is followed by construction which, if carried out in accordance with the drawings and the specification, produces a building with a structure that has an acceptable level of safety, serviceability, durability, etc.

In an existing building, the structural performance is ascertained through the process known as structural appraisal, in which information is obtained, by various means, about the physical facts of the structure and this information is assessed so as to check that the structure is adequate for the intended use of the building. Structural appraisal was first fully described in 1980 in the report by The Institution of Structural Engineers, entitled: 'Appraisal of Existing Structures'. Much of the following is based on this report, the second edition of which was published in October 1996.

2.1 THE APPLICATIONS, SCOPE AND PROCESS OF STRUCTURAL APPRAISAL

2.1.1 The Need for Appraisal

It is usually necessary, and always advisable, to check the adequacy of the structure in an existing building in the following situations:

- When defects of general design or construction have been discovered or are suspected as a result of apparent substandard behaviour (e.g. large deflections).
- When the building contains a type of construction or a particular material, that has been found to cause problems in other buildings.
- When significant deterioration of structural members has been discovered or is considered a probable risk (e.g. rotten floor beams in a dilapidated building).
- When there has been accidental or other damage to the structure in the past (e.g. cutting into structural members to accommodate service pipes or ducts).
- When refurbishment for continued use is considered to include improved services (e.g. ventilation ducts or computer wiring).
- When significant works, adjacent to or under the building, are proposed; for instance, demolition of adjacent house in a terrace, tunnelling under the building or excavation of a deep basement close by.
- Whenever a change of use is considered.
- When there is going to be a change of ownership or tenancy.

2.1.2 The Scope of Structural Appraisal

Appraisal usually encompasses the following aspects of structural performance:

- *Overall stability and robustness* (Is the building slowly keeling over or is it susceptible to disproportionate collapse in case of minor damage to a structural member?)

- *Stability and load-carrying capacity of structural components* (Is there an acceptable margin of safety under loads that are likely to be imposed by the continued use or by the intended new use and/or upgrading of the building?)
- *Serviceability* (Is there going to be excessive deflections or vibrations of floors and/or significant cracking of finishes during the expected remaining life of the building?)
- *Fire resistance* (Will the structure stand up long enough to allow occupants to escape to safety *and* give firefighters reasonably secure access?)
- *Durability* (Will the owner/tenant be faced with disproportionate repair bills, in order to maintain structural integrity?)

Impermeability (weather-proofness) and *appearance* are often included in the structural appraisal.

2.1.3 The Process of Appraisal

Structural appraisal is a process which usually encompasses the following: document search, inspection, measurements and recording, and structural analysis; sometimes it includes testing of materials and occasionally load testing of entire structures is involved.

The process is cyclical, as shown on the flow charts, Figs 2.1–2.3. The gathering of information alternates with the processing of it. At any one stage the amount of information, to be collected, should be limited to that which at that stage needs to be assessed. For example, there is no point in taking samples of bricks and testing them for strength if a simple 'back-of-envelope' calculation shows that stresses are well below those permitted for the weakest type of brick likely to have been used. Conversely, no more analysis should be done, than is justifiable by the information available at that stage; there is no merit in an elaborate elastic analysis of a roof truss, if the joints have not been surveyed in sufficient detail to enable their strengths and stiffnesses to be estimated.

Figure 2.1 illustrates the path of initial appraisal, which often can be complete in itself, if the situation is one in which structural adequacy 'by force of habit', as discussed in Sec. 2.4.3, is evident. Figure 2.2 illustrates the path of a full appraisal, when a clear-cut assessment of the load-carrying capacity can be carried out, following the usual design methods and criteria. This is often, but not always, the case and when this approach does not yield a satisfactory answer, the approach to the assessment of the margin of safety has to be reconsidered, as outlined in the following section.

2.1.4 Approaches to the Assessment of Margins of Safety

There are important differences between the basis for assessment of the margin of safety, that is usually adopted in design of new structures, and the approach, which is appropriate when appraising an existing structure. Whilst the full range of procedures and methods, used in appraisal, will be dealt with in Chapter 3, a brief resumé of the history of design procedures may be helpful at this point.

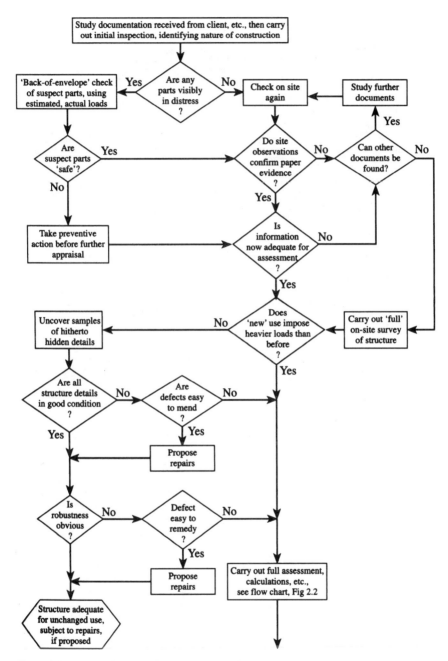

Figure 2.1 Paths of initial appraisal (*after 'Appraisal of Existing Structures', 1996*).

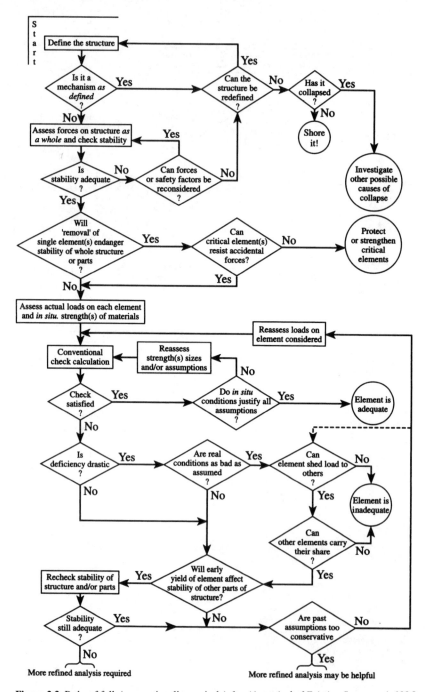

Figure 2.2 Paths of full, 'conventional' appraisal (*after 'Appraisal of Existing Structures', 1996*).

2.2 THE HISTORICAL DEVELOPMENT OF DESIGN PROCEDURES

2.2.1 Rules-of-Thumb and Early Regulations

The earliest structures were 'designed' and built by the same man, who had learnt by experience that certain geometric proportions, e.g. the ratio between span and depth, led to a satisfactory structure, whilst others did not. When buildings got more ambitious and builders became 'specialists', as opposed to the 'do-it-yourself' farmer-builders, design rules developed among the bands of itinerant tradesmen who were employed for the building of temples, castles, cathedrals, etc.

These design rules were generally in the form of prescribed sizes and proportions of the spaces to be enclosed *and* of the structural elements forming the enclosure. Some of these, e.g. the layout of Roman legionary forts, found their way into army manuals or were recorded by authors such as Vitruvius. Later, in the middle ages, rules were handed down verbally, often as carefully guarded secrets of the forerunners of the trade guilds, where the mediaeval master-builders-cum-architects had to learn them during their apprenticeships.

The structural rules-of-thumb developed by trial and error (several of the great medieval cathedrals suffered collapses, when their constructors crossed the frontiers of their technology). They were still in the form of span-to-size ratios for floor beams of specified timber qualities and height-to-thickness ratios for walls of specified masonry construction. Gradually most of them were recorded and incorporated in early building regulations, some of which survived until the middle of the twentieth century.

2.2.2 Load Assessment and Prototype Testing

As the 'art' of architecture became divorced from the 'science' of engineering, with its development of calculation methods for structures and with the emergence of the 'new' materials such as cast and wrought iron during the industrial revolution, there came a new approach, based on establishing the force in a structural component, caused by its loading and testing prototypes to failure (as well as, in some cases, testing each component to a force slightly above its working force and checking its deflection against that of the prototype, under the same load, so-called 'proof testing'). Out of the prototype testing came the first definition of the 'factor of safety' as being:

> The force, producing failure in the tested prototype, divided by the force, to which the component would be subjected in the actual structure.

This method was occasionally extended to include load tests on reduced scale models of entire structures. Such a test was carried out in 1845 by William Fairbairn for Robert Stephenson's design of the tubular Britannia Railway Bridge over the Menai Straits.

2.2.3 Standard Strengths and 'Blanket' Safety Factors

With the subsequent advancements in the understanding of theory of elasticity came the ability to calculate stresses in structural members (subject to certain assumptions, which are now known not always to apply). Such calculated stresses could be compared with the strengths of the structural materials, as ascertained by tests on standard samples and stipulated in standard specifications, or government norms, for the supply and use of these materials.

This was clearly much more convenient than testing whole components and all that remained to establish a complete design procedure was to determine the basis of comparison between calculated stresses and tested strengths. This was done by the introduction of permissible stresses. These were derived from the tested and/or standardized strengths by division by 'factors of safety' agreed by committees of learned experts, design practitioners and representatives of vested producer interests.

The resulting design procedure, which persisted unchallenged until the 1940s, was as follows: from the dimensions of the designed structure and the loads that it was required to carry, the engineer would calculate the stresses in the structural members, assuming linear-elastic behaviour (Hooke's Law, see Sec. 1.2.1). He would then adjust the cross-sectional dimensions of the members such that the calculated stresses would be just below the permissible stresses laid down in the norm or code of practice for that type of member (e.g. beam or strut) in the material envisaged.

The factor of safety had now become defined as:

The ratio of the standard tested strength of the material divided by the permissible calculated stress under working load.

This factor was intended to take account of the variability of the material, the inaccuracies in the design assumptions and various other factors affecting the strength of the structure and, in addition, it was sometimes modified to cover serviceability considerations, e.g. to limit deflections.

2.2.4 Partial Safety Factors ('γ factors')

Whilst convenient for the design office and for the checking authority, the application of a blanket factor of safety might, for one structure, be unnecessarily restrictive, whilst for another it provided an inadequate margin of safety. Growing dissatisfaction with this state of affairs led to the development of the 'partial factor' approach, in which separate partial factors were applied to the loading, to the inaccuracies in design and construction, and to the strength of the material.

Unfortunately, there is, to date, no practical and rational way of determining these partial safety factors, because to do so involves comparisons of death and injury with damage to property. New design codes were therefore 'calibrated' against the old ones, and these endeavours to 'harmonize' the sets of partial factors with the old 'blanket' safety factor did sometimes result in less than sensible numbers.

2.2.5 Brief Summary

- *Antiquity—late eighteenth century*: Rules-of-thumb, e.g. the thickness of a wall should be not less than its height divided by a certain number.
- *First half of nineteenth century*:

$$\text{Factor of safety} = \frac{\text{failure load of prototype in test}}{\text{working load on component in structure}}$$

- *Second half of nineteenth century—1970s*:

$$\text{Factor of safety} = \frac{\text{strength of standard test specimen}}{\text{calculated stress in structural member}}$$

- *1970s*: The load effects *multiplied* by the partial 'load' factor 'γ_L' must be equal to or less than the strength of the structural member *divided* by the partial 'strength' factor 'γ_M'.

2.3 THE UNCERTAINTIES COVERED BY THE SAFETY FACTORS USED IN DESIGN OF NEW STRUCTURES

When making drawings and writing specifications for a structure to be built, the designer is trying to predict future events. He or she is, however, not master of the destiny of the structure, once it comes to the construction: the dimensions of the structure and the non-structural elements, that it has to support (such as screeds and partitions), will vary from those specified or assumed, the materials used for the structure may be stronger than the minimum specified, but they may also, by the time they are placed in the structure, be weaker than in the standard control test specimen (e.g. the concrete test cube). Once completed, matters can get even worse; owners and users have been known to let workmen do quite unspeakable things to innocent structures.

The designer is thus faced with a future uncertainty about the loads, that his structure has to carry, the effective dimensions of the structural members and the actual strength, in situ, of the structural materials. The designer's provision against each of these uncertainties consists of a component of the safety factor(s): the greater the uncertainty, the greater the corresponding component. The designer is, however, usually absolved from fixing the values of these individual components; they are all combined in the overall safety factor(s) prescribed by the norms, regulations or codes of practice. In fact, even if the designer were willing to exercise engineering judgement, by weighting one or the other component of the safety factors, he was in the past effectively prevented from doing so in a rational way, as no breakdown was given of the overall factor. In recent decades, a breakdown of each of the partial safety factors into components, covering specific uncertainties, is in fact being provided by national and international standardization bureaux.

One last point on design of new structures: building structures (as opposed to bridges and similar 'purer' structures) often provide a multitude of paths for the loads to reach the foundations. To explore which proportion of the total load goes down which path, would be very time-consuming and hence, in most cases, too expensive for the design office. Normal design practice is therefore to ignore all but the main path. The result is that the client gets a marginally increased load capacity for a negligible extra cost of material, if any. This practice is legitimate (and generally universal) for new construction, but if it becomes a habit, it may blind engineers to such alternative paths, so that they unnecessarily compensate for, what is seen as a shortfall of load capacity in an existing structure.

2.4 FACTS TO BE RECOGNIZED IN APPRAISAL OF EXISTING STRUCTURES

As mentioned in Sec. 2.2, the design of a new structure contains an element of unfulfilled prophecy. When appraising an existing building, the structural engineer is however confronted with a set of physical facts which he must take into account in his appraisal.

2.4.1 Facts to be Taken into Account

1. The physical presence and condition of the building; it is there, it has performed its function(s) for so many years, it may show such and such signs of 'fair wear and tear' and it may or may not show signs of past or present structural distress.
2. On closer examination, the structural system may turn out to be quite complex with, in places, several unorthodox but perfectly effective load-carrying assemblies providing multiple load paths.
3. The structural members are there and their dimensions can, in principle, all be verified (although there may be difficulties in practice) and the actual strengths of the materials, in situ, are there and can, in principle, be tested.

2.4.2 Consequences for the Assessment

Fact 1 can be, and should be, used as a qualitative check on any numerical assessment. If the building stands up and the calculations show that it should have fallen down, the calculations or the assumptions on which they are based must be wrong. What may jokingly be referred to as 'structural adequacy by force of habit' can be a real condition; it is however not without its problems and these are discussed in Sec. 2.4.3.

When qualitatively assessing the soundness of the structure, one must not be misled by a superficial display of 'architectural merit'. Where there are elaborate or expensive finishes, the structure may be 'cheap and nasty', and even when the original structure was sound, any subsequent alterations might not have been. Generally, the more hectic the building activity at the time of construction, the worse the structural quality.

Fact 2 is likely to be a challenge, but unless the obvious primary system can be shown to be adequate on its own, the challenge has to be faced by the engineer, even if it involves 'non-linear analysis' of a three-dimensional structure: to do otherwise is tantamount to condemning a perfectly good structure out of laziness. 'Linear-elastic' analysis assumes that Hooke's straight-line stress–strain relationship applies regardless of the stress intensity; non-linear analysis takes account of the curvature of the real stress–strain diagram and of any yielding of the material prior to fracture; see Sec. 1.2.1.

Fact 3 allows the elimination or reduction of some or all of the design uncertainties described in Sec. 2.3 and thus leads straight to Sec. 2.5. Figure 2.3 illustrates the path of improved appraisal.

2.4.3 Adequacy of Existing Load-carrying Capacity for New Use

If a structure has carried a certain load for a considerable period in the past, without showing any signs of having been overloaded and if it has not suffered any significant deterioration (due to rotting of timber, corrosion of metal, etc.) nor been damaged by past alterations, it is clearly capable of carrying the same load for a further term of use. Similarly, if the loads imposed by a previous use were greater than the loads to which it will be subjected in a proposed use, the structure should be safe for the new use, again subject to neither deterioration nor past damage.

Often, the problem is to establish what the intensity of the past loading has, in fact, been. Prior to the introduction of loading requirements into building legislation, some guidance on design loads was given in various textbooks of the time. The figures varied and were invariably well above today's requirements, as were the loading requirements of the London Building Act 1909 (the first to introduce such stipulations in Britain), shown in Table 2.1.

Experience in appraisal of buildings from that period shows, however, that only a few builders followed these recommendations. It is not unusual to find that, by today's criteria, the structure, as built, cannot carry the loads stipulated in today's regulations, without the stresses exceeding permissible values, even after review of the safety factors according to Sec. 2.5, and as for the loads recommended at the time of its construction, they are out of the question. This has led to some building control authorities becoming reluctant to pass a proposed use of a building on the strength of a past 'design load' being more than that of the new use.

Where, however, actual loading has been observed within living memory, such approval should be easier to obtain. Old photographs of interiors that show, albeit indirectly, the loadings imposed by the use of the building at that time, have been known to convince a building inspector.

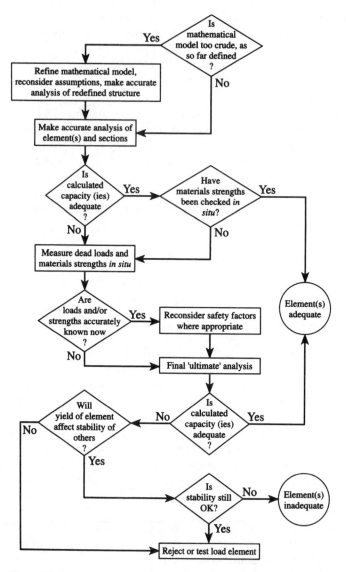

Figure 2.3 Paths of improved appraisal (*after 'Appraisal of Existing Structures', 1996*).

2.4.4 Realistic Assessment of Imposed Loads from New Use

A further fact to be borne in mind is that the imposed loadings, stipulated for certain uses, are not always realistic. The British Standard Code of Practice 6399, Part 1 stipulates an imposed load of 2.5 kN/m^2 for offices for general use. During the 1980s and 1990s, it was however usual for developers to demand that floors in speculative office buildings should be designed for 5 kN/m^2

Table 2.1 Early loading requirements

	kN/m^2	lbs/ft^2
Domestic	3.4	70
Offices	4.8	100
Shop or workshop	5.4	112
Warehouse (minimum)	10.7	224
Roofs (less than 20° slope)	2.7	56
Other roofs (wind included)	1.3	28

(Wind loading to be taken as 30 lbs/ft^2 (1.4 kN/m^2) on the upper two-thirds of the building face.)

imposed load. This was intended to allow any part of an open floor to be used for filing.

Surveys have been carried out by the Building Research Establishment, English Heritage and Stanhope Properties/Ove Arup & Partners between 1970 and 1992. These have shown that the real imposed office loading varies from about 0.4 kN/m^2 in an open plan financial adviser's office to about 1.6 kN/m^2 in a fairly close-packed design office with many plan-chests. Loading in the libraries and the conventional filing rooms, that were surveyed, varied from 1.8 to 2.6 kN/m^2. Against this, archives stored in moveable racks were found to impose loads in the region of 10 kN/m^2. Realistic loads, likely to be imposed by the proposed new use rather than stipulated values, should therefore be used in check calculations of existing structures. Archiving should, sensibly, be confined to basements and/or specially adapted rooms.

2.5 REVIEW OF SAFETY FACTORS IN APPRAISAL

2.5.1 Reduced Factors Do NOT Mean Reduced Safety

The mention of the possibility of reducing safety factors, when appraising an existing structure, often produces a 'shock-horror' reaction from engineers, not acquainted with the underlying reasoning. What has to be emphasized is that it is only *the number in the calculation*, that is reduced, *NOT the real margin of safety*.

2.5.2 Reduced Uncertainties Justify Reduced Safety Factors

As explained in Sec. 2.3, some components of the safety factor(s) *in the calculations* are there to cover certain uncertainties in the as-built structure. If one can reduce or eliminate some of these uncertainties, one is justified in reducing or eliminating the corresponding components of the safety factor(s) used in the calculation, because the real margin of safety will remain.

The Institution of Structural Engineers' report 'Appraisal of existing Structures' discusses this at some length and suggests numerical values; the reader is referred to pp. 23–27 in the 1996 edition. A caveat is however in place here; as explained later under Sec. 3.5, the practical difficulties of procuring enough test data, to justify a reduction of the safety factors on the strengths of the materials, can be great indeed and gaining enough data on the existing dead loads can be disruptive.

THREE

PROCEDURES OF STRUCTURAL APPRAISAL

Some 'tools of the trade': Drawings, ancient and modern, old reports, tape measure, folding rule, plumb line, levelling instrument, penknife, callipers, sketch pad, camera. These are only a selection of the many aids to structural appraisal.

The procedures which together make up an appraisal are mentioned in Sec. 2.1.3. In this chapter they are each described in some detail. (They may not all be required in every appraisal and some will only occasionally be employed.)

3.1 EXAMINATION OF RECORDS AND OTHER DOCUMENTATION

The first requirement for a structural appraisal is a complete description of the structure, including any defects. The most complete record is, of course, the building itself, but this is in most cases too complicated to grasp as an entity. This is important because in most cases one cannot make a sound assessment of the causes of defects if they are observed in isolation. What is needed before a safe diagnosis can be made is not only a complete pattern of symptoms, but also an overall picture of 'the patient'.

3.1.1 Drawings

What is needed for the overall picture is something that can be studied in the relative calm of the office. Drawings are therefore essential. For the purpose of structural appraisal, drawings should show sections and elevations, generally of small enough scale that the whole of the building can be shown on one sheet. These sections and elevations should include all significant structural features, and any observed defects should be superimposed after the inspection.

The production of such drawings can be time-consuming and expensive if a measured survey has to be carried out first. It is therefore worth a lot of effort to try and locate existing drawings, whether from the original design and construction or from subsequent repairs or alteration works. Whilst such drawings can be very helpful, they will, however, rarely show the structural essentials and will often include features that mask, or confuse, the structural issues. Drawings for the purpose of structural appraisal will therefore usually have to be produced by the engineer (see Sec. 3.4.1).

Ground conditions can be crucial for the appraisal; the geological map of the area should therefore be studied.

3.1.2 Original Specifications, Accounts, etc.

The original construction specifications and/or accounts may give useful indications of the nature and origin of materials or even on forms of construction. For instance, a contemporary document stated that the church of All Souls', Langham Place in London had to be founded at extra great depth due to poor ground conditions and that the pier loads were spread along the strip foundations by inverted arches; this was an essential information for the project to create a new meeting hall underneath the church.

Documentation from the original or later works may also suggest the locations of potential sources of trouble and thus help to concentrate and limit the amount of investigative opening-up.

Other documents that may also be helpful are descriptions, whether contemporary or later, of the construction practices of the period when the structure was built. Whilst these should also be taken with a pinch of salt, they may help to explain apparent idiosyncrasies and puzzling features and again point at the likely troublespots.

It is therefore worth going to the archives of the building owner (present or past), the building control authorities, the local museums and archaeological and/or historical societies and persuade the custodian to look for such information and, if found, to allow it to be copied, if at all permissible. The search may end up being fruitless, but if it unearths a reasonably accurate set of plans and elevations, it may save several man-weeks of work on producing a complete measured survey. (A list of British sources of information is given in the Institution of Structural Engineers report 'Appraisal of existing structures'.)

With regard to the materials and forms of construction likely to be found, Table 3.1 shows the approximate historic periods of the availability of various building materials and the use of different forms of construction. Whilst not complete, nor exact, it may help to give a first indication of the nature of the structure when its age is known—and vice versa. (The density of the shading indicates the extent of use.)

3.1.3 Documentary Evidence Must Be Supplemented

The preceding paragraphs do not mean that a site inspection becomes superfluous. For one thing, buildings are not always constructed fully in accordance with the original drawings. Original variation orders, given on site, are not always recorded and a building 200 years old may have been subjected to several unrecorded alterations during its life. A site visit, to check how far reality corresponds to the documents, is therefore essential, but that is a far less onerous task than a complete measured survey *ab initio*. It will also give a first impression of the condition of the building and the nature of the site. The latter may come as a surprise, as many nineteenth-century architects drew elevations that did not show the slope of the site.

3.2 INSPECTIONS

Documentary information is, as mentioned, rarely adequate for a structural appraisal and it is the responsibility of the person in charge of the appraisal to inspect the condition of the structure.

3.2.1 Permits, etc.

Before attempting any internal inspection, one must make sure that one has the owner's and/or the tenant's authority to enter and that the occupiers are informed

Table 3.1 Periods of usage of forms of construction

Material and/or form of construction	Period of availability/major use in UK	Notes
Wood based	1600 1650 1700 1750 1800 1850 1900 1920 1940 1960 1980 2000	
Timber framed with clay or brickwork infill: home-grown hardwood		
Timber frame with brickwork, or tile-cladding,softwood		
Timber frame and inner leaf with brickwork cladding		
Roof trusses and floors on masonry walls: home-grown hardwood		
Roof trusses and floors on masonry walls: softwood		
Bolted, and/or glued, laminated beams and frames		
Plywood – web and similar beams		
Natural stone masonry		
Loadbearing walls and piers: solid or with rubble cores		
Subsequent facing to timber frames and loadbearing brickwork		
Bonded facing to brickwork backing		
Cladding to steel, or reinforced concrete, construction		
Brickwork and blockwork		
Loadbearing walls: clay bricks in lime mortar		
Loadbearing walls: clay bricks in cement* mortar		* Includes cement/lime mortars
Loadbearing walls: calcium-silicate bricks in cement* mortar		
Loadbearing walls: concrete blocks in cement* mortar		
Infill to timber frames: clay bricks in lime mortar		
Infill/cladding to steel, or concrete, frames: clay bricks, cement* mortar		
Infill/cladding to steel, or concrete, frames: calcium-silicate bricks		
Infill/cladding to steel, or concrete, frames: concrete blocks		

Material and/or form of construction	Period of availability/major use in UK	Notes
Cast iron		
Beams		
Roof truss components		
Columns		
Components for special structures: SG cast irons		
Water and gas mains, drain pipes, etc.		
Wrought iron		
Tie rods, straps and bolts in timber structures		
Tie rods, chains, cramps and dowels in masonry		
Rolled strip, L and T sections		
Rolled I - and [sections		
Rivetted, built-up sections		
Wire for bridge suspension cables (chains used earlier)		
Corrugated sheet (superseded by steel from 1890–1900)		
Structural steel		
Plates and tie rods		
Rolled L -, T -, I - and [- sections (Ls up to 610×190)		
Rivetted, built-up sections		
Structural tubes		
Rolled 'parallel-flange' I sections up to 1100 mm high		
Rectangular hollow sections		
Welded, built-up sections		

Period of availability/major use in UK scale: 1600, 1650, 1700, 1750, 1800, 1850, 1900, 1920, 1940, 1960, 1980, 2000

41

Table 3.1 Periods of usage of forms of construction

Material and/or form of construction	Period of availability/major use in UK	Notes
	1600 1650 1700 1750 1800 1850 1900 1920 1940 1960 1980 2000	
Other steel products		
Corrugated sheet, galvanized: see wrought iron		
Profiled sheet, plastic-coated		
Tubes for plumbing, etc.		
Mild steel reinforcing bars, smooth		
High tensile reinforcing bars, cold twisted square		
High tensile reinforcing bars, cold twisted ribbed		
High tensile reinforcing bars, hot-rolled smooth		
High tensile reinforcing bars, hot-rolled ribbed		
Wire for suspension cables and for pre-stressing tendons		
High Tensile pre-stressing bars		
Non-ferrous metals		
Lead: plumbing		
Lead: dowels in masonry, roofing and guttering		
Lead: underlay for heavy bearings on masonry		
Lead: damp-proof coursing in masonry		
Aluminium: alloys in structural sections		
Aluminium: sheeting, flat and corrugated for roofs and wall cladding		
Copper: roof sheeting		
Copper and copper alloys: plumbing and roof drainage		

Material and/or form of construction	Period of availability/major use in UK	Notes

Period of availability/major use in UK axis: 1600 · 1650 · 1700 · 1750 · 1800 · 1850 · 1900 · 1920 · 1940 · 1960 · 1980 · 2000

Cement and concrete based

Mass concrete: lime or pre-Portland cement based

Mass Concrete: Portland cement based

Clinker aggregate concrete in filler-joist floors

Reinforced concrete: early patent systems

Reinforced concrete: conventional design

Pre-tensioned concrete

Post-tensioned concrete

Lightweight structural concrete, blocks and screeds

Miscellaneous

Rammed earth 'Cob'

Wattle and daub

Asphalte roofing, guttering and waterproofing

Bituminous felt roofing

Asbestos-fibre-reinforced cement sheet materials

Glass-fibre-reinforced cement (cladding and permanent formworks)

Other fibre-reinforced concretes (screeds and ground floors)

Polymer-fibre-reinforced concrete

Organic polymer based

Glass-fibre-reinforced plastic

Polymer sheeting

about the purpose of the visit. If the inspection is prompted by anxiety over structural adequacy, it is however best not to say so outright at the beginning, if asked by the occupiers; it does not help anybody to create unnecessary panic before the inspection has confirmed that there is cause for concern.

It is as well to remember at this stage that any physical interference with the building fabric, whether for purposes of the eventual restoration, or merely for structural examination or testing, may require permission from several authorities, particularly if the building has been 'listed' (as in Britain) or otherwise registered as being of historical interest. The appendix given at the end of this chapter contains detailed information on this.

3.2.2 Inspection Equipment

Having dealt with the formalities, it is time to check the equipment. It should be remembered that the purpose of inspections of buildings is to observe and record physical facts in such a way that a considered appraisal can be made in due course after study of additional information (if any) and/or some structural analysis.

The two most important items on the equipment list are therefore a pair of open eyes and an open mind. Imagination is likely to interfere with the faculty of observation and is therefore best left outside the door at this stage.

Next on the list come the means of recording the observations: a clipboard with a pad of A4 (30 × 21 cm) or similar size paper, or a similar size book with stiff covers. Squared paper will be a help to some when drawing sketches, while others prefer blank paper on the grounds that old buildings are rarely square and the grid may tempt one to draw them as if they were. For working out-of-doors when it is raining, a clear plastic 'flip-sheet' to protect the paper, in between making notes, can be a great help, and pencil, as opposed to some pens, does not smudge when the paper gets wet. Soft (B or 2B) pencils and ditto rubber facilitate corrections. Wooden pencils are less prone to mechanical breakdown than the plastic push-button types, but do require some implement for sharpening. If the push-button or propelling type is preferred, two should be carried and the supply of spare leads should be checked before leaving base!

A pocket dictaphone can save a lot of time making notes and enables recording of immediate impressions during the initial 'walkabout'. When it comes to checking whether everything relevant has been recorded, it is however much more difficult to do this from the tape than from hand-written notes and sketches, and it is even more awkward to add the extra information in the place in the sequence where it makes sense. When access will not be possible at a return visit, e.g. if the scaffold is being taken down, a video camera can be useful, but 'panning' must be done *very* slowly. All such devices, and particularly their batteries, must be checked before leaving.

A sturdy pocket knife of 'Swiss Army' type with a sharp-pointed bit for stabbing suspect timber and a screwdriver bit for gouging mortar joints (and a blade for sharpening pencils!) should be carried.

For the purposes of a structural inspection, when the geometrical survey either exists or is to be done by others, a pocket measuring tape about 3 m long, with lock, is usually sufficient (16 mm wide ones are more rigid than 12 mm ones and hence easier to use in 'cantilevering'). A 2 m folding rule will stand up on its own, better than a tape.

A ball of string is useful for measuring bowing of walls and deflection of floors; with a moderately heavy object, e.g. a bunch of keys, it will also act as an improvised plumb line. It should be strong but thin, so as to define a sharp line from which to measure.

A torch (with fresh batteries, a spare set and spare bulb!) is usually indispensable; a pencil-thin beam is no advantage, a wider diffuse light can be achieved by a piece of tracing paper under the lens. High-powered battery lights, such as are sold for video recording, can be useful in large spaces. Head-band mounted lights leave both hands free.

A large version of a dentist's mirror is useful for looking into inaccessible places such as spaces between floor joists. Where disturbance has to be kept to a minimum, a 'borescope' may be used. This is a miniature periscope, usually about 300 mm long and between 7 and 12 mm in diameter, with a built-in light, usually powered from a belt-carried battery. It has an eyepiece for direct observation, but can be attached to a camera for recording purposes. There are models available for viewing at 90°, 45°, and straight ahead. Requiring only a hole 10–16 mm in diameter, it enables observation of spaces between floors and ceilings, as well as in wall cavities.

Connections in exposed roof trusses over churches and other tall spaces are difficult to inspect at close quarters as adequate ladders are rarely available. A pair of binoculars can therefore be of great help. Binoculars are specified by two numbers: the magnification $\times$ the diameter of the object (front and larger) lenses. Great magnification is rarely required; what matters, when inspecting badly lit spaces, is light transmission. This is dependent on the diameter of the 'exit pupil' which is calculated by dividing the diameter of the object lens by the factor of magnification; for example a pair of binoculars designated 8×32 will have an objective 32 mm in diameter and a magnification of 8 times, and its exit pupil will therefore be $\frac{32}{8} = 4$ mm. This is adequate for most uses, but as the pupil of the human eye can expand to 6 mm or more, a pair of binoculars 7×40 will be better for inspection purposes. Light transmission depends on other factors as well, and some makes of pocket-size binoculars may well prove to be adequate. What is important, however, is the ability to focus on close objects: if you rest your ladder on the tie beam of one roof truss, the suspect joint that you want to inspect may be only 3–4 m away; binoculars that will only focus down to 5 m are then of no use.

Photographs are very good as *aide-memoires* and can be the best way of illustrating certain points in a report. The camera should ideally have a wide angle lens for general views and a telephoto lens for picking out details, or, alternatively, a close-focus zoom lens. As most cameras nowadays rely on batteries for one or more essential functions, spare batteries should be carried. Auto-focus, if incorporated, should ideally be capable of being manually overridden (the camera may not realize that you want maximum definition of the detail on the second roof truss, not the closest one). Alternatively, aim the little box, in the centre of the viewfinder, on the bit that is wanted in focus, push and hold the shutter button gently to lock the focus, then frame the picture before pushing the button home. As with all equipment, one must make sure that one knows how it works, before leaving home. It is often better to utilize the available light with a long exposure, using a tripod and cable release, rather than using flash: The illumination from the flash decreases with the square of the distance; it is therefore extremely difficult to get pictures with anything approaching even lighting. In large dark spaces, one can alternatively open the shutter on time exposure and walk around firing the flash on the distant parts.

Different films have different advantages: Colour-negative films allow quick and inexpensive processing and printing, and have large exposure latitude (they allow a fair degree of over- or under-exposure). Colour-slide films usually give better definition and colour rendition and are generally preferred by printers, for high-quality reproductions, but correct exposure is critical. Black-and-white films give the best definition and allow the easiest and least expensive monochrome reproduction for report illustrations. For all films, higher speed rating (ASA No.) gives coarser grain and reduced sharpness; for ordinary inspection purposes, 400 ASA print film is adequate.

Digital cameras have the advantage of immediately showing what has been taken and are often very good at 'seeing in the dark'; the cheaper ones will however produce poor sharpness of any prints. Where evidence is wanted in legal cases, traditional film will have to be submitted.

Identification of the photographed items can be a problem back in the office, particularly when confronted with prints showing several similar but different details. Cards about 10×15 cm and a broad felt-tip pen with which a location code can be marked are helpful here if the card is included in the picture (fixed with drawing pin or adhesive putty). If the cards are coloured and the pen is a contrasting colour, the notes will not only be easier to read than black on white, but will also enable subsequent colour printing to be given correct colour balance.

3.2.3 Health and Safety Precautions

Old buildings are often dirty and sometimes hazardous, so old clothes and/or overalls should be worn and a supply of tissues should be carried. Safety shoes or boots may be ungainly, but will protect toes against falling objects and are unlikely to be penetrated by nails protruding from lifted floor boards. Wearing 'trainers' is taking unjustifiable risks, unless the building is in use and occupied. 'Hard hat' should be worn if the building is empty.

Whilst inspection of an occupied building can obviously be hampered by the presence of the occupiers and furniture and by restrictions on removal of finishes to expose the structure, unoccupied buildings, whilst free of such impediments, can on the other hand be hazardous to enter and move about in. For this reason one should not go unaccompanied on a first visit and on follow-on visits, one should always leave a message of where one is going and what time one expects to be back. If one has to go alone, a mobile phone, *with fully charged battery*, must be carried *on the person*—a mobile phone in your bag on the ground floor is not much help when you are on the second floor with a broken leg, having fallen through some rotten boards on the third floor! (There is a further argument for inspections not to be carried out solo: two pairs of eyes are better than one.)

Some of the dangers that may be encountered in badly maintained buildings are listed below:

- rotten floors, protruding nails from lifted, overturned floor boards, fragile roofings;
- missing handrails to stair wells (easy to fall into, when stepping back for a better look);
- exposed sprayed asbestos insulation, torn asbestos pipe lagging (fully encapsulated asbestos is *not* dangerous);
- basement pits containing asphyxiating or explosive gases (e.g. carbon monoxide, methane);
- wasps nesting in roof spaces;
- rats (carry Weil's disease), dust from disturbed pigeon droppings (carries Salmonella);
- Abandoned cats and dogs.

Some further safety hints:

- Ladders and scaffolds: only one person on any one ladder at a time; check that scaffolds are complete. Current British Health & Safety rules limit the height of ladder that may be used and as scaffold towers are getting easier to obtain, they should be used where necessary.
- Sash cords can break suddenly: prop open the sash window before looking out.
- Flat roofs/parapet gutters: if there is a harness system, use it; if not, keep away from the edge!
- Death traps from past investigations: watch out for loose or missing floor boards, open trial pits, etc. and make sure that *you* do not leave similar hazards behind!

Some of the governmental health and safety rules, applying to these hazards, may appear irksome, but it is foolish (and illegal!) to ignore them.

3.2.4 Inspection Procedures and Methods

There are two important objectives to a structural inspection: the first is to obtain an overall view of the structural condition of the whole of the building; the second is to make sure that all significant details are recorded. The first is sometimes best achieved through an initial 'walkabout' inside and out, preferably unaccompanied by any owner's or tenant's representative, who would be too keen to point out the details that worry *him or her*, but which may not be significant for the whole. The second requires a very systematic procedure of observation and recording.

To avoid missing anything, one should proceed in one direction when inspecting the exterior and similarly within each space of the interior; whether clockwise or anticlockwise is a matter of personal choice. From each viewpoint one then observes and records from top to bottom, covering as far as possible a well-defined area, e.g. an exterior bay, defined by two successive buttresses to the wall. It is helpful to head the notes for each bay in identical fashion, e.g. roof ridge, roof slope, gutter, top storey, next storey and so on, or for interiors: ceiling, cornice, wall over windows, wall between windows, windows, wall under window sills, skirting, etc.

Having done the bay-by-bay recording, it is advisable to step back, literally *if possible without danger*, and certainly mentally, and let the eye follow any horizontal features, distortion of which may have been missed but could be of structural significance. One should look for defective gutters, drainpipes and gullies and for evidence of changes to the surrounding surfaces, e.g. new paving.

The main problem, internally, is that in an old building most of the structure, if not all, is often hidden by finishes. These make it nearly impossible to judge the soundness of walls and floors and selective removal of finishes may be a prerequisite for final assessment. Wallpaper and/or plaster may hide cracks across the length of a wall and vertical cracks at right angle junctions between walls. It may also hide inadequate lintels over openings. Ceilings may conceal woefully inadequate joist bearings, and whilst in the case of bearings on walls these may be more conveniently inspected by removing floor boards, this is not necessarily so with bearings of joists on main beams. It is worth looking for floor boards that may have been lifted in the past for installation of wiring, etc.; they will be easier to lift.

As wall cracks do not always show up on photographs, their positions, widths and directions should be recorded on sketches or on prints of the 'as existing' drawings. Straight vertical cracks should be closely examined to ascertain whether they are, in fact, the result of an unbonded junction between two walls. This is sometimes found at the junctions between facades and cross-walls.

'Bonding timbers' are sometimes found in Georgian and Victorian brickwork, particularly in rough work such as rear walls. They are frowned upon by building control officers, who may insist on their removal, because they may be sources of dry rot. Their location, condition and size should be recorded, if at all practicable (this may have to wait until the building is vacated for refurbishment). Timbers may also have been embedded to enable wood panelling to be fixed; their presence, or absence, should, if at all possible, be investigated by selected removal of a small area of panelling.

Some buildings, apparently of Georgian or Victorian brick construction, incorporate walls from an earlier period and these may in fact turn out to be half-timbered. This can often only be ascertained after removal of plaster, but it is a possibility to be borne in mind. Both sides of walls should be checked for plumb: walls of early origin may consist of two skins with rubble infill. They may split and the two skins will then lean or bulge away from each other.

If cracks in walls appear as if they might be due to settlement, it may be advisable to expose enough brickwork to enable any waviness or dipping of the coursing to be ascertained. Past settlement may reveal itself through out-of-true window and door openings. If doors and windows open freely, the settlement is most likely no longer going on. It may however start again if the permanent foundation loads are increased as a result of adaptation of the building or, if the building is on clay soil, there is renewed foundation movement, due to changing moisture content of the soil (see Sec. 8.1.5). It is also possible that the doors or windows have recently been 'eased' and may bind again, if the settlement continues (look for recently planed, hence unpainted edges).

Excessive springiness, or a gap between the bottom of the skirting board and the floor, may be indicative of inadequate boards and/or joists. Floors should therefore be walked on all over, but cautiously at first. In dilapidated buildings, that have been empty for some time, it is advisable to examine the floor construction from below, before putting one's weight on them. Stabbing with a penknife may then help to give an indication of the presence and structural significance of rot and/or insect attack in boards, joists, beams, wall plates, etc.

Roof timbers and particularly their joints should be carefully recorded, unless it is obvious (e.g. from extensive visible 'dry rot') that complete replacement is necessary. Many traditional timber details are, in the light of current knowledge, unsound engineering and will not stand even a modest increase in load from replacement of slates with tiling or new ceilings with insulation and/or services.

Adjacent buildings should be examined, from the front, from the back and sides, and if at all possible, inside. Much valuable information may be gained from their condition and past behaviour, as ground conditions usually do not change abruptly from one building to the next.

Reference has been made above to removal of plaster, etc. This is often not appropriate for a first inspection and particularly not for an occupied building. It is included here, as it may well form part of later investigations together with specialist testing for rot and beetle attack on timber, exploration of foundations, etc., but if the building is unoccupied and derelict, limited areas of plaster and/or floor boards can often be fairly easily removed.

3.2.5 Verbal Information

Before leaving a site, remote from one's office, it can be beneficial to arrange to meet any local archaeologists and art historians as these may have unearthed useful bits of information that have not found their way into the official documentation. Similarly the proverbial 'oldest inhabitant of the village' may remember the last time the building was restored or altered and may convey information not otherwise available. Needless to say, any such verbal information must be treated with circumspection, but it should not be ignored because, unlike the 'responsible person', the one-time labourer has nothing to lose by blurting out the truth.

3.2.6 Recording

The findings of the inspection must be recorded so as to be available in reliable form for the subsequent desk work. Sufficiently descriptive field notes are tedious to write and worse to read; annotated sketches communicate the observations far more efficiently, but they must have a unique location reference (or if they show a detail/defect occurring in several locations, each of the locations must be clearly specified). If plans of the building are available, however rudimentary, it can be a great help to superimpose a common reference grid on to these and locate all observations and/or detail sketches by grid 'co-ordinates', e.g.: 'foot of roof truss, grid D-7'.

It is important to record the total number of structural items as well as the number of defective ones: '13 rafter feet rotten' takes on an entirely different significance depending on whether the total number of rafter feet is 14 or 70 (93 per cent as opposed to 19 per cent).

Notes and sketches of textures and colours are difficult to make fully descriptive and photographs may provide a better record, provided that they are identifiable. Each detail view should therefore include a card with a bold felt-pen location reference, e.g.: 'D/7-8, 2 floor, W.face'. It is helpful also to make a note of the exposure numbers for each viewpoint.

Field notes tend to be scruffy at the best of times and the observers' handwritings may after a time be indecipherable to themselves as well as to others. It is therefore advisable to make a fair copy, whilst the observations are fresh in the mind; the original sketches *must* however be preserved.

Fair copies of field notes and 'sanitized' sketches are however not the same as a proper appraisal report. This will be dealt with in Sec. 3.6.

3.2.7 Further Investigations

Depending on the condition of the building and the intended future use, the initial visual inspection may be sufficient in itself. It may however have been limited to superficial examination by problems of access, etc. and a 'closer look' at some features may be required. There are various aids to this

- The 'Borescope': A more robust variant on the instruments used by the medical profession for visual examination of internal organs. It enables inspection of cavities through small (10–16 mm diameter) holes. Judging distances can be difficult.
- 'Tele-Hoist'/'Cherry-picker'/'Simon Platform'/mobile access towers: All provide an elevated observation platform, which enables cost/time-effective inspection at high level. Vehicle access is needed, and if it is in a city street, the closure of a lane, or the street, may have to be arranged with the traffic authorities. 'Zip-up'-towers serve a similar function indoors. There are moving platforms, designed to be used for tall spaces indoors, provided that the floor can carry the concentrated loads from the 'feet'.
- Abseilers are usually engineers/technicians with rock-climbing as their hobby. To earn their income, they will climb spires or lower themselves down elevations of tall buildings, taking with them a video camera and a 'walkie-talkie'. In this way they can make observations and collect samples, as directed by an engineer on the ground, who is watching the monitor screen.

As often as not, a need for further investigations will come to light during the initial and/or the further inspection and appropriate recommendations will have been made in an interim report. In broad terms, these will come under the headings of measurements and/or monitoring, of defects, detailed structural analysis, testing of materials' properties and possible degradation, and probably, investigations of foundations and subsoil.

3.2.8 Staging of Investigations

Occasionally a building may be vacant and cleared of all contents before any investigation work starts. Usually, however, the building is occupied, and it is not clear what further use it can be put to. The owner will at that stage want to spend as little money as possible and suffer the least disruption from the investigation. On the other hand, the results from the investigation are required for the feasibility studies and for obtaining finance, especially if an application for grant-aid, to subsidize conservation, is to be made. The strategy of the investigations therefore has to be planned to take account of this.

In theory, one would carry out all the investigations first, so that a complete project for any necessary works could be worked up and a firm cost established. This, however, ignores several facts of life. Some of these are:

- The extent and nature of further investigations may depend on the choice of the future use of the building.
- No investigation within the limits of practicality can guarantee that no surprises will be uncovered during the execution of the works.
- An investigation, such as a detailed survey of defects on a stone masonry façade, requires extensive (and expensive) scaffolding, in order to satisfy the health and safety requirements. The scaffold has to be taken down, pending the final decisions, only to be re-erected later when the works are to be carried out. Such doubling of the cost of temporary works can more than offset any hoped-for advantage of a 'fixed-price' contract.

It is therefore usually best to proceed in a step-by-step manner, in which the evaluation of each stage of investigation indicates what the next step, if any, should be.

3.3 MEASUREMENT OF DEFECTS

Measurements of defects are aids to the appraisal of the structure, they are not an aim in themselves. Quantifying a defect, relative to others in the same building and relative to similar ones in other buildings, helps the assessment of its significance, particularly if the appraiser wishes to consult somebody, who may be an expert, but who has not carried out a complete inspection. There are several different kinds of measurements, and they can serve many different purposes, all contributing to the appraisal, so it is important to be clear about what one wants to measure and why.

3.3.1 The Purpose of Measurements

Before embarking on any programme of measurement it is most important to clearly establish in one's mind the purpose of the measurements. There are, for instance, measurements that will quantify past defects as an aid to structural analysis and there are measurements to show whether harmful movements are still continuing and so require remedial measures: These two kinds of measurements require entirely different accuracies and hence different instruments. There are measurements that show the diurnal and seasonal response of the structure to the climatic environment and whilst the instruments for these may be similar to the ones used to detect continuing movements, the measurement locations and the frequency of readings will be different.

If measurements are seen as an aid to decision-making on remedial action, all the possible outcomes of the measurements should be considered in advance and the line of action to follow from each one should be stated in principle. If, when this has been done, the line of action is the same for all the possible outcomes, then the measurements will not serve their perceived purpose, but will only postpone the decision.

3.3.2 Defects to be Measured

The most obvious structural defects in an old building are the existing cracks in the fabric. They are easily observed, but when measuring them, it is important to ascertain the 'structural gap' as opposed to the appearance on the surface, because past repairs may make a 50 mm gap appear as a 5 mm crack. Using a good quality ruler, it is fairly easy to estimate the width of cracks with an accuracy of ±0.15 mm, which is good enough. Alternatively there are crack width gauges: pieces of clear plastic with printed lines of different width, to compare the cracks with.

From the point of view of the diagnosis of an existing situation the most important measurements are those of differences of levels, which may be evidence of past differential settlement, and tilts of vertical features. These latter may be consequences of settlement, but they could indicate an inherent tendency of the structure to 'spread out' or 'drift'.

More often than not it needs to be established whether or not observed existing defects are still deteriorating, whether due to the original cause or due to some recent change in environment; alternatively, some 'engineering intervention' is envisaged and it is desired to monitor the influence, if any, of this. In these instances one needs to measure the changes of levels and tilts as well as any changes in crack widths. These measurements of ongoing movements require different techniques from those appropriate to measurement of existing defects, caused in the past.

Finally there is a phenomenon which is increasingly being blamed for deterioration of old buildings: vibrations. These may be generated by outside sources, such as traffic, or by internal functions, such as ringing of bells in churches, or they might even be imagined rather than real, due to a high noise level. Measurements of frequency distributions can help to indicate the source and possible means of reducing the effects, but only in rare instances can a measured amplitude be used to prove that damage is being caused.

3.3.3 Measurement of Past Defects

For deformations which have occurred between the original date of construction and the time of measurement, only moderate accuracy can be achieved, whatever methods are employed. This is due to the inherent dimensional inaccuracies of the original construction. For this reason, fairly crude methods of measurements suffice: an ordinary auto-levelling instrument will do, but it should have the best possible telescope to assist reading in poor light. As one may have to take levels in triforium galleries in churches and in other hard-to-get-at places, the instrument should be as light and as compact as possible. When taking and recording levels it should be remembered that any one section of the building which was built as an entity in one period would have been finished to within ± 6–12 mm on horizontal features such as string courses. Anything built 20–40 years or more afterwards may however have been affected by intervening differential settlement.

For plumbing, simple plumb lines or theodolite-plumbing will be chosen on the basis of which is the easier to do, remembering that plumb lines are not easily fixed to great heights, and that they require a calm day to be of any use externally. Where there is more than one storey, and particularly where the work above a floor is of a different vintage from that below, fairly accurate (± 25 mm) correlation of pier section shapes are however necessary to ascertain possible eccentricities.

Ordinary rulers, divided in millimetres suffice for measuring offsets of plumb lines and steel or glass-fibre/plastic tapes suffice for length measurements. The measured deviations from level and plumb should be indicated, possibly to exaggerated scale, on drawings which also show cracks and other defects (see Fig. 3.1). This 'description' of the structure as existing in reality (as opposed to the idealized representations sometimes found in works on architectural history) can then form the basis for a structural analysis, if such is deemed desirable (see Sec. 3.4).

Rectified photographs of walls (in which true verticals and horizontals are recorded as such, without any 'inward tilting uprights') can be an effective way of identifying in-plane distortions of the walls. Photogrammetry can be used to find distortions normal to the plane of the wall, but this is a far more involved process, and hence much more expensive.

3.3.4 Measurement of Ongoing Movements

It will, at times, be desirable to know whether cracks or deformations are still 'live', or, having resulted from some cause in the past, are now static and therefore require nothing but 'cosmetic' treatment. (This knowledge can also be useful in counteracting any 'panic', when it shows that there is no runaway

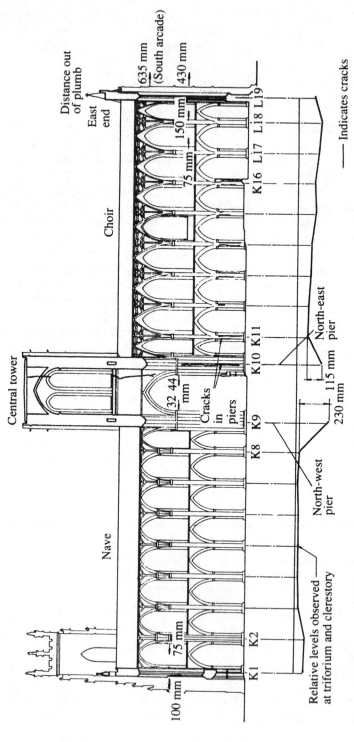

Figure 3.1 Example of drawing, showing measured and observed defects (*from Dowrick and Beckmann, 1971*).

movement going on.) Likewise, one may have carried out structural repairs, which are compromises for economic and other reasons and would like to know whether they have been effective in arresting movements. In this instance one is dealing with slow movements (250 mm settlement in 500 years averages 0.5 mm per year) and to get meaningful answers within a reasonable time, one must employ measuring methods of high accuracy.

Level measurements should preferably be based on a deep datum benchmark which can be assumed to be unaffected by movement of the structure or the soil underneath it and likewise be independent of structures in the vicinity. At York Minster, such a benchmark took the form of a steel tube capped with a stainless steel dome, with its lower end grouted into the sandstone at the bottom of a bore-hole, sufficiently deep to ensure that no movement was likely at that depth. The lining tube of the bore-hole was left in place but terminated clear of the grout so that it could settle with the soil without dragging down the inner tube which forms the bench mark. The space between the two tubes was filled with a rust-inhibiting compound.

The levelling instrument should be fitted with a tilting plane-parallel glass disc in front of the lens making it capable of reading to an accuracy of 0.2 mm or perhaps even 0.05 mm. A self-levelling 'automatic' instrument may have advantages because it is less likely to drift out of adjustment than very accurate bubble levels with their opposed-screw adjustment. The levelling points should be either grouted-in sockets into which a removable ball-headed bolt can be fitted for supporting a precision levelling staff (see Fig. 3.2), or be short lengths of non-tarnishing metal scale, permanently plugged and screwed to the fabric, preferably at re-entrant corners, where they are less likely to be damaged.

When checking variations in out-of-plumbness with the accuracy required, plumb bobs may have to be furnished with vanes and suspended in buckets of oil or water to dampen oscillations, and measurements must be taken with a micrometer between the plumb wire and round-headed metal studs permanently fixed in the masonry. Whilst cheaper in prime cost, this method is expensive in labour when a number of points have to be checked and optical plumbing should be considered.

Optical instruments are available which will read to an accuracy of 1 mm out-of-plumb in a 100 m height (see Fig. 3.3), but for the smaller heights in question here, the accuracy will be affected by the precision with which one can focus on the target. High level targets should be fixed by stout three-legged brackets out of reach of ordinary maintenance ladders. They should have graduated scales printed on glass and be capable of illumination from the ground. The low level bulls-eye target should be set below floor level, preferably in a pocket in a mass of concrete large enough to remain undisturbed by floor renovation, etc. For ease of operation it is worth setting it to coincide to the centre of the high level target at the beginning of the measurements so that no zero corrections need to be made to future readings.

Figure 3.2 Grout-in levelling socket and ball-head bolt (BRE pattern).

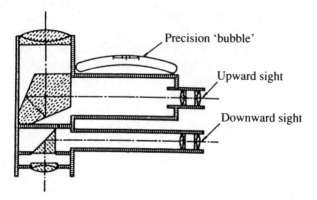

Figure 3.3 Principle of optical plumb (*from Beckmann, 1972*).

Lasers are nowadays commonly used in the construction industry for level-ling and plumbing. They have very high accuracy of aim, but the finite width of the laser beam of light means that they may not provide the reading accuracy required for our purposes. There may also be a possible hazard with older instru-ments if measurements have to be taken whilst the building is open to the public (eye damage). There is however rapid developments going on in this field and the latest information should be obtained prior to specifying instrumentation.

An alternative way of checking changes in slope would be to employ preci-sion spirit levels with graduated tilting screws such as used occasionally in mechanical engineering. These have plane machined bases which have to be placed on special machined reference plates fixed on brackets. This method will, however, only measure the change in slope at each reference plate position and the change in shape of the member will have to be reconstructed by geometric integration, thus losing some degree of accuracy.

The same applies to the more sophisticated electrolytic level, which is a small sealed glass phial with two pairs of electrodes, partially submerged in a conduct-ing liquid, and which works by recording the change in resistance, as tilting makes the liquid rise on one set of electrodes and fall on the other. This instrument is only some 50 mm long, so it is easy to fix directly to the structural member. A number of these can be permanently connected, by means of slim (3–4 mm diameter) multi-core cables, to a central computer, which converts the changes in resistance to angles of tilt and can print these out at set time intervals (alternatively, each electrolevel can be pro-vided with a 'tail' with a multi-pin connector to allow periodic reading with a roving hand-held 'black box'). These devices are extremely accurate, but very expensive and they may require some calibration with mechanical instruments. One example of their use is the monitoring of the movements of the Leaning Tower of Pisa.

Similar instruments have been used for measuring movements in soil due to excavations, etc. They rely on a 'torpedo' moving on a track installed in a pile or in a borehole lining tube. For use in a building, a similar track would have to be permanently fixed to the parts of the structure for which one wanted to monitor the variations in tilt. The advantage of these instruments is that they allow continuous monitoring of the tilt over the total height of the column or wall. They thus offer a more accurate 'integration' of tilts into overall out-of-plumb at any point of the height, such as may be desired if checking bowing of a heavily loaded pillar.

Linear dimensions can be checked with Invar tapes which are practically unaffected by temperature variation. It may be possible, if they are permanently installed on pulleys with tensioning weights, to read to an accuracy of say 0.2 mm with a vernier device. However, if they have to be taken down and re-hung every time, the inaccuracy of the measurements is likely to double. For permanently installed Invar tapes, improved accuracy can be obtained by connecting one end of the tape, running over a pulley, to a linearly variable displacement transducer (LVDT) (see below under crack measurements). This will also allow remote reading, a great advantage if one is checking the spreading of a vault or dome 20 m above the floor.

Invar rods are used in geodetic surveying to measure lengths of base lines with very great accuracy, but they require special supporting trestles and would therefore be inconvenient to use in buildings, particularly at high levels. Electronic distance meters, which work by measuring the travelling time of light, are extensively used in surveying. Their accuracy is however in the order of ± 1.5 mm at the distances found in buildings. High-precision triangulation can be extremely accurate, but that requires special targets to be fixed to the structure.

The traditional way of checking movements at cracks has been the use of tell-tales which have taken the form of either glass slips cemented on either side of the crack, or specially z-shaped pieces of pipe clay with a waisted portion in the centre. Experience at York Minster indicated that failure of glass tell-tales sometimes did occur where they were glued to the structure and might in some instances have been produced by shrinkage of the polyester putty used to fix them. Even if the cracking of the glass is due to structural movement there is no practical way of ascertaining the magnitude or the direction of any further movement with adequate accuracy. This also applies to the pipe-clay variety.

Sometimes an initial way of qualitatively ascertaining whether a crack is widening is observation of its length. Unless a crack extends from one side of a component to the other, it is virtually impossible for it to get wider without getting longer. Pencil lines across the ends of the crack, with the date written on them form a datum against which a later, longer, crack length, and hence larger width can be judged.

Where the main direction of a suspected movement can be foreseen, one can use a pair of small oblong brass or aluminium plates which are placed parallel to the expected movement and each with one of its long edges touching that of the other. These plates are cemented and/or rawplugged and screwed, one to either side of the crack. If a line is scribed at right angles to the touching long edges, future movement of significant size can be directly measured from the off-set in the scribed line.

A development of this principle is marketed under the trade name of 'Avongard'. The device consists of two strips of acrylic plastic, one overlapping the other. One has a black two-dimensional millimetre grid engraved on a white background, the other, which overlaps it, has a pair of cross-hair lines engraved in red on the clear plastic (see Fig. 3.4). This allows movement across the crack and along it to be

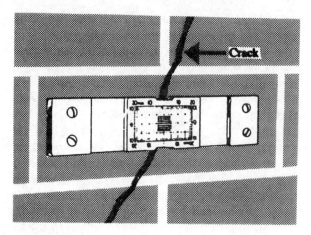

Figure 3.4 'Avongard' crack-measuring device (*from 'Avongard' catalogue*).

estimated to ±0.15 mm, *provided* that the slips are firmly attached to either side of the crack; this means that they have to be screwed to plugs which grip the masonry *behind* any plaster. The screw holes in the plastic strips, as supplied, may have to be enlarged to achieve this. One recent version of this device has a small protruding 'tooth' on each strip, arranged so that measurements of variations in their distance can be taken with a calliper gauge. There are also variants, which allow movements normal to the surface to be measured.

At York Minster a demountable mechanical strain gauge was used to measure the variation in distance between the centre punch marks of three stainless steel discs glued to the structure in the pattern of an equilateral right-angled triangle, so that two studs were situated on one side of the crack and one on the other. Measurements were taken of the variations in distance between the single stud on the one side of the crack and the two on the other; a simple calculation then produced the movement parallel to, and at right angles to the crack. The instrument (trade name: 'Demec') (see Fig. 3.5) is fairly expensive, but easy to use. The gauge length usually used is 100 mm and a setting template, which is part of the outfit, ensures that the studs are set at the correct centres. (On very irregular masonry, a 200 mm gauge length may be preferable.)

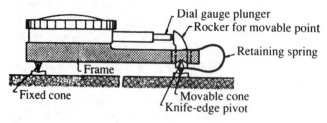

Figure 3.5 Principle of Demec gauge (*from Beckmann, 1972*).

The Demec gauge has a dial, each division of which corresponds to a movement of about 0.0015 mm (varying depending on each instrument's individual calibration). This accuracy, which stems from the original laboratory purpose of the instrument, is excessive for measurements of building movements, but makes reliable readings fairly easy to take, even for the inexperienced. The instrument is also fairly robust and as all readings are referred to an Invar reference bar with factory-punched centres, any accidental off-setting of the zero reading on the dial is of no consequence.

A Precision calliper gauge may be used across two screws in lieu of the Demec gauge but it is slower and more dependent on the operator's skill, particularly in installing the screws with their shanks parallel to each other and perpendicular to a common surface. Such calliper gauges can be read to an accuracy of ±0.05 mm, which is a quite adequate precision, but the reliability of the readings depends on the skill of the operator in applying the gauge parallel to the line connecting the centres of the screws.

A rather less accurate instrument is the magnifying glass with a graduated reticule commonly sold as a 'crack microscope'. The reticule is available divided into tenths of a millimetre, and provided that initially a sharp pencil line is drawn on the plaster (or on the smooth stone) around the circular base of the instrument and the instrument is always applied concentrically with the marked circle, a fair degree of accuracy can be achieved.

The use of crack measuring instruments, so far described, requires close access. This can make readings very laborious, if ladders have to be moved from location to location. If crack locations cannot be reached by ordinary extending ladders, then repeated measurements become impractical. A range of electronic devices have therefore been developed to allow remote readings to be made whenever desired. The most common of these devices is the LVDT in which the variation of the distance between its fixing pins causes a proportional variation of the inductance of a coil in the turnbuckle-like body of the instrument. This variation of inductance is then in turn, read by a 'black box' situated at a convenient position and connected to the LVDT by a slim electric cable. Accuracies of ±0.1 mm are easily obtained and the black box can be set to read several transducers at predetermined intervals and to record the readings automatically. The latest version, allows the 'black box' to be removable, being connected to all the transducers by a multi-pin plug and socket. The standard LVDT is about 150 mm long and its body about 15 mm in diameter. A further development uses linear potentiometers (variable resistors). These have a body about 50 mm long and 5 mm square and are therefore unobtrusive.

These and similar electronic devices have to be installed and wired up by specially skilled engineers and technicians. Whilst very knowledgeable about their gadgets, some of them are however woefully ignorant of building construction and the behaviour of building structures. It is therefore essential, when such measurements are commissioned, to specify very clearly the reason why the measurements are wanted, the positions where they are wanted, the accuracy required and the frequency of the readings as well as the expected duration of the monitoring programme.

Seasonal fluctuations of climatic environment are likely to cause some structural movement, e.g. masonry will expand with rising temperature and in consequence crack widths will tend to reduce. It is therefore important that each set of crack width readings is accompanied by a reading of ambient air temperature near the crack location and preferably also a reading of relative humidity. In the case of electronic measurements it may be tempting to attach a thermocouple, so as to measure the actual temperature of the masonry. It should however be remembered that it is the temperature in the main bulk of the masonry that determines the over-all movement—not the surface temperature. (Conversely, strong variations in surface temperature, such as those produced by prolonged solar radiation, can produce variations in the crack width on the surface, which do not reflect the movement of the wall as a whole.)

There is always a possibility that movement will occur as a result of remedial measures being carried out. Such movements may be indicative of certain unforeseen reactions to the building operations of the fabric or the foundations and should therefore be checked in a similar way to that described above. They do, however, tend to be more rapid and in consequence the instrumentation can be slightly cruder than that described above for checking ongoing movement, but it must nevertheless be considerably more accurate than that which suffices to establish what movement has taken place in the past.

All measurements of continuing movements require permanent reference points, if not permanently installed instruments. This can lead to difficulties in measuring in buildings which are used and occupied: In public buildings with not enough supervision, reference markers within reach may end up as 'souvenirs'; and in smaller churches the misguided zeal of voluntary cleaning ladies has been known to lead to removal of Demec studs.

Another difficulty may be caused by tourist popularity: In famous cathedrals and similar buildings it is virtually impossible to take levels in the daytime during the tourist season, e.g. between March and November in the northern hemisphere.

3.3.5 Recording and Interpretation of Measurements

It is essential that readings are recorded in logbooks, kept specifically for the purpose, in such a way that a newcomer can, by studying the introductory notes in each book and the setting out of the results, immediately see what the figures represent. It is tempting to let 'George' maintain his own system of booking, because 'he knows what he is doing', but George may walk under a bus tomorrow, and unless his hieroglyphics are decipherable, all his good work will be wasted.

Similarly, 'hard copies' must be produced of all electronic readings and, like the logbooks, must be kept by the person responsible for the building. There are admittedly advantages in electronically logged data, in that they can be sampled for different purposes by the computer, but computer systems are not 'virus'-immune and data storage discs, whether magnetic or CD-ROMs, are not accident-proof. Ideally, therefore, duplicate copies should be produced of all readings, whether on paper or on 'disc', and one copy kept in a safe in a different place from the 'working copy'.

Even the best kept note-book in the world is however useless on its own, when it comes to interpreting the readings. Basically, any series of readings should initially be plotted against time. Only this will enable one to distinguish between the three features, which here, for convenience, will be called 'noise', 'seasonal fluctuations' and 'movement'.

'Noise' is the fine waviness of the plot caused by inherently unavoidable random inaccuracies in instrument reading. 'Seasonal fluctuations' are the movements about a steady average position caused by changes in temperature and humidity.

If one were to take hourly readings with a sufficiently accurate instrument, one would get a plot showing the diurnal variations. If the readings were continued over a year one would see the diurnal variations superimposed on annual movements, but over several years there would be no total movement in one direction or another from 'seasonal fluctuations'.

'Noise' is common to all observations where an instrument displays readings, which are more accurate than the measurements being taken. 'Seasonal fluctuation' is mainly observed in the widths of superstructure cracks. The range of both these confusing phenomena must be established before any significant 'movement' can be deduced from the readings. It follows therefore, that when it is intended to measure the effects of a proposed 'engineering intervention' or of construction works on an adjacent site, it may be necessary to commence the readings 12–18 months in advance, in order to establish the base pattern of seasonal fluctuations.

To establish seasonal variations, it will usually be enough if readings are taken once a month at the same time of the day. In the case of suspected ongoing, but slow, movements, readings should initially be taken monthly, unless it is clear that, say, six-monthly or annual readings will suffice. Again, the results should be plotted as soon as possible, as the graphs can indicate if and when the frequency of the readings can be reduced, *or if they may have to be increased* (see Fig. 3.6).

Where there is concern about the possible effects of a potentially damaging external disturbance, such as a deep excavation on an adjacent site, it may be advisable to take readings every week, or even daily and the same applies in the case of engineering 'surgery' to the structure itself. In these circumstances the readings must be plotted immediately, so that the shapes of the graphs can be examined for early warning signs of impending serious damage and avoiding action, if necessary, can be taken in time.

Once the external disturbance has ceased or the internal structural works have been completed, it should be possible to gradually reduce the frequency of reading, e.g. monthly for six months, then half-yearly for 2 years, annually for another 2 years and then every 5 years or 10 years, or a similar interval that corresponds to the inspection/maintenance cycle of the building. Plotting may not need to be instantaneous for such readings, but it must nevertheless not be neglected.

The graphs should be kept up to date and they should be reviewed at regular intervals, preferably at the time of the quinquennial (or similar regular) inspection, in order that any incipient failure of remedial works or any effects of changes in internal or external environment can be detected in time.

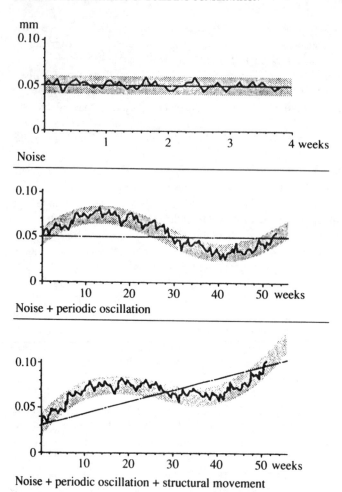

Figure 3.6 Plots showing 'noise', 'seasonal fluctuations' and 'movement' (*from Beckmann, 1972*).

Instead of plotting the movement of individual points against time, it is some-times useful to plot a profile of a series of points, along or across the building, each time readings are taken. Such a profile will show the movements of the points relative to each other, something that is particularly useful in case of settlements. For instance, if all the levels of the measuring points on the building appear to be going down relative to the benchmark, it means that either the building is settling uniformly, or that the benchmark is rising and in neither case is the superstructure likely to suffer. (Such uniform settlement was observed at York Minster between 1984 and 1989; it was considered to be caused by a general consolidation of the ground due to fluctuations in the groundwater regime (see Fig. 3.7).)

Settlement profiles and similar graphical representations can be helpful in arriv-ing at an overall picture of the various movements of the entire fabric of the building.

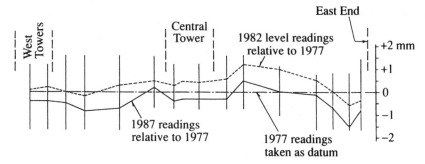

Figure 3.7 Plot of settlements along south arcade of York Minster 1977–87.

Monitoring, which concentrates on one part of a building, can be misleading, unless correlated with measurements of other parts and with relevant elements of the history of construction and subsequent occurrences. At St Paul's Cathedral in London, the out-of-plumb of the South Transept had been measured in 1933. It was measured again in 1970 and was found to have increased by about 12 mm. At the same time, some cracks, hitherto not included in the monitoring, were noted. The rate of movement, 12 mm in 37 years, was out of all proportion to the movements recorded elsewhere in the cathedral over that period. These discoveries caused great concern that settlement of the transept had started again, possibly as a consequence of traffic vibrations. Another theory was that wartime destruction of the buildings to the South of the cathedral, which had not been rebuilt, had exposed the South wall of the transept to greater temperature fluctuations than before, due to solar gain.

In 2001, closer engineering analysis of all the measurements from 1933 showed near-perfect proportionality between the out-of-plumb of the transept piers, that had been measured then, and the differential settlements of the dome piers relative to these transept piers. As the settlement of the dome piers had been found not to have increased, settlement as the cause of the increased out-of-plumb and of the cracking could be ruled out.

Looking for an alternative 'one-off' cause, it was then remembered that in 1941 a heavy bomb had exploded in the North Transept, causing great damage there and elsewhere in the cathedral. This would have created an intense pressure rise inside the building. Whilst most of the damage elsewhere would have been due to the following suction wave, the South Transept wall, being right opposite the exploding bomb, would have been severely hit by the pressure before it had vented itself by blowing out the windows. This could easily account for the 'step increase' in the lean of the transept wall. Subsequent thermography and vibration measurements eliminated thermal effects and vibrations as significant influences.

It was the engineering correlation of the 1933 readings of out-of-plumb with the settlements, together with the evaluation of a historic event 60 years ago that showed that the sudden increase in out-of-plumb was a one-off occurrence. This demonstrated that extensive (and expensive) structural remedial works were not required.

3.4 STRUCTURAL ASSESSMENT

Appraisal of an existing structure requires an 'analysis' in the broadest sense of the word, in which structural calculations form only one part of the whole process. Drawings, defining the structure, are fundamental to the whole process, but there are other basic considerations to be made before launching into numerical calculations.

3.4.1 Consolidation of Information

Once all the available information has been gathered, the structural engineer should consolidate it, by hand, on to a set of drawings of the structure, 'as existing'. Doing it by hand, as opposed to by computer, reduces hardware distractions to broken pencil points and allows the brain—the most sophisticated 'software' known—to concentrate on the structure. The process of working out precisely what has to be drawn will usually answer a lot of questions.

The aim of these drawings is to show the structural essence. Anything that is not part of this is best omitted. Many structural 'near misses', requiring last-minute redesign, can be traced to the failure of the engineer to draw a crucial cross-section. It follows that elevations and sections are usually far more informative than plans. The elevations and sections may have to be constructed 'from scratch', as they often have to be taken where previous records have not included them. This is particularly so for internal structural walls, extending through several storeys, where one needs to draw the elevation in order to see how beam bearings and door openings relate, so as to (hopefully) find paths for the loads to follow on their way to the foundations. Features on the far side of the wall, such as intersecting walls and floors, should, of course, be included in broken lines. External walls are usually best shown from the inside, leaving out bay windows, etc., that might make the wall look more solid than it is.

A line and level survey is often necessary. It will be carried out by specialist professionals and will nowadays be presented as CAD drawings. These will contain far more information than the engineer needs. The brief for the surveyors should however include sections and elevations, so that the Engineer can derive his drawings from theirs by 'stripping away' the non-structural information and adding the structure, where needed. This will save much time and effort in producing the structural appraisal drawings.

Even if the project only includes part of a building, it is essential for all but the most minor works to draw the whole structure. For instance, when dealing with a terraced house, remember that the structure is the terrace—not just the individual house.

3.4.2 Overall Stability of the Structure

Modern buildings of any size usually have structures, that are separate from the 'architectural' elements, such as cladding. Their overall stability is usually provided by one or more 'cores' of reinforced concrete walls, or diagonally braced bays. The other parts of the building are secured to the core(s) by 'engineered' connections.

In cathedrals and large churches, transverse stability is often inherent in each bay, due to the interactions of arches, vaults and buttresses. This is, however, not always the case, longitudinally.

In ordinary traditional buildings, overall stability was provided by the walls and floors acting together as a three-dimensional cellular structure. Wall openings for doors and windows, and floor openings for stairs were limited in size, so there was no need to consider stability explicitly. The connections between the elements came about 'naturally' without any need for 'engineering design', and this may disguise their structural significance.

When buildings are to be adapted for further use, there is a temptation to remove walls and make new openings. This will tend to reduce the stability. Any refurbishment proposals must therefore start with an assessment of the stability of the existing structure, pinpointing any possible shortfall, followed by identification of where improvement is required, both in the 'as existing' state and after adaptation.

3.4.3 Structural Analysis for Assessment of Safety

In the construction industry today, structural analysis is normally used for calculating sizes of members in structures to be built or to check stresses against 'permissible' stresses in existing structures. In existing buildings the sizes are given and, when considering the safety of the structure, the stresses should be evaluated from first principles because the 'ultimate strengths' or the safe working stresses, laid down in codes and regulations, are often determined from considerations other than those of safety, e.g. limiting initial and long-term deflection, preventing cracking, etc. In an old building one should accept the existing deflections, and there is therefore no need to limit one's stresses below what is necessary to maintain an adequate factor of safety and to limit further deterioration.

One should also approach with an open mind the problem of what constitutes failure and how it happens. In masonry, failure under compressive load will take the form of shearing, bursting, or splitting, but very rarely crushing of the stone. In fact, stresses in traditional masonry structures, including most Gothic, slender, flying buttresses, tend to be very low compared with the crushing strength of the stone.

As far as timber roof structures are concerned, one need rarely concern oneself with the members, as it will invariably be the joints that are the weak points (with the exception of cases with local rot or beetle attack away from the joints).

Very few ancient buildings have 'text-book' structural frames and an accurate elastic analysis can therefore become complicated, or even impractical, but for purposes of safety one can resort to fairly simple limit state assumptions. What is important, however, is to analyse how the structure *actually* works, as opposed to how one would assume that it should work, according to conventional structural design theory (*or* how it might have been intended to work). The emphasis in the initial stages should be on simplicity. If equilibrium can be established between the external forces acting on a part of a structure and the internal forces in the structure, and if the internal forces can be mobilized without requiring an excessive stress level anywhere, then the structure will be safe, regardless of whether the assumed stress distribution is in fact the correct elastic one.

The partial safety factors, prescribed by codes of practice for limit state design of new structures, are usually not appropriate for these calculations and the best estimates of the *actual* loads should be used *unfactored*, together with estimated *actual, unfactored*, strengths to give approximate factors of safety.

Lintel and post structures will usually not pose any special problems; timber trusses can, however, be structurally highly indeterminate with a large number of 'redundant' members, but provided the joints are capable of a certain amount of 'give', one can assume that all parts which are capable of contributing to the load-carrying capacity of the truss will in fact do so.

The analysis of a simple masonry arch can often be carried out as a simple thrust-line construction. Due to the usually low stress levels, all that is required for safety is to find a thrust-line which will balance the forces on the arch whilst remaining within a distance away from the boundary of the arch equal to between 5 and 10 per cent of its thickness. The thrust-line is constructed (see Fig. 3.8) as an inverted string polygon to the forces and the horizontal thrust exerted by the arch can be read off the diagram (see also Sec. 4.6.1).

One can extend this exercise to construct the thrust-line extended from the arch down through pier and buttress right to the foundation. Here the position of the thrust-line can become much more critical; few soil strata will allow much unevenness of ground pressure before they suffer differential settlement which will result in the foundation rotating together with whatever it supports.

The East end of York Minster is a prime example of this. Here the eccentricity of the load on the buttresses and the end wall foundations led to a rotation which in turn aggravated the eccentricity to the extent that the east wall was about 0.63 m out-of-plumb in 1967 (see Fig. 3.9).

If the results of the initial calculations are not clear-cut, a more elaborate and/or refined analysis may be required. The Institution of Structural Engineers' report: 'Appraisal of Existing Structures' contains useful guidance on the principles to be applied and the procedure to be followed (see also Sec. 2.4).

The ISE report states by inference that, where the actual strength of the structural material has been ascertained by sampling and testing, the use of a *lower* factor of safety *in the calculations* will still result in an *undiminished real margin of safety* (see also Sec. 2.5).

As will be discussed under Sec. 3.5, the number of samples that have to be tested to establish the actual strength with adequate confidence may, however, be more than is practical.

Refinement of the mathematical model of the structure, used in the calculations, may however still be worthwhile, particularly if it originally was rather simplified; for example, tie beams in timber roof trusses often have considerable bending strength, which is ignored in simple truss calculations; inclusion in the calculation of the bending stiffness may show that most of the roof load is comfortably carried in bending by the tie beam and only a minor proportion by truss action.

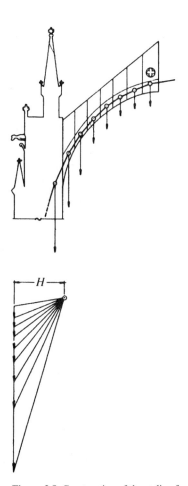

Figure 3.8 Construction of thrust-line for flying buttress (*from Beckmann, 1972*).

There is among engineers today a pre-occupation with the calculation aspect of structural design and this is often reflected in their attitude to structural appraisal. It should, however, not be forgotten that if a building structure is carrying its own weight and the loads imposed on it *without any signs of structural distress or significant deterioration of the materials of the structure* and if it has done so for 50 years or more, it can in all probability go on carrying those loads for the foreseeable future *as long as the loads are not increased and the structure is not tampered with or suffers serious deterioration from rot or rust, etc.*

3.4.4 Structural Analysis to Explain Observed Defects

It is of paramount importance for the correct design of remedial works, that the mechanism causing the defects is clearly identified; for example: are cracks in masonry arches caused by differential settlement of foundations or by spreading

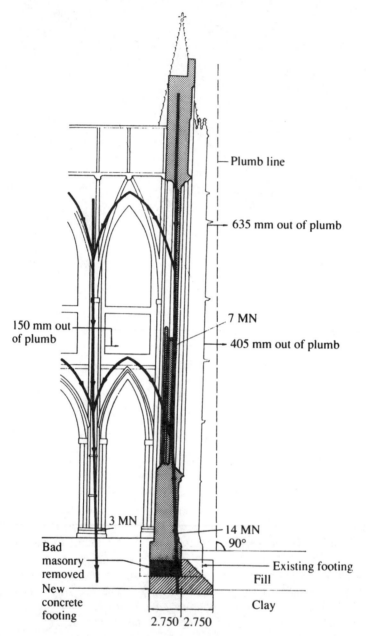

Figure 3.9 Thrust-lines for the east end of the arcade and the buttress of the choir end wall of York Minster (*from Dowrick and Beckmann, 1971*).

of abutments?—and if the latter, is it an inevitable consequence of the original structural geometry, or is it due to rotating pier foundations or removal of buttressing after completion of the original construction? Structural analysis of various kinds can help to answer such questions.

It has in the past been postulated that if a masonry arch stands up for 5 minutes after the scaffolding has been removed, it will stand up for 500 years. This statement assumes that the supports of the arch in question are immovable in space and time and that foundation settlements are negligible. In the real world, foundations do settle appreciably and differential settlements do give rise to cracks, and so does the spreading of abutments of arches due to the horizontal thrust. But how does one distinguish between the effects of the two different causes?

There is a kind of structural analysis which can be done 'by inspection' which will indicate the likely causes of certain cracks. For instance, if there are cracks on the inside of an arch near the crown, in a symmetrical pattern, the supports are likely to have moved out horizontally. On the other hand, if there are 'anti-symmetrical' cracks at the quarter points, i.e. on the inside of the arch at one quarter point, and at the outside at the other, the cause is likely to be differential settlement of the supports. Similarly, if a wall leaning outwards is concave on the side of the lean, the cause is likely to be horizontal thrust from roof or vault, whilst if it is convex on the side of the lean the most likely cause is an eccentric foundation which started rotating during construction.

Sometimes old buildings display quite severe deformations for which no cause can be found in the present structure. It is occasionally possible to solve these problems through 'analysis by eye': St David's Cathedral in Dyfed, Wales, has a very low-pitched roof 'truss' which is supported on walls which lean out alarmingly. There is, however, no indication of movement between the main horizontal beam of the roof 'truss' and the walls. As a truss-beam of this kind does not exert any horizontal thrust, it must mean that the lean of the walls was caused by an earlier roof in conjunction with weak foundations and that with the replacement of the roof, the cause of lean was eliminated and the movement arrested. Inspection of the east and west elevations of the tower does in fact reveal remains of drip mouldings corresponding to an earlier roof with about 60° slope; such a roof, if inadequately tied, would in fact have thrusted outwards at the eaves.

When one has to deal with multi-bay arcades surmounted by triforium arches, and possibly clerestories above, where the spandrel infills over the arches contribute significantly to the strength and stiffness of the whole system, it usually becomes too difficult to make adequate deductions directly from observations and even the determination of force paths and stress patterns may defy simple hand calculations.

One might at this point consider the use of structural models and, on the face of it, photo-elasticity is an attractive technique as it is possible to load a three-dimensional model in a heated chamber and let the model cool off whilst loaded. When this has been done, all strains are 'frozen' into the model which can then be cut into slices which can be analysed on a photo-elastic bench. However elegant this method is in principle, the making and loading of the model is a difficult and expensive task, and the actual photo-elastic analysis is a fairly lengthy mathematical procedure if quantitative results are required. If one could make a model of a fairly soft rubber and coat it with a brittle lacquer, one could get a very easy visual indication of regions of high stress, but there are great difficulties in simulating gravity loads on a model like this.

On balance, therefore, the best choice is a mathematical model which can be analysed with the aid of a computer. In some early particular examples of arches with spandrel panels, the mathematical models, that were used, conveniently took the shapes of rigid frames with finite elements to model the spandrel panels. Today, a full finite element model would probably be preferred. An objection to this method is that the loads and the material properties are not known with an accuracy, commensurate with that inherent in a computer analysis. It must however be remembered that, in this instance, the purpose of the computer is not to produce extreme numerical accuracy but to enable one to manipulate a large mass of data. Stress levels are usually not crucial, so reasonably good approximate figures for loads will be adequate for the purpose. Likewise, as one does not rely on the analysis to produce exact magnitudes of individual deformations, as long as the pattern is in proportion, the elastic properties of the material need not be specified with very great precision, as long as one can be reasonably confident that no part of the structure is loaded far beyond the linear elastic range of the material.

The Central Tower complex at York Minster was treated this way as follows. An analysis was carried out for a structure consisting of the Transept arcade and the Central Tower pier walls in that plane, modelled as a plane frame with finite elements to represent the spandrels. Gravity and wind loads were first considered and it was found that the calculated movements and stresses did not explain the deformations and cracks observed in the fabric (see Fig. 3.10).

A similar analysis was then carried out assuming that the Central Tower had settled relative to the rest (as was in fact observed). Here again it was found that the calculated results of such a settlement were in conflict with the observed and measured distortions of the fabric, *not only in magnitude* but, and this was significant, *in the directions of the movements* (see Fig. 3.11).

The engineers were then faced with the puzzle of how the transept piers had come to lean the way they did. It was only the realization of the existence of an early English tower, built about 1250 and a casual reference to its fall in 1407, recorded in the minutes of the chapter meetings, and pointed out by the local historian, Dr Gee, that presented the solution to this riddle.

There was still the remaining problem of the large cracks in the tower spandrels under the lantern and whilst analysis had shown small tensions under the windows due to gravity loads, the correlation was not satisfactory. Re-examination of the records of the measurements of the existing defects revealed that the northwest pier showed signs of having settled about 150 mm more than the remaining three at the level of the lowest pier pedestal course, whilst the difference at gallery level was only 75 mm.

A space frame model was now analysed on the computer assuming these differential settlements to have taken place within the tower itself and this time the magnitude and disposition of tensile forces checked with the actual position and extent of the cracks (see Figs 3.12 and 3.13).

This example (described in detail by Dowrick and Beckmann, 1971) shows how knowledge of the history of original construction and subsequent alterations/rebuilding can help diagnostic structural analysis. It is clearly unrealistic to

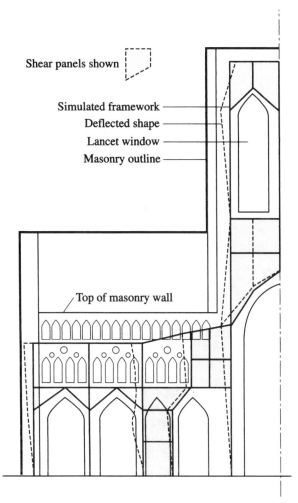

Shear panels shown

Simulated framework

Deflected shape

Lancet window

Masonry outline

Top of masonry wall

Figure 3.10 Mathematical model and calculated deflections under gravity loads of the Central Tower and South Transept of York Minster (*from Dowrick and Beckmann, 1971*).

expect structural engineers to decipher minutes of chapter sessions written in mediaeval Latin, but they should engage in dialogue with people, such as historians, who can.

3.4.5 Analysis of Dynamic Effects

Vibrations are often blamed for structural deterioration and much effort is sometimes expended in trying to prove or disprove a claim by measurements of the level of vibration. There is, however, very little information on the actual deleterious effect on a traditional structure, that a measured, absolute, level of vibration causes. Hence the best that can be done is usually just a comparison with levels

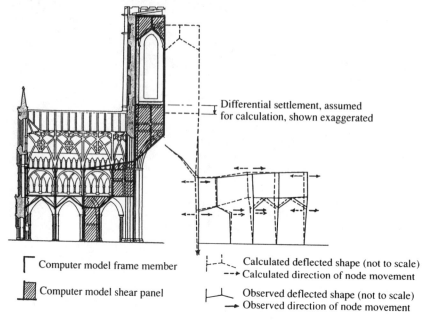

Figure 3.11 Calculated v. observed deflections under assumed relative downward displacement of the Central Tower of York Minster (*after Dowrick and Beckmann, 1971*).

of vibration which have been shown by experience to be harmless, or harmful as the case may have been, in another building of more or less similar construction.

It is sometimes possible, by assessing or measuring the inherent vibration characteristics of a structure, to predict whether that structure will be particularly susceptible to vibrations of known characteristics from a new source.

Before describing methods of assessment of vibration characteristics, it is worth recalling certain basic tenets:

1. When a body vibrates, it undergoes periodic movements in alternating, opposite directions. It is therefore subjected to reciprocating accelerations, which result in the body having periodic reciprocating velocities.
2. The frequency of a vibration is the number per second of complete movements (from position at rest to one extreme, through rest to the other extreme and back to rest). For a given value of maximum displacement, the velocities and accelerations increase with the frequency of the movements.
3. A structure, like any other supported body, will have certain natural frequencies, at which it will vibrate, if it is given an impulse and then left alone. A simple example is the string of a musical instrument which, when plucked, will vibrate with a frequency that depends on the length, the tension and the mass of the string.
4. All practical structures have, in addition to their natural frequency, another property which is of importance to their response to vibration; this is their

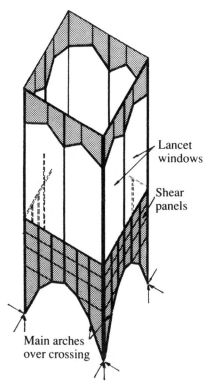

Lancet
windows

Shear
panels

Main arches
over crossing

Figure 3.12 Three-dimensional mathematical model of the Central Tower of York Minster (*from Dowrick and Beckmann, 1971*).

internal damping. This is a resistance to movement, which for most building structures has its origin in the internal friction in the building material. It is quantified by the 'logarithmic decrement', which is the natural logarithm of the ratio between two successive amplitudes of oscillation.

When an ordinary structure is subjected to a fluctuating force, the effect can be described in terms of the static value of that force, multiplied by a 'Dynamic Amplification Factor'; this factor depends on the ratio between the forcing frequency and the natural frequency and on the damping.

If r is the ratio between the forcing frequency and the natural frequency, and if δ is the logarithmic decrement (due to the damping) then the dynamic amplification factor (DAF) is:

$$\mathrm{DAF} = \frac{1}{\sqrt{(1 - r^2)^2 + (\delta r/\pi)^2}}$$

The DAF can be seen to attain its maximum value when the forcing frequency approaches the natural frequency i.e. $r = 1$; this is called resonance, and

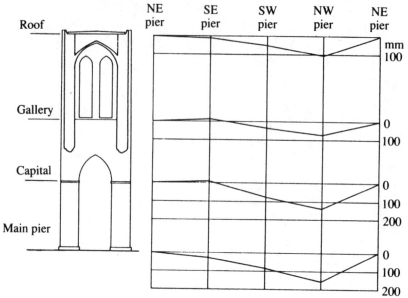

Differential levels of the four corners of the central tower measured at various heights

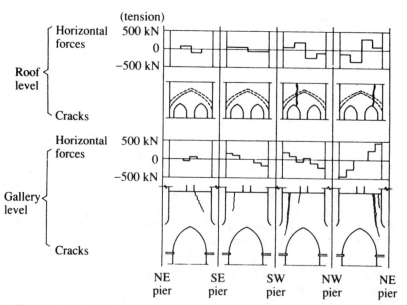

Observed cracks and horizontal forces calculated assuming the differential settlements shown above.

Figure 3.13 Results of three-dimensional analysis of 'internal' differential settlements of the Central Tower of York Minster, compared with the crack pattern (*from Dowrick and Beckmann, 1971*).

the amplification then depends solely on the damping. It is therefore often as important to establish frequencies as to measure amplitudes.

The frequency of the disturbing, imposed, vibration may have to be measured or may be provided by the manufacturer of the 'machinery', e.g. the frequency of the swing of a bell will be given by the bell foundry. The natural frequency of the structure or of a structural element may, however, in some cases be ascertained by calculation.

For simple structural elements, ready-made formulas for the natural frequency exist. For example, for a simply supported beam of constant cross-section, the natural frequency is:

$$f = \frac{\pi}{2l^2}\sqrt{\frac{EI}{m}}$$

where f is the natural frequency in hertz (cycles/second), l the span of the beam in metres, E the Young's modulus of the beam material in N/m^2, I the moment of inertia of the beam section in m^4 and m the mass per unit length of the beam in kg/m.

This formula can be simplified to:

$$f = \frac{18}{\sqrt{y}}$$

where y is the static elastic deflection in millimetre under dead load.

Structures with more complex geometries or with cross-sections that vary along their length require more elaborate calculations. For those of fairly slender configuration, such as church spires, it may be possible to use a mathematical model of 'sticks and lumps' in which the total height is divided up into segments, each of which is represented as a corresponding height of vertical beam with a constant stiffness equal to that of the mean stiffness along that segment and the masses of the segments are considered as concentrated in 'lumps' at the junctions between the segments. Such mathematical models are amenable to relatively simple computer analyses. For greater refinement, or for more complicated geometries, finite element analysis may be possible.

It must, however, be remembered that calculation of natural frequency requires knowledge of Young's modulus of the material and this is rarely available with any great accuracy. The damping inherent in traditional forms of construction is an even more elusive quantity and this being so, very elaborate mathematical analysis aimed at establishing the dynamic amplification factor, is hardly ever appropriate. Measurements may however be of some assistance.

When calculations are considered appropriate, they should be carried out with two sets of values of these parameters; the values being chosen so as to represent both limits of the probable range, so that the effect on the results of these uncertainties can be evaluated.

3.4.6 Analysis of Effects of Remedial Works and Additions

When carrying out remedial works or structural strengthening, one sometimes has to use modern materials to act together with the old, and even if traditional materials are used, they will not have been under load for centuries, as the existing ones have. Nevertheless, the two will have to work together, compositely.

Whilst calculations of ultimate load-carrying capacity can usually be based simply on the sum of the individual strengths, checks must be made to ensure that one component will not be overstressed, before its companion carries its share of the load. Such calculations must therefore allow for the probably different elastic moduli of new and old construction.

It may also be necessary to check the effects of different values of coefficients of thermal and moisture movement, creep and shrinkage. The exact values of these materials' properties are unlikely to be ascertainable, so values representing reasonable upper and lower boundaries of the range should be used to evaluate the influence of these uncertainties.

Whilst a measure of variability in the calculation parameters has to be allowed for in this way, it is however imperative that these compatibility calculations are based on a mathematical model that will simulate the real behaviour of the original structure together with the repair or strengthening.

It is frequently necessary to temporarily remove a portion of a structure in order to carry out repairs or strengthening; for example if the foot of a timber truss is badly decayed, a length of rafter and tie beam will have to be cut out and new timbers spliced on. In such a situation, the problem of the stability of the structure during the operation has to be addressed and in all but trivial cases, such as superficial replacement of small areas of facing masonry on massive walls or piers, calculations should be carried out to ascertain if temporary support need to be provided and if so, what form it should take and to what degree it may need to be pre-loaded so as to produce active rather than passive support. In such calculations allowance must be made for possible joint movement between scaffold members as well as for their elastic deformations.

3.5 TESTING

It is natural to want to defer a conclusion, even an interim one, of a structural appraisal until tests have been carried out and the results have been studied. Most tests are, however, expensive and all of them are disruptive to the occupants of the building. *Before* commissioning a regime of tests, one should therefore write two draft conclusions; one for the case that the test results are *favourable* and one for the case that they are *unfavourable*. If the two courses of action, arising from the two conclusions are similar, the tests will serve no useful purpose and should not be undertaken. Similarly, no tests should usually be commissioned, if their cost is likely to approach or exceed the cost of reasonable remedial work. An exception to this rule arises when the testing is likely to show, in a way that calculations cannot, that an existing structure of historical interest will be adequate for future use.

There are several types of test that can aid the assessment of the load-carrying capacity of a structure.

3.5.1 Static Tests as an Aid to Structural Appraisal

1. *Laboratory tests on samples of materials taken from the structure*: These will indicate the strength of the material in the sample, either directly as in a tensile test on a metal sample, or indirectly, as in a chemical analysis of a mortar sample. The results of the test are then used in calculations to demonstrate the load-carrying capacity.

2. In-situ *non-destructive (or minimal damage) tests to estimate materials' strengths*: These must always be calibrated against a number of laboratory sample tests, but can then be used in calculations to demonstrate load-carrying capacity.

3. In-situ *non-destructive flaw and void detection techniques*: For masonry, these are commonly based on pulsed radar reflection and absorption. An alternative is thermography. For cast iron, ultrasonic techniques are used. These techniques are useful when the load-carrying capacity may depend on the presence and extent of voids and imperfections, e.g. badly constructed rubble fill inside a masonry wall. They rely on highly sophisticated equipment that needs to be operated by specialist technicians who are very skilled in the use of the instruments, but sometimes ignorant of building construction. Instrument read-outs must therefore, in the case of masonry, be calibrated against a number of visual examinations of opened-up holes. For cast iron, mechanical calibration tests are required.

4. *Strength tests on samples of construction or assemblies*: These may be appropriate if the load-carrying capacity cannot be satisfactorily derived by calculations, based on the results of laboratory tests. For instance, in the case of masonry made with materials no longer in current use or in the case of a connection of unusual three-dimensional geometry in a cast iron structure, it may be possible to carry out load tests on a few samples of construction, either in-situ, as for wall panels, or in a structural testing laboratory, as for the iron connection. These tests will cause at least some disturbance to the fabric of the building (and to its occupants!), because of the need to isolate or remove the sample, even if they are not pursued to destruction.

5. *Load tests on whole structures or large parts of structures, such as a substantial area of a floor*: These tests are aimed at direct demonstration of capacity to carry a pre-determined load (such as dictated by an intended use) and can be used if, after inspection local materials tests and calculations, the load-carrying capacity is still in doubt.

The tests outlined above, under 1–4 will be described in more detail in the chapters dealing with the various forms of construction, e.g. masonry, iron, etc. They are also, together with load testing, 5, discussed in the ISE report 'Appraisal of Existing Structures' (pp. 29–34 and Appendix 7 of the 1996 edition) and in Sec. 3.5.3.

3.5.2 Tests of Dynamic Effects

Whilst normally not directly assisting the assessment of load-carrying capacity, tests to ascertain the response of a structure to dynamic influences can sometimes be helpful when decisions have to be made on, for instance, whether bell ringing is likely to cause serious structural deterioration in an old church tower. The instruments commonly used for vibration measurements are electronic transducers which work in a similar way to gramophone pick-ups and microphones. Accelerometers and 'velocimeters' have an internally suspended mass which remains stationary while the body of the transducer follows the movements of the structure to which it is attached. Accelerometers are usually fairly compact, light and easy to fix in any position; they are most sensitive to higher frequencies, but will function down to 5–10 Hz (1 Hz = 1 cycle/s). Velocitymeters are larger and heavier and may require a surface to stand on. Displacement transducers are useful for measuring relative movements between structures, or between two parts of a structure separated by a crack.

The output from the transducers is fed through an amplifier and can either be directly read as a mean value within each frequency band (e.g. 'octave') or it can be recorded on magnetic tape or light-sensitive paper for subsequent analysis. The recording technique is particularly appropriate when studying the effects of transients such as sonic booms, blasting, etc.

The measurement techniques are thus well established but, as mentioned before, this is not the case when it comes to assessing the significance of a measured level of vibration. Very little information exists about which levels of vibration create damage in different types of buildings. What little, that does exist, is mostly derived from experiments on domestic buildings of modern design and of modern materials. It is therefore rarely applicable to large historic buildings, which often tend to have heavy masses and be built of materials with different susceptibilities to those which form the basis of the existing official criteria.

On the other hand, seventeenth- and eighteenth-century buildings of originally domestic nature often have plaster ceilings which can be considerably less robust than modern ones. Having measurements taken of the vibration levels in an old building therefore rarely helps, when it comes to deciding whether, for instance, restrictions on traffic are necessary to protect it.

Even less information exists on the cumulative effect of vibrations. In some old buildings constructed of two skins of ashlar with a rubble-and-mortar filling, there is a possibility that certain vibrations may gradually shake out mortar particles from between the rubble and thus create voids in the masonry. Proper experiments, to find out more about this, are likely to be expensive and time-consuming. They would entail the building of two identical specimens using similar materials and techniques to the original construction, then subjecting one to a certain level of vibration continuously for, say, 5 years and subsequently demolish both and examine their interiors for any difference.

A test of similar nature was set up by the Royal Aircraft Establishment to investigate the possible effects on stained glass windows from sonic booms produced during the proving flights of 'Concorde'. The results from the 'Synthetic Boom' tests were however not generally applicable, nor entirely conclusive. The measurements of the response of structures to the actual booms during over-flights did, however, show fairly clearly that the effects of sonic booms from commercial aircraft were not significant compared with the natural loads such as gravity, snow and wind, which act on the main structural elements. Conversely, one of the large stained glass windows on York Minster was so delicate that an accelerometer, attached to one of its panes registered the barking of a dog in the street below.

Measurements can sometimes help in ascertaining whether a general state of dilapidation is caused or aggravated by vibrations from a particular existing source or whether a new source of vibration is likely to cause damage. An example of this is bellringing in church towers, particularly when the bells are rung in the English fashion where the bells swing through 360°. The inertia forces from these low-frequency oscillations (0.3–1.0 Hz) are large, and if in resonance with one or more of the natural frequencies of a slender tower might well give rise to overall damage, and alternatively, if the tower is stubby, with a high own frequency, the bell frame may have worked loose in the tower and be acting as a battering ram on the walls.

For slender towers and spires, it may be possible to deduce the basic modes of oscillation and to calculate the natural frequencies, but for squat towers with complex buttressing from other parts (e.g. nave and transepts) this is not practicable. For structures of this kind, the dynamic characteristics can be ascertained by attaching accelerometers and/or velocimeters to, the structure and then subject it to a controlled amount of dynamic input.

In the case of a church tower, one would at first swing one or two bells for a short time and then stop. The instruments would then measure the natural frequency and the damping at low levels of energy input. This would establish whether simple resonance would be a problem. One could then proceed to swinging more bells and, by recording the response, find out if the natural frequency and/or the damping changed for the higher energy input. This would indicate that mortar joints might be opening up, due to the swaying movements of the tower, caused by the inertia forces from the swinging bells.

Displacement transducers with one end fixed to the bell frame and the other to the tower wall could also be installed to measure any relative movement between the bell frame and the tower.

Apart from any possible 'battering ram' effects from the bell frame, the most important fact to be ascertained is whether there are any possible resonances between the natural frequencies of the structure and the basic frequency or any of the harmonics of the driving vibration. The amount of damping in the structure at the various frequencies may also be significant and this can, as mentioned, also be derived from the measurements.

Similar measurement techniques are also sometimes used to assess the sensitivity of a structure to earthquakes and typhoons. As in this case there is no source of vibration, such as a swinging bell, available in the building, a vibration generator with eccentrically mounted weights on a rotating shaft is bolted to the structure, usually at roof level. The instrumentation for measuring the response of the structure is similar to that used for the bell towers.

3.5.3 Advantages and Limitations of Testing

There are some general aspects, which have a bearing on practically all types of tests. Tests can produce improved knowledge of physical and chemical properties of materials and thus reduce the uncertainties in the structural appraisal and hence allow the use of lesser factors of ignorance (= safety factors) in the calculations. It must, however, be remembered that each test result only applies to the sample or the part of the structure that has been tested. The properties of all materials and components are not uniform, but vary from place to place within the structure. One cannot therefore know whether a sample is representative of the best, the worst or the average of the structure as a whole.

The way that a test result is used in calculations of load-carrying capacity will depend on what the tested sample represents, i.e. a different safety factor would be used in the calculation, depending on whether the test result represented the mean or the worst of the material in the structure. It is currently a common practice to base structural calculations on a value for the strength that will be equalled or exceeded by 95 per cent of the material. This 'characteristic value' of the strength depends on the mean value and on the variability: the greater the variability is, the lower the characteristic value will be, relative to the mean value.

If a finite number of samples are tested, the mean value can be simply calculated. It is similarly possible to calculate a measure of the variability of the results, known as the 'standard deviation' (this is a sort of average of the differences between the individual test results and the mean value). The statistical branch of mathematics then enables the characteristic value to be calculated as:

> The mean value, minus the standard deviation multiplied by a coefficient, κ, that depends on the number of samples that have been tested; κ is the larger, the fewer tests that have been performed.

$$\text{Mean value } \bar{x} = \frac{x_1 + x_2 + x_3 + \cdots + x_n}{n}$$

where $x_1, x_2, x_3 \ldots x_n$ are the individual test results and n is the number of samples.

$$\text{The standard deviation, } s = \frac{\sqrt{[(x_1 - \bar{x})^2 + (x_2 - \bar{x})^2 + \cdots + (x_n - \bar{x})^2]}}{n - 1}$$

The characteristic (95 per cent confidence) value $= \bar{x} - \kappa s$, where the coefficient κ has the following approximate values:

n = number of samples:	3.0	4.0	6.0	10.0	15.0
κ:	3.37	2.63	2.18	1.92	1.7

It follows that for materials with high variability, such as old masonry and early cast iron, where the standard deviation may be as much as 25 per cent of the mean value, it may be necessary to test more than 10 samples, in order to arrive at a characteristic value that is more advantageous than the approximate values quoted in contemporary references.

The advantages that can be gained for the structural appraisal by testing of the materials may thus be limited by the difficulty in extracting sufficient samples. Before any testing is put in hand, it should therefore be carefully considered whether it will be practicable to extract the number of samples that may be required in order to obtain a more favourable assessment.

In some circumstances, it may however be possible to locate areas of the weakest material by means of non-destructive techniques. In this case, a practical *minimum* strength (which can be used with a similar, or possibly lower, safety factor to that applying to the characteristic strength) can be assessed from tests on a small number of samples, e.g. 3.

3.5.4 Practical Points on Commissioning Tests

No tests should be commissioned, unless their results will help to answer an essential question that cannot be answered any other way, nor if their cost is likely to exceed that of reasonable remedial works (see introduction to Sec. 3.5).

The reasons for this are that, as mentioned in Sec. 3.5.3 above, tests are usually expensive and often disruptive. Laboratory tests require taking of samples, which can entail removal of finishes and cutting operations in several locations, with the attendant nuisances of dust and noise. When the rough samples have been taken, they have to be carefully cut to the correct size and shape for the testing machine. Laboratory technicians have to have special skills and are therefore well paid and their equipment is expensive to maintain and calibrate. In-situ tests may call for even higher skills of the testing technician, whilst requiring the same provisions of access, etc. as for taking of samples.

Load testing of whole structures or structural elements, e.g. an area of floor, has significant limitations. Its results only apply to the area tested and it cannot demonstrate the actual factor of safety, if it is to remain non-destructive. The temporary scaffolding, etc. necessary for safety reasons during the test and the process of applying the load are expensive; load testing of columns may cause progressive collapse, unless very elaborate arrangements are made.

Test house technicians are usually good at their job, but sometimes ignorant about structures. It should therefore be clarified, well in advance, whether the proposed method of testing will provide results that are relevant and useful in the process of appraisal (it is no use testing the compressive strength of bricks, if the mortar properties cannot be ascertained).

Test reports have an air of authority because they appear to be dealing with immutable physical facts. Unfortunately, some test houses have a habit of including in their reports some comparisons between the results of the tests and current standard specifications for new materials. This can be distinctly unhelpful, especially when one is trying to demonstrate, to an inexperienced building control officer, the adequacy of an old structure that needs appraisal from first principles. Test reports should therefore be specified to be strictly factual, leaving any interpretation to the person responsible for the structural appraisal.

3.6 REPORTING

Every structural appraisal must have a report as its end result; that is what the client pays for. The report forms the basis for, and should justify, the decisions which determine the future of the building.

3.6.1 Terms of Reference, Limitations, Facts v. Deductions

When writing reports on structural appraisals of buildings, one should consider all the potential readers, e.g. owners, tenants, financiers, etc. and their, probably limited, familiarity with technical issues and technical jargon. At the same time, one should also bear in mind that there may be technically knowledgeable readers of the report, and they need to be given the technical evidence, on which the conclusions are based. This may call for the technical details to be included, but as an appendix.

It is important, at the beginning of the report, to define one's responsibility by quoting who authorized the appraisal, the terms of reference and the physical, administrative, financial and possibly political limitations on the inspection and subsequent investigations.

Whilst field notes may include speculation as to causes of defects, one should when writing the report keep observed facts and hypothetical interpretations clearly distinguishable. One way of achieving this is by allocating separate sections of the report to each.

The adoption of a set format for reports may be helpful in achieving these aims. The format, set out below, has been found useful by experience. As it stands, it applies to a full appraisal and items (8) and (9) will rarely be relevant to an initial report, dealing only with the inspection. The remainder will, however, be generally applicable.

3.6.2 Report Format

1. *Synopsis*: One or at the most two pages of plain, succinct language, summarizing the gist of the contents for the really busy 'top man', who will make his decision on the basis of this.
2. *List of contents*: For the top man's assistant, so that he can quickly find the section dealing with a particular aspect of the report, when the top man asks about it.
3. *Terms of reference*: Who instructed you to do what, and when? Refer to the date and the sender of the letter and quote the part of the letter, which describes the expected extent of your work. Also quote any subsequent modification of your

brief with date of letter or phone call. This is useful to both parties; it helps you to make sure that you are answering all the right questions and it makes it clear to the reader what you were asked to do—or not do.

4. *Documents examined*: List the documents made available to you and by whom (e.g. client's or solicitor's letters, reports from others sent by client, etc.) and describe those which you would have liked to have but did not get (e.g. structural drawings) with an indication of why not. This will indicate to the reader from what basis of information your work was carried out and may justify why certain questions remain unanswered. It is also a check list to ensure that you have looked hard enough for information.

5. *Description of the structure*: Just because you have bumped your head against every collar beam in the roof space, it does not follow that your reader has a clear picture of what you were investigating Make it brief and pictorial, include a potted history of its construction and use and refer to diagrammatic drawings and perhaps photographs in the report. Avoid the 'catalogue' style. At this point it is worth remembering that drawings, made for the engineer's or the constructor's use, hardly ever make good illustrations. When they are reduced to A4 page size, fine lines break up and lettering becomes too small to read. The drawings mentioned in Sec. 3.4.1 are usually rough and ready but serve the engineer's purpose; they will however look too scruffy in a report. Report drawings of the structure should show the essentials to the non-technical reader and should do just that. They should be purpose-made, with bold lines and explanatory 'labels'. Isometric presentation may be helpful to the 'lay' readers.

6. *Inspections*: Who looked at what, how and when? It may be important to make it clear that an adequate number of inspections were made by adequately qualified people. The dates may be important in developing situations. Any limitations on the effectiveness of the inspections should be indicated, e.g. 'inadequate opening up, hence only superficial visual examination possible'. Follow with a concise description of what was seen on each occasion. If necessary refer to an appendix with a schedule of individual observations, but do not discuss the inference or significance of the observations in this section. Well-presented drawings (i.e. 'tidied-up' field sketches) of observed crucial details can help to describe what was seen on the inspection(s). These will be examined by the 'technical readers', so the 'tidying-up' must not 'improve' the evidence. Clear, well-printed, photographs (with film originals) are often the best evidence.

7. *Additional (verbal) information (if any)*: What you were told, by whom, when, and how credible does the information appear to you.

8. *Sampling and testing (if any)*: Who took how many samples of what? When and where were they taken? To whom were they sent? When, and for what kind of testing? Why? What were the results? It may be important that samples are taken in the presence of representatives of other parties, hence name both the 'taker' and the 'watcher', if there was one. It is important that mechanical tests and chemical analyses are carried out by reputable independent laboratories, so name them. Give a brief précis of the results and refer to an appendix with photocopies of the laboratory certificates. Do not discuss yet.

9. Calculation checks (if any): What relevant and necessary information was obtained from drawings, specification or original design calculation? Which necessary parameters have had to be assumed or deduced? What type of calculation was carried out and which criteria have the results been judged against? Summarize the findings and refer to an appendix containing the actual calculations. Do not draw any conclusions yet.

There may be occasions when it is advantageous to present the activities covered by items 6, 7, 8 and 9 in chronological order, even if it means that each item may be split into two or more parts, e.g. 'Initial inspection'—'General design check'—'Detailed inspection of . . . '—'Detailed stress analysis of . . . '—'Sampling and testing of . . . ', etc.

Up to this point the report has described facts which can be, or could have been, checked. What follows is interpretation and/or opinions which can more easily be challenged. The report will be safer to use and easier to defend in a possible conflict situation if 'facts' and 'opinions' are clearly separated.

10. *Discussion of evidence*: As the heading indicates, this is the item under which one discusses the importance of each of the findings described under 6, 7 and 8 and, particularly, their relevance to the questions raised in the brief.
11. *Conclusions*: If items 9 and 10 have been well written, a reasonably intelligent reader will by now have arrived at the correct conclusions unaided. This item need therefore only contain a brief paragraph stating in plain language the (by now hopefully obvious) answer to each of the questions in the terms of reference.
12. *Recommendations*: A brief description of the course(s) of action which are recommended as the logical follow-up to the conclusions. Stick to the broad principles and describe them in clear, plain language, intelligible to the lay reader (e.g. the top man's assistant). Details, even if commissioned in the terms of reference, should be banished to an appendix with a reference. Remember also that in the present litigious climate, it may be prudent not to proffer 'remedies' of defects, but merely 'mitigations'.

The many references to 'plain language' are prompted by the fact that one may have become so used to dealing in one's daily life with technical minutiae, which are most conveniently described in jargon, that one forgets that the end result is rather simple bits of hardware, which the layman can understand if they are described in plain words. It takes effort, but it can be done and, in the case of reports, it is worth it.

APPENDIX TO CHAPTER 3

Permits for Investigations, etc.

On most new building projects, structural engineers do not have to obtain planning consents for construction work. Indeed, clients sometimes do not appoint an engineer until they have secured planning permission. The situation is different with existing buildings, and totally different with buildings of special architectural or historic interest. The architect may still take the lead in obtaining all the additional consents necessary, but the engineer needs to be aware of the context in which he is working. In some very sensitive buildings, even investigation work needs consent.

In Britain, to determine what consents will be needed, one needs to know:

- If it is a Scheduled Ancient Monument.
- If it is in a river or watercourse under the control of the Environment Agency and if there may be bats' resting places in the building.
- If it is a Listed Building, and if so what grade.
- If it is an unlisted building but in a Conservation Area.
- If it is a building belonging to the Crown.
- If it is a Church of England Building used for worship.

Scheduled ancient monuments are usually unoccupied buildings, ruins or sites with buried archaeological remains. Almost any investigation work, even digging a trial pit, will require consent. Scheduled monument consents are issued centrally by a government department (at the time of writing, that of Culture Media and Sport) acting on the advice of English Heritage. The local authority may be consulted, but they are not part of the approval process.

Working in or adjacent to rivers and watercourses can involve the Environment Agency, who may be interested in whether the works may cause pollution, or may damage wild life (flora or fauna) on the foreshore. If work is likely to disturb bats' roosts or resting places, a licence from the appropriate department (currently Environment, Farming and Rural Affairs) is required.

In listed buildings, normal maintenance does not require what is known as *Listed Building Consent*, but more disruptive investigations do if they involve operations that would not form part of a maintenance programme. Alteration work *of any sort* needs listed building consent. It is important to recognize that the description in 'the listing' does not define what is of interest in the building. The whole building and its curtilage are listed, and what is mentioned in the listing is merely there to describe the building for identification purposes. There is no such thing as a 'listed façade'.

If a building is not listed, consent will not be needed for investigations, *but* if it is in a *conservation area* and work involves partial or complete demolition, consent is needed. This is irrespective of the contribution that the building, or the part of it, to be demolished makes to the quality of the conservation area. Under the current guidelines, it is *the merits of the replacement building* that determine whether consent to demolish will be granted.

Listed Building Consent and *Conservation Area Consent for Demolition* are normally granted by local authorities in consultation with English Heritage. Planning permission is a separate consent, which is wholly determined by the local authority.

The situation is slightly different if the building is the property of the Crown. Because the Monarch is superior, in the hierarchy, to Parliament, under whose legislation consents are granted, the Monarch cannot, by virtue of his or her status, ask for consent. However, for all practical purposes the process is similar to that for non-crown buildings. This is because of an understanding reached in the 1980s and set out in Department of the Environment Circular No. 18 of 1984. Under '18/84' the Crown presents proposals more or less as if it were a normal building owner. So if the Crown site is a scheduled ancient monument, the Crown will apply to the DCMS for 'scheduled monument clearance', and if it is a listed building the application will go to the local authority. The local authority will consider the case as any other, but without the public consultation part. Applications are determined in the normal way, and the Crown normally accepts the decision, though it is not legally bound to do so.

Church of England buildings consecrated for worship (but not other church buildings) are subject to local authority control where the work requires planning permission. All but the most minor work to Parish Churches (whether they are listed or not) requires a 'Faculty', which is a legal consent issued by the Chancellor of the Diocese. Faculty jurisdiction extends far further the listed building legislation; for example, a faculty is required to introduce a carpet into an unlisted building!

This church system was established centuries ago, before listing of buildings was introduced, and because of this, and its more extensive coverage, it has been allowed to continue in parallel with the secular system. To avoid duplication of applications, the granting of a faculty takes the place of listed building consent (but not of planning permission). Diocesan chancellors are lawyers, so they will seek technical and conservation advice from others on the merits of proposals. Where a building is listed they will normally want to see the result of a consultation with English Heritage and the local authority, as well as with the relevant amenity society. They will also be guided by the recommendations of Diocesan Advisory Committees, made up of professionals, who vet proposals before they go to the chancellor. There is a similar process for cathedrals, except that consent is ultimately granted by the Cathedrals Fabric Commission for England, advised by the Fabric Advisory Committee for the cathedral in question. This church system is generally referred to as the 'Ecclesiastical Exemption'.

It follows from the above that not only the architect but also the structural engineer need to know the status of the building, in order to plan the approach to the project. Those granting listed building consent are also increasingly seeking reassurance that the proposal before them can actually be realized in practice. Engineering proposals demonstrating the need for the work and the feasibility of doing it are therefore often required when consent is being sought. Similarly, work to important historic buildings is often funded by grants from bodies such as the Heritage Lottery Fund, grant-giving trusts and from generous individuals. They also need early reassurance that the work will receive consent and that the proposals have been well thought through.

This means that the structural engineer, together with the architect, must not only devise wise proposals based on a sensible amount of investigation, but they also need to put this over in a convincing manner to assessors who, whilst experienced in building and conservation matters, may not be engineers. It is not sufficient for the engineer just to put in enough for another engineer to understand. What goes into this sort of report needs careful thought so that the argument is made concisely, and that the reasons why things are proposed (or not proposed) are clearly stated.

Regardless of the status of the building, *consent for investigations* must be obtained from the appropriate owner (who may not be the client) and it must be agreed what is to be done about making good afterwards.

Other countries have their own frameworks of regulations for the purpose of conserving historic buildings. Few will have the ramifications, arising from the ancient status of the Monarch and the Church of England. Against that, their political government structure may introduce other complications, such as a dichotomy between federal and state administration and funding, and the degree of regulation will be affected by the extent to which government interference with property owners' freedom is accepted by the population.

FOUR

MASONRY

Bootham Bar, York: Mediaeval limestone masonry, largely built as two skins of dressed stone with a core of rubble and mortar (see Sec. 4.4.3). Being originally built as a fortification, the heavy construction of this gate-tower has ensured that it is still intact despite the quite severe weathering of the surface.

Masonry, whether of natural stone or of fired clay bricks, is the most durable of the traditional forms of construction. It is also one of the most versatile, in terms of structural form, having been used as walls, pillars, arches, vaults and domes. Each of these has its own particular structural behaviour and is susceptible to its own particular structural weaknesses and potential defects, requiring their own particular remedial measures. Masonry is often exposed to view, so repairs and strengthening measures have to be visually acceptable as well as functionally efficient. Due to the mechanical properties of the units (bricks or blocks of natural stone) and their usually weaker joints, masonry is the form of construction that exhibits the greatest differences in the way it responds to imposed forces of different kinds (e.g. compression or tension). The general strength properties are therefore dealt with first.

4.1 GENERAL STRENGTH PROPERTIES AND CAUSES OF DAMAGE

Masonry is generally relatively strong in compression, has a moderate resistance to sliding along the bed-joints (shear) and is usually weak in tension. Masonry structures are therefore generally designed to work in compression and sometimes in horizontal shear. Damage to masonry structures usually takes the form of tension and/or shear cracks caused by imposed deformations, e.g. differential settlement, or by excessive lateral forces, e.g. from arch thrusts; or alternatively it may suffer from incipient bursting due to buckling of ashlar or expansion of rubble fill in structures that consist of two ashlar faces with a core of rubble and mortar.

All unintended gaps in masonry are usually referred to as 'cracks'. For a proper assessment of causes and significance, one should however distinguish between the following categories: (a) cracks through units, (b) opening of bed-joints and butt joints and (c) 'sliding' along bed-joints. Whilst all three may have structural implications (a) is more likely to lead to a risk to people from falling bits of masonry.

4.1.1 Cracks Due to Imposed Deformations

The cracks caused by imposed deformations are usually not dangerous in themselves, but with the movements caused by temperature fluctuations, debris in the form of sand grains and mortar particles tend to fall into the cracks when they open and prevent them from closing in the second half of the temperature cycle. In this way, a ratchet action develops which will gradually wedge the cracks open to an increasing degree. Similarly, if water can get in, the expansion caused by frost will gradually wedge the cracks wider and wider (see Sec. 4.4.3).

4.1.2 Cracks Due to Excessive Lateral Forces

The cracks caused by excessive lateral forces, in the plane of the wall or pier, are usually accompanied by significant deformations and are a symptom of structural deficiency somewhere or the other and unless the cause is eliminated, the cracking is likely to get progressively worse until such time that instability results. Cracks due to pressure from rubble filling are also a sign of a continuing process, and unless this process is halted, the structure will eventually burst and collapse.

4.1.3 Cracks Due to Eccentric Loading

One type of cracking that sometimes causes unnecessary structural concern is that due to eccentric loading. If the load on a wall or pier is not acting along the central plane, it will subject it to bending and the stress will vary across the thickness of the wall (Fig. 4.1*b*). If the load acts one-sixth of the thickness from the central plane, the stress will vary linearly from nothing at the face furthest from the load to a maximum at the other face (Fig. 4.1*c*). If the load moves outside the 'middle third' of the wall thickness, the conventional theory of bending stresses postulates that tensile stresses should develop near the face away from the load (Fig. 4.1*d*). In coursed masonry, particularly in older work built with ordinary lime mortar, there is negligible tensile capacity across bed-joints, so tensile stresses cannot develop. Instead, the stress drops to zero over part of the section and this may be accompanied by a slight opening of the bed-joint(s). It is sometimes thought that this indicates failure of the wall or pier. This is not so; what is overlooked in this argument is the fact that equilibrium is possible between the load and a triangular stress distribution across the uncracked part of the thickness of the wall (Fig. 4.1*e*). It should however be borne in mind that, once this stress pattern is reached, the compressive stress increases more rapidly with increasing eccentricity. On the compression face of the wall, there may therefore be some slight spalling of arrises along the bed-joint. This is not necessarily a cause for alarm, as long as the peak stress is below the safe limit, and it can be avoided by recessing the pointing.

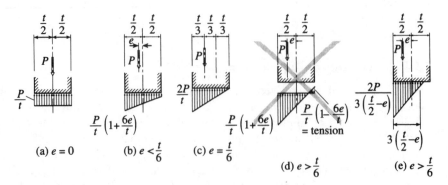

Figure 4.1 Stress distributions due to eccentric loading.

4.1.4 Cracks Due to Embedded Ironwork

In the past, stone masonry was sometimes built with embedded iron cramps or 'staples', which were intended to hold together two adjacent stones in areas where tension might occur, temporarily or permanently.

Similarly, where arch or dome thrust needed to be restrained (see Sec. 4.6), tie rods were provided within the thickness of the masonry. In both cases, the iron components were usually placed in horizontal rebates formed in the top surfaces of the stones and the ends were bent down into vertical holes.

In the case of dry-jointed masonry (see Sec. 4.4.1), rainwater is almost certain to find capillary paths to the iron and in combination with atmospheric oxygen cause rusting of the ironwork. This hazard is aggravated in exposed coastal situations, where salt-laden spray may be carried inland by wind. This is because the chlorides in the salt accelerate the corrosion (cf. the chloride effect on reinforcement corrosion; Sec. 7.3.2).

Embedded iron cramps can, if anything, cause more damage in mortar-jointed masonry than in dry-jointed. This is because the mortar, which originally was alkaline and therefore protected the iron, has been carbonated by carbon dioxide from the atmosphere and hence has lost its protective function, whilst it still confines the corrosion products. As rust takes up between 6 and 10 times the volume of the iron from which it is formed, very great expansive forces are created by the confinement resulting sometimes in spectacular displacements of large stones and/or gaping cracks.

To protect the ironwork against corrosion, it was often wrapped in thin sheet lead. Alternatively, molten lead may have been poured around the iron, fixing it in the holes and rebates at the same time as providing protection. However, the lead wrapping may not have been watertight to begin with or the mortar joints may have weathered with time and allowed lime-laden rainwater to get to the lead and corrode it and form pinholes through which water could get at the iron. In either case, the net result is the same: rust forms and spreads under the lead. The apparent effect is to make the ironwork expand (and, in the process, split the lead wrapping, if any). Whilst this has been known to lift the entire structure above the ironwork, with the bed-joint just above the iron gaping open, it is more common to find that cracks have formed on the face of the stones (usually along straight horizontal lines) and whole slivers of stone may have spalled off. Cracks from this cause can often be identified by the stones below being stained by rusty water leaking out from the iron.

The location of embedded metalwork, even when it is not (yet) causing problems, can usually be found by the use of certain metal detectors. There are, however, some bricks and some sandstones which contain a significant amount of iron; this has been known to 'confuse' the instruments, so that they register metal where none is present and vice versa. They should therefore be 'calibrated' by a small local opening-up.

Similar problems can be caused in steel-framed buildings with masonry cladding (see Sec. 4.8).

4.1.5 Defects in Brickwork Cladding on Reinforced Concrete-framed Buildings

Brickwork cladding to concrete-framed buildings is supported on edge beams, usually one-storey height at a time. As will be discussed in Sec. 7.1.4, concrete, when subjected to a compressive stress, will go on shortening for a considerable time, albeit very slightly, even when that stress is not increased. This is called 'creep'. If the cladding is built tight, between the floor below and the edge beam above, the effect of creep is to make the storey height of the frame too small for the brickwork; the cladding is being 'pinched' and may then buckle.

The problem is sometimes exacerbated by bricks being laid too soon after being taken from the firing kiln. In that situation, they will still take up moisture from the atmosphere and expand after being built in.

4.1.6 Problems with Cavity-ties

Cavity brickwork is found as cladding on multi-storey, framed buildings, as well as in load-bearing walls of dwelling houses.

On older houses, inadequate corrosion protection of the ties, together with the use of 'black ash' mortar in the outer leaf, can lead to the ties rusting. The ties, then used, were cut from 25×4 mm flat steel, twisted in the middle and split at the ends; they introduced a substantial iron component in the bed-joints, and the rusting sometimes cracked the joints.

During the 1960s, ties made of galvanized wire, twisted into a butterfly shape, were introduced. Corrosion can again be a problem with these, but the main danger in this case is that they rust away almost completely, leaving the outer leaf with no connection to the backing structure.

The most common defect with this type of cavity wall is, however, the failure of the bricklayer to put in the required number (usually specified as one for every 45 cm in every third course). Spacings of more than double the specified are not uncommon and have in some cases led to the outer leaf being sucked off a façade during high winds.

4.2 THE NECESSITY OF DIAGNOSIS OF CAUSES OF DEFECTS

It follows from the above that repair or strengthening must be preceded by ascertaining the real cause of the observed defects, and if the cause is still active, it must be eliminated. Otherwise the repairs will be short-lived. There is a parallel in the field of medicine: any treatment must be preceded by a correct diagnosis.

The techniques used for repair or strengthening must depend on whether the cracks are 'live', i.e. still moving, or 'dead', i.e. being due to a cause, which has been eliminated in the past. If the cracks are live, the cause should be treated, if at all possible, and the subsequent repairs to the masonry will be largely cosmetic.

If treatment of the cause is impractical, one may have to consider strengthening the masonry in such a way that it can resist the cause of the cracking.

It may not be practical to strengthen the masonry to resist the cause of past live cracking so that no further cracking takes place at all. In this case, one should aim at strengthening in such a way that subsequent cracking is substantially reduced, if not eliminated. This means a strengthening which is capable of being 'stretched'. Adding tensile capacity has the effect of distributing the cracks so that the individual cracks become fine and hence unobtrusive.

If the cracks are dead one should consider future environmental influences such as temperature and moisture fluctuations, earthquakes and also possible changes in ground conditions, due to activities on adjacent sites, before deciding finally to apply cosmetic repairs. A sixteenth-century mansion had suffered repeated incidences of settlement, separated by long 'static' intervals. The cause was traced to a layer of the subsoil, which softened on the rare occasions when the groundwater rose to particularly high levels. The remedy was a system of land-drains to deal with the peak water levels.

4.3 ASSESSMENT OF STRENGTH

The conservation of a building often requires a change of use or an adaptation that will impose increased loading on parts or the whole of the masonry. In that situation, the strength of the masonry will have to be assessed.

4.3.1 Codes of Practice, Empirical Formulae and Local Experience

Codes of practice are intended for new construction (some of them state this explicitly). The current British masonry codes base their design stresses on a combination of the following: the specified strength of the bricks or stones, usually verified by testing; the strength of the mortar, as implied by the specified mix proportions and sometimes verified by testing; and the control of materials and workmanship, particularly the bond.

This procedure is often unsatisfactory for existing masonry for the following reasons:

(a) Even if the age and the source of the bricks and hence their likely strength were known and the mix proportions of the mortar have been verified by chemical analysis of a small sample, old lime mortars that have gained strength slowly over a long period are sometimes significantly stronger than the new mortars of identical mix composition, on which the code stresses are based.
(b) The quality of the bond may not match the code assumptions (although this does not necessarily mean that it is inadequate).
(c) The code stresses are intended for relatively thin (100–350 mm) walls and therefore incorporate substantial safety factors to allow for possible future uncontrolled cutting of chases, etc.; this is not appropriate for thick walls nor

for buildings where alterations will be subject to proper supervision, as should be the case for any conserved building.

A practical, if conservative, alternative is found in the empirical design stresses, adopted by building inspectors in cities like London. These were based on experience, accumulated in the past, and have been passed down to today's officials.

For one-off buildings of modest size, it may be the best solution to accept such 'traditional' working stresses and deal with local areas of potential overstress by incorporation of suitable spreader beams/padstones or by the provision of new supporting piers, not bonded into the existing masonry, so that they can be removed as part of a future restoration back to the original layout. Alternatively, if it can be shown that higher stresses are already present in (other) parts of masonry of the same quality, however undefined, and that these have not led to structural distress, then these higher stresses should be permitted. This has been successfully argued in a number of cases.

4.3.2 Testing of Material Samples

If an initial visual inspection indicates that the units (bricks or stones) may be stronger than those on which the empirical working stresses, mentioned above, are based, it may, in theory, be possible to extract a suitable number of units (usually 12) for compressive testing, together with a quantity of mortar (about 1 kg) for chemical analysis, so as to justify higher working stresses. Some of the difficulties of this approach are mentioned in (a), (b) and (c) earlier. Further obstacles are:

(d) It may be difficult to extract an adequate number of units, particularly large stones, and similarly a large enough sample of mortar.
(e) It is difficult to take out the test samples without damaging, and hence weakening them, and the extraction may significantly weaken the masonry, particularly in the case of large stones.
(f) Old bricks, even from the same wall, often show such great variation in strength, when tested, that it is difficult to arrive at a significant strength with any confidence.
(g) Façade walls are often constructed with a better quality brick on the face, bonded to a lower quality brick in the backing work, in which case the strength of the *masonry* depends on the two separate strengths together with the mortar quality and the bond in a combination that is impractical to quantify.

If, despite all difficulties, one has succeeded in sampling and testing the units and the mortar, there remains the problem of translating the test results into permissible stresses. The British Standard Code of Practice BS 5628 provides one set of 'design strengths' (to be used with suitable γ factors) for different combinations of unit strength and mortar composition; the Highways Agency's Design Manual for Roads and Bridges BD21/01 includes a graph, which gives more 'favourable' figures for certain types of stone masonry; and the Draft Eurocode DD ENV 1996-1-1 gives yet different formula. The latter is to some extent helpful, as it advises that masonry strength is proportional to unit strength to the power of 0.65, but the mortar strength only raised to the power of 0.25, i.e. mortar strength may not be crucial.

4.3.3 Testing of Masonry Samples

The uncertainties, outlined above, may suggest that it would be better to establish the strength of the masonry directly, by testing samples of masonry, rather than samples of units and mortar. This would require the cutting out and transporting to a structural testing laboratory an undamaged sample of wall.

Such a sample has to be quite big for the following reasons: As explained in Sec. 1.2.2, the plates, that transmit the compressive force of the testing machine to the sample, will exercise a lateral frictional restraint to the ends of the sample, which will enhance the strength near the ends. Such restraint does not exist in the parent wall or pier, except very near the top or bottom. To obtain an unenhanced, and hence realistic, strength from the test, the sample should be at least twice as high as it is wide. Furthermore, as failure of the parent wall may be governed by splitting, parallel to the faces, as well as by vertical cracking at right angle to the faces, the sample should be at least as wide as it is thick. For a 220 mm thick brick wall, this requires a sample to be six courses, i.e. 450 mm high by 220 mm wide. Even using diamond sawing, the cutting out of such a sample is not easy and it would need very elaborate clamping and crating, to ensure that it would not be damaged in transit to the laboratory.

Laboratory testing of masonry samples is therefore in most cases not a practical proposition.

4.3.4 In-situ Strength Testing

In some cases, it may be possible to measure the strength of masonry in situ by isolating a small pillar and loading it with a flat-jack. The method was originally used by P.P. Rossi of Bergamo, who employed a specially developed flat-jack, inserted in a cut-out mortar joint to measure the stress in the stone masonry with a minimum of damage. Flat-jacks are capsules of very ductile metal; commercially available ones usually have a circular plan shape, with a cross-section that is originally dumb-bell shaped. When inflated with hydraulic fluid, similar to that used in car brakes, they can exert very great forces—up to 15 N/mm^2 of their plan area, whilst they expand a small amount at right angle to their plane—usually up to 25 mm.

A procedure, adapting Rossi's principle so as to use a commercially available flat-jack to measure strength, would be as follows: A pillar of the required size and proportion is freed (but not removed) from the wall by means of diamond sawing or 'stitch drilling' (overlapping horizontal diamond cored holes). At the top of the pillar, a slot to accommodate a spreader beam is formed by similar techniques. A spreader beam is bedded against the top of the slot and a spreader plate is bedded on the top of the pillar, with the flat-jack centred between spreader beam and spreader plate, as shown on Fig. 4.2.

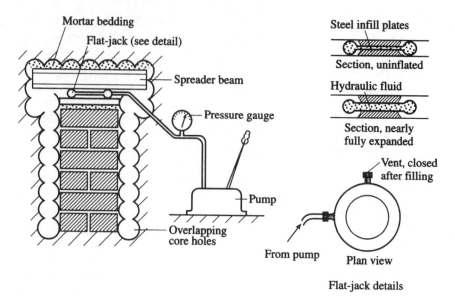

Figure 4.2 In-situ strength testing of masonry wall.

The hydraulic fluid is now pressurized by the pump until the pillar fails. The maximum pressure on the gauge, multiplied with the effective area of the jack, gives the failing load of the pillar. It is obviously essential to ensure that there is sufficient weight of masonry above the upper spreader beam to balance the maximum jacking load.

The deformation characteristics ('elastic modulus' and transverse deformation) can be measured during the early stages of the test by suitable instrumentation, such as 'Demec'-gauges (see Sec. 3.3.4). If such instrumentation were to be installed prior to the cutting free of the pillar, one could even measure the elongation when the load from the wall above is removed by the cutting of the slot. When this elongation is brought back to zero by the inflation of the flat-jack, the force in the jack, divided by the area of the pillar will indicate the initial stress in the wall.

This method, whilst direct, is anything but non-destructive and it disturbs a substantial area of wall. For this reason, the split-cylinder test, developed at Karlsruhe University under the direction of Professor Wenzel, and briefly described below, may provide a more convenient method. It is well researched and a practical application on sandstone masonry is described by Egermann (1997).

4.3.5 Cylinder Splitting Tests and Ultrasonic Techniques

As mentioned in Sec. 1.2.2, a so-called compressive failure of a brittle material is, in reality, a transverse tensile failure. In masonry, this effect is intensified by the tendency of the mortar to have a greater transverse deformation than the units, for the same axial stress.

In the original Karlsruhe research project (SFB 315), horizontal core samples were taken from brickwork and placed horizontally between the plates of the testing machine so that the cylinder was subjected to loading along two diametrically opposite lines, parallel to its axis. This method of testing (known in the field of concrete research as the Brazilian Cylinder Splitting Test) produces an almost uniform tensile stress on the plane joining the two lines of loading. The following relationships were being investigated:

(a) The ratio between the vertical failure stress of small pillars, tested in the conventional way, and the splitting stress of cylinder samples (cores) incorporating a bed-joint at mid-height (at right angle to the direction of loading).
(b) The ratio between the splitting stress of jointed cylinders and that of horizontal cylinders taken entirely through the bricks.
(c) The ratio between the splitting stress of unjointed horizontal cylinders and the 'compressive' failure stress of cores taken vertically through individual bricks (Fig. 4.3).

Experience has indicated that it is possible to assess the strength of masonry, made from prismatic units laid in reasonably regular courses, by taking and testing a relatively small number of horizontal cores. The method may, however, not work on brickwork made from bricks with indentations ('frogs') on the bedding surfaces, as the cores, to be of reasonable size, would 'bite' into the 'frog'.

Another line of research, pursued at Karlsruhe, has attempted to link the ratio of the velocity of ultrasonic pulses through brickwork and the velocity through individual bricks to the ratio between the strength of brickwork and the strength of its parent bricks. The principle of ultrasonic pulse velocity testing is described in Sec. 7.2.4. At the time of writing, the method has, however, not been proven practical for the assessment of the strength of masonry.

Impulse radar can, however, be very effective for testing the integrity of masonry. It can find hidden voids and it can determine the thickness of masonry,

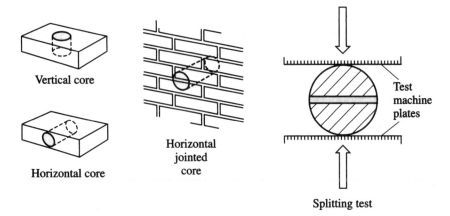

Vertical core

Horizontal core

Horizontal jointed core

Test machine plates

Splitting test

Figure 4.3 Cylinder and core testing.

when only one face is accessible, as for instance in basement walls below ground.

For random block masonry and random rubble walling, there appears at present to be no prospect of anything to replace judgement based on experience.

4.3.6 Stability and Robustness

As mentioned in Sec. 1.5, stability of masonry walls and pillars may be inherent in their height-to-thickness ratio. More often, however, the stability is achieved by a cellular wall layout on plan, that provides mutual support of intersecting walls, or by arches and spandrels providing lateral support at the top of pillars. The cellular wall arrangement does not have to be all in masonry; in some houses, one finds 'trussed partitions', that is, substantial timber structures, that span between party walls (walls, usually at right angles to the façades, which separate adjacent properties) and which support floors. These will often form adequate parts of the cellular layout. 'Braced partitions', being triangulated without necessarily spanning between masonry walls, can likewise contribute to stability. Floors, even of timber, can also be made to act as horizontal diaphragms and in this way brace façade walls to cross-walls, and vice versa.

In domestic buildings (including some palatial ones), lack of stability is occasionally due to poor original design, but it is more often found where a building has, in the past, been converted to other uses, such as shops or restaurants, or where individual rooms have been combined to form a big conference space, or similar. In such conversions, internal walls were often removed and replaced by beams, supported on slender columns, and windows and doors in façades were enlarged to create wide, glazed, shop fronts. This will have robbed the remaining walls of the lateral support that they received from the internal walls, and the restraint from the substituted beams may not be enough to stop walls from leaning if, for instance, subsequent excavation of a service trench were to disturb their (often shallow) foundations (see Figs 1.21a and b).

There are some open-plan, multi-storey buildings, such as textile mills, where the original use precluded a cellular wall layout. In these, stability of the long external walls depends on the restraint from the ends of the floor beams, built into the walls and on the floor structure acting as a horizontal diaphragm, spanning between the end walls. This fact must be borne in mind when considering conversion schemes, which involve replacement of (parts of) floors, and hence temporal or permanent removal of floor structures. In the case of timber floors, the diaphragm action depends on the connections between floor boards and beams.

For most buildings of masonry construction, robustness is allied with stability; there must be enough strength in the connections between the restraining elements and the walls that they restrain, to resist the, normally small, destabilizing forces. Often, traditional details provide enough connection strength to provide a good measure of robustness against the effects of earthquakes and explosions. Mechanical anchorages are desirable, but the mere bearing of floor joists on walls can make a crucial difference. In Budva in Montenegro, after the 1979 earthquake, it was seen that in many houses the entire height of an external wall had fallen, showing the floor joists being parallel to the wall, whilst walls with the floor joists bearing on them were left standing.

It is important to recognize this action and ensure that it is maintained during any refurbishment, or conversion-works. Over-zealous temporary propping of a floor has been known to deprive a wall of the lateral restraint from the floor joists, with subsequent collapse as a result.

Roof rafters and trusses do not in themselves have any stability against tipping over, but rely on the stiffness of the roof-cladding (sometimes suspect in old buildings) or on bracing in the planes of the ridge. Stability and robustness are in this case not only necessities for the roof, as such; most gables rely on the roof for their wind resistance.

4.3.7 Qualitative Assessment of Seismic Resistance

Whilst conventional elastic, dynamic, analyses may predict the behaviour of traditional forms of masonry structures during mild earthquakes, it cannot explain what happens during more severe movements. Many structures, which on the basis of such calculations would be deemed unsafe, have nevertheless survived centuries of earthquakes.

There are several explanations for this; for one thing, once the movements reach such a magnitude that mortar joints begin to open, the damping characteristics increase to a significant, but incalculable degree. Secondly, the structure may, through cracking, transform itself into a hinged assembly of rigid bodies, with walls and pillars rocking about their bases. A structure will generally be able to develop several mechanisms of this kind, provided that the individual elements remain intact and do not suffer internal disintegration. Experience indicates that surviving structures were originally built of good materials used with good workmanship, such as bonding, etc. A qualitative assessment of building techniques and workmanship, if practicable, may therefore form the basis for kinematic analysis of the possible hinged mechanisms and as such be more valuable than strength-testing of materials and masonry samples.

A possible exception to this may be posed by relatively modern masonry structures, built with strong mortars. The seismic resistance of these may depend on the in-plane horizontal shear strength of the masonry at low level, where it is subjected to the vertical load from the structure above. In this case, it may be useful to conduct an in-situ shear test, by cutting a pocket either side of a masonry unit and pushing it sideways by means of a hydraulic jack.

4.4 CHARACTERISTICS AND PROBLEMS OF LOAD-BEARING WALLS AND PILLARS

This is the most widespread application of masonry and the one of which the greatest amount is exposed at eye level, where minute cracks are easily seen and provoke comment, concern or consternation, depending on their size. There are essentially two kinds of wall and pillar: dry-jointed and mortar-bedded. European mediaeval masonry is predominantly mortar-bedded and may either be through-bonded or consist of two leaves of dressed stone (or coursed brick) with a rubble-and-mortar core. A third variety, usually of later origin, has two leaves of dressed stone, fully bonded to a core of bonded brickwork.

4.4.1 Dry-jointed Masonry

In this type of masonry, which is generally only found in buildings from specific cultures and periods, each block of stone was cut and dressed to very precise dimensions and then laid directly on the blocks below and against the adjacent block without any intervening layer of mortar. Some of the final dressing of the upper bedding surfaces of the blocks, already laid, must almost certainly have been done in-situ, in order to produce a plane surface for the next course.

However meticulous the dressing of the stone (it is often impossible to insert a thin knife blade anywhere in the joints), it is inherent in this form of construction that the blocks only bear on each other over a small proportion of the bed-joint surface. In consequence, the bearing stresses at the contact points are very high indeed. This does not create any problems away from the edges; the lateral components of the stresses, as they spread out from the contact points, cancel out each other; there are no stresses to cause splitting (Fig. 4.4*a*).

If, however, a 'hard' point occurs at the edge of the bedding surface, i.e. at the face of the wall, then the inclination of the stresses, as they converge towards the contact point, produces a horizontal force, which is only resisted by the tensile strength of the stone; if that is overcome, a thin sliver is split off the face of the stone (Fig. 4.4*b*).

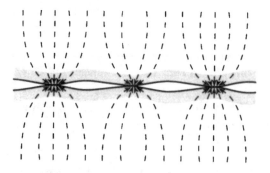

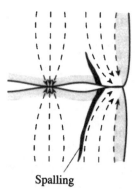

Spalling

(a) Stress flow at bed joint, remote from edges: horizontal components balance, no splitting stresses

(b) Unbalanced horizontal components push outwards at edge causing spalling

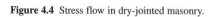

Figure 4.4 Stress flow in dry-jointed masonry.

Another possible cause of splitting is rusting of iron cramps, that may have been used during the original construction to tie together individual blocks in one course (whilst some ancient wrought iron is so pure that it does not rust in a reasonably dry, clean, atmosphere, it will corrode if exposed to salt, for instance from wind-driven sea spray).

4.4.2 Solid, Through-bonded, Mortar-jointed Masonry

In all mortar-jointed masonry, the mortar in its initial, plastic, consistency flows into the irregularities in the surfaces of the stones or bricks so as to provide uniform bearing, when it hardens. In this way, the high local stresses at the contact points in dry-jointed masonry are avoided. It can be said that the original function of mortar is to keep the stones or bricks apart—not, as it is often thought, to glue them together! The mortar also performs the function of preventing weather penetration through the joints to the inside of the building.

The term 'through-bonded' has been used to denote masonry in which the units, the coursing and the bonding remain basically the same through the thickness of the wall (Fig. 4.5) in contrast to the rubble-cored walls that will be discussed in Sec. 4.4.3. The through-bonding provides a connection of some tensile strength between the two faces of the wall. Failure under vertical load of such a wall or pier will therefore usually manifest itself by vertical or near vertical cracks originating at butt joints and going through the units or zig-zagging along the joints.

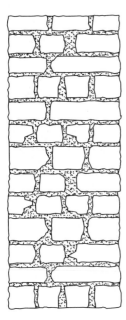

Figure 4.5 Section through solid, 'through-bonded' wall.

The alternative failure mode by splitting parallel to the faces of the wall will require cracking of the bonding stones or sliding of the bonding stones along the mortar joints and the only outward symptoms will be a slight bulging of the wall or a visible 'pulling in' of the bonding stones from the face. This bulging-splitting failure is, however, far more likely to occur with rubble-core walls, which have only a few bonding stones scattered through the core, randomly disposed over the area of the wall.

In some seemingly well-bonded brick walls, with better bricks showing on the face and inferior bricks used in the backing work, the original builders may sometimes have saved on the more expensive facing bricks by 'snapping' them in half and laying each half to appear as if it were a header, whilst bonding the face to the backing work with a minimum of real headers. If, for some reason these few headers fracture, problems, similar to the one described above, are likely to arise.

The strength of the mortar joints is important for the performance of the masonry. If they are too weak, the consequence could be premature sliding-splitting under high vertical loading; this is, however, likely to be a rare condition. If the mortar is too strong, compared to the strength of the stones or bricks, and a deformation, such as differential settlement, is then imposed on the masonry, the result may be cracking that goes through the units, rather than follow the joints. Such cracks are much more difficult to repair in a visually satisfactory manner (once the cause of the deformation has been dealt with), than cracks that follow the joints. This has led to problems with some repairs from the late nineteenth and early twentieth centuries.

The mortar quality also influences the weathering performance of exposed masonry. If it is too weak and porous, it will be eroded too quickly. If it is too strong, it will almost invariably be too impervious (adding a small proportion of Portland cement in substitution of lime will only increase the strength of the mortar marginally, but it will reduce its permeability significantly). It may then trap water that has got into the units, e.g. from driving rain, and prevent it from draining down; such water, lodging in the units over each impervious bed-joint, is likely to exacerbate frost damage.

Another cause of disfigurement or damage that can occur, when new bricks of certain types are used for exposed repair work, is a chemical reaction between salts in the bricks and the lime in the mortar; rainwater, absorbed by the bricks, dissolves the salts (mainly sulphates) in them and the solution drains down into the mortar beds, reacts with the lime or cement and the reaction products then re-crystallize immediately behind the surface as the water evaporates. The re-crystallization is accompanied by a volume increase which causes the face of the bricks to spall off.

Cracks, due to embedded iron, have been mentioned in Sec. 4.1.4. 'Best Practice' was to set the cramps at least 6 in. (150 mm) back from the face and run molten lead around them. The lead provided a physical, rather than galvanic, protection, at the same time as making a slightly resilient connection. Because the lead envelope was never perfect, water would get to the iron, if the pointing of the mortar joints was neglected over a period, and the rusting would start. The rust is spongy and retains moisture and thereby accelerates the corrosion.

Another source of potential problems, mainly found in Georgian and early Victorian brickwork, is horizontal timbers, embedded either for the purpose of levelling up apprentices' wayward coursing or as fixing grounds for wood panelling. Their propensity to harbour dry rot is mentioned in Sec. 3.2.3 and the possibility of their shrinkage leading to buckling of the wall as a whole is illustrated under Sec. 4.5.7.

Where gutters are hidden, thermal expansion can create problems. The upstands or parapets will have a gutter flashing, tucked into a bed-joint on the inside, and there will often be a lead damp-proof course in one of the bed-joints near the façade face. These features, which are required to protect the masonry further down, provide horizontal planes that are weak in shear. Solar heating will make the parapet or upstand expand and slide on the damp-proof course, etc., but the resistance against sliding may be enough to stop the parapet sliding back when it cools down and wants to shorten; this will make it crack. A similar phenomenon is sometimes observed where the parapet stays intact, but the near-horizontal coping on top of it 'walks off the end'.

4.4.3 Rubble Core Masonry

As the name suggests, walls in this type of masonry were built with two 'skins' of mortar-jointed stone blocks, carefully laid, between which a filling or 'core' of mortar and stone rubble was placed (Fig. 4.6). Piers and columns similarly have

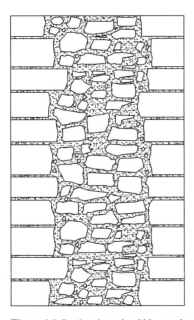

Figure 4.6 Section through rubble-cored wall.

an outer shell of block masonry surrounding a rubble core. In some instances, bonding stones may bridge the core, but more often the only connection provided is through random stones protruding from either skin into the core. This lateral connection is obviously much weaker than that found in through-bonded masonry and this weakness can give rise to additional problems.

During construction, the core filling would be brought up a short way below the skins and would support its own weight. Its modulus of elasticity would, however, be significantly lower than that of the skins, so that any subsequently imposed load would be largely carried by the skins. If the skins are thin compared to their height and if they are not well connected to the core, they may buckle. In the case of heavily loaded columns or piers, the stresses in the shell of block masonry may become so high that splitting failure becomes a possibility.

The tendency for load to be shed from the core to the skins or shell can be aggravated by water ingress. The mortar in the core filling is often porous, the lime content is low, and due to the thickness of the wall, it may not have carbonated completely. If for whatever reason, such as the roof missing for some years following a fire or defective pointing of mortar joints on the weather face, rainwater enters the core, it will dissolve or leach out lime and loosen the fine particles; these then get washed down, leaving behind cavities which grow until the rubble collapses into them. This process will be greatly assisted by the expansion caused by freezing of the trapped water during winter months.

The net result is that the core filling slumps, leaving the skins standing as separate thin walls. This can happen at the top due to roof or gutter defects, but it can also occur at lower levels, particularly where stone replacement has been carried out in the past. The reason for this is that, whereas during the original construction the core material would automatically fill the space between the skins, so as to leave no voids behind the stones of the skins, it is extremely difficult to achieve such a complete filling when a stone is removed and another put in its place. Whilst stone masons are very skilled at cutting out defective stones and keying in new ones, some of them did in the past look down on pointing, considering it to be semi-skilled work, and it may therefore have been neglected. If a leaking joint communicates with a void behind a stone, rainwater will accumulate in the void, dissolve the lime and/or freeze and cause the void to grow. In some cases, the frost expansion may by itself cause bulging of the wall or column (Fig. 4.7).

Another problem can be caused, if substantial loads are imposed on the core filling. Having, by its nature, very little lateral strength or stiffness, the core will respond to vertical load by exerting horizontal pressure on the skin(s) in the same way as earth exerts pressure on a retaining wall. In this case, bulging might presage bursting.

Another example of core expansion was that which in 1967 was found in the choir piers of York Minster. These had been re-faced after the 1829 fire. Careful inspection of drilled core holes showed that the cracks were confined to the fairly thin Victorian re-facing and did not penetrate into the remains of the original fourteenth-century work. The explanation was thought to be that the re-facing was carried out with stone fairly fresh from the quarry and that some of the inner

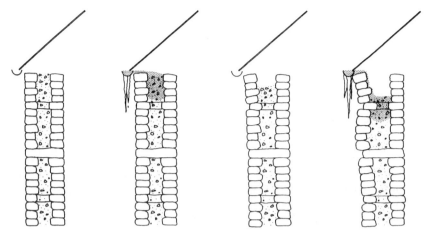

Figure 4.7 Deterioration of rubble-cored wall due to frost expansion (*from Beckmann, 1985*).

masonry (which otherwise was of excellent workmanship) had been left in its partially calcined state. The combination of the expansion due to moisture absorption by the desiccated core and the drying shrinkage of the new facing stones would then produce vertical cracks in the re-facing stones.

It should, however, be noted that not all rubble cores are weak and crumbly. At St Paul's Cathedral, a passage way had to be created between two parts of the crypt by tunnelling through the masonry at the bottom of the south-west tower. This was found, in the end, to require the use of pneumatic hammers!

4.4.4 Slender Monolithic Column Shafts

A point worth bearing in mind is that when a slender column consists of a single piece of stone, a potential buckling failure will be initiated by *tensile* failure of the stone (see Sec. 1.2.4). This type of failure will occur *without any warning signs*. This is in contrast to a column, made up of several 'drums' of stone; in that case, the joints will tend to open in a similar way to that shown in Fig. 4.1 and failure will be preceded by crushing of mortar and/or spalling of stone at the compressed face of the column.

4.5 REPAIR AND STRENGTHENING TECHNIQUES FOR WALLS AND PIERS

4.5.1 Compatibility of Materials

For all types of construction, repair materials must be compatible with the original materials. This applies to the stone or brick as well as the mortar. Failure to observe this has led to problems in the past. For instance, on York Minster, during a certain period, some masonry repairs were carried out with a stone which, when placed below the original magnesian limestone, suffers badly from 'chemical erosion'. This is because the rain washing over the magnesian limestone dissolves magnesium salts which, when they wash over the masonry lower down, decompose the cementing minerals in that stone.

The quality of the repair mortar will influence the weathering performance of exposed masonry. If it is too weak and porous, it will be eroded too quickly. If it is too strong, it will almost invariably be too impervious (adding a small proportion of Portland cement in substitution of lime will only increase the strength of the mortar marginally, but it will reduce its permeability significantly). It may then trap water that has got into the units, e.g. from driving rain, and prevent it from draining down; such water, lodging in the units over each impervious bed-joint, is likely to exacerbate frost damage in the form of spalling-off of the face of the bricks or stones.

The other objection to the use of modern strong mortars is that they prevent the slight sliding of stones or bricks on the bed-joints which, when walls are subjected to imposed deformations such as differential settlement, usually only lead to minor cracking along the joints. With such sliding movement being prevented, cracks will form through the stones or bricks and such cracks are obviously far more disfiguring and more difficult to repair than cracked joints.

The current consensus is that repair mortars should have a composition as close to the original as possible. A possible problem with this approach is that some of the original limes may have contained impurities that actually improved the properties of the mortar and which are absent from the lime available today. For instance, a modest inclusion of clay in the raw lime could make it 'hydraulic', when burnt, i.e. capable of hardening in the absence of air. Mortars made with hydraulic lime tend to be stronger than those made with pure lime in the same mix proportions. A similar improvement of mortar properties was obtained by the Romans by the incorporation of crushed brick, or other 'pozzolans' into otherwise plain lime mortar. These aspects are dealt with to some degree by Ashurst and others, who go into some detail about 'cosmetic' and weathering repairs (see Ashurst and Ashurst, 1988).

Structural repair and strengthening techniques are conveniently considered under the headings of the defects they are intended to remedy.

4.5.2 Repair of Static Cracks

Minor cracks, of which the original cause has been removed or has ceased (e.g. shrinkage) and which are situated where extreme temperature or moisture movements are not likely, may only need filling and pointing so as to exclude water ingress. Pointing also has the beneficial side effect of improving the strength of the wall by restoring its full cross-sectional area and by retaining the (possibly weak and crumbling) bedding mortar. Careful pointing, retaining the original appearance of the wall, is therefore an integral part of any masonry repair.

When there are cracks that are of some consequence for the integrity of the masonry, and when the main cause of these has been eliminated by other structural intervention, so that only slight subsequent movement is expected, they can be treated by 'stitching' as follows: short lengths of masonry courses are taken out and new stones, bridging the cracks, are mortared in. The mortar used for such stitching should be of a composition as near as practicable to the original: if it is too strong, any slight future movement will crack the stones, rather than the mortar. Great care must be taken to avoid leaving voids behind the stitching, in which water can collect (see Sec. 4.4.3). A possible safeguard against this would be to incorporate temporary injection nozzles into the mortar joints to allow grouting of the space behind the stone after the mortar has set. If grouting is carried out soon after the placing of the new stone, it must obviously be done with low pressure to avoid displacement of the stone.

4.5.3 Provision of Tensile Resistance

In walls which are cracking due to tension stresses, which are produced by differential settlement or by thrusts from roof structures, and if these root causes cannot practically be completely eliminated, it is beneficial to provide tensile resistance against further opening of the cracks. This can be by means of tie rods which, whilst inside the building, are external to the masonry. This is frequently seen in church towers, that have been 'strapped up' in the past. Alternatively, they can be placed in holes, drilled longitudinally within the thickness of the wall, as was done in the central tower of York Minster (Fig. 4.8).

Another method is to form horizontal rebates in the masonry and insert reinforcing bars, which are mortared into the rebates to provide horizontal tensile strength. A variation of this, which is appropriate to relatively thin brick walls, is to rake out every third or fourth horizontal joint and embed thin stainless steel reinforcing rods or stainless steel strand, such as used in the standing rigging of yachts, in the re-pointing mortar. Theoretical and experimental research at Bath and Karlsruhe Universities has demonstrated that this is an effective alternative to foundation underpinning, when remedying settlement cracking in brick façades (see Cook et al., 2000). For joint reinforcement, as well as for tie rods, the future forces in the 'uncracked' wall should be assessed, as far as practicable, by calculation or otherwise, and used as the basis for the 'design' of the reinforcement.

Where the cracks extend from, and are most serious at, the top of the wall, the remedy may be a complete capping beam of reinforced concrete. This device is most appropriate for thick walls where the beam can be formed in a chase, cut in the top of the core of the wall, so as to leave the face masonry intact on both sides. Such beams are particularly effective in towers, where they form complete ring beams. In a bell tower, the reaction forces, created by the swinging of the bells, can be evenly distributed, from the bell frame over the entire perimeter of the tower walls, by such a ring beam.

Objections are sometimes raised against the introduction of alien materials, such as concrete and particularly steel, into masonry. The use of more 'sympathetic' forms of construction, such as the introduction of courses of tiles, is advocated as an alternative. Two or three courses of tiles with staggered butt joints will

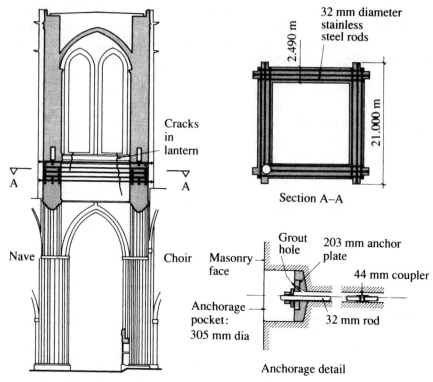

Figure 4.8 Tensile strengthening of the spandrel walls of the central tower of York Minster (*from Dowrick and Beckmann, 1971*).

undoubtedly provide a measure of tensile resistance (tiles, being more thoroughly fired, will have a much greater tensile strength than bricks), but, leaving aside the question of the violation of authenticity by tile courses inserted where originally there were none, there are two structural objections to this technique. Firstly, the damage to the original masonry, in carrying out this work, may be greater than in the case of the deprecated 'high-tech' techniques. Secondly and more significant, while tile courses may be quite strong in tension, their failure mode will be brittle; they will allow hardly any movement before they suddenly crack and lose all their strength and thus become of no further benefit.

In contrast to this, steel, whether in the form of tie rods or as reinforcement in a concrete beam, will yield if overstrained. During yielding, it will elongate and thus allow some movement, but at the same time, it will continue to provide tensile resistance.

A structurally more valid objection to the introduction of steel tie rods, etc., is that the different elastic moduli of steel and masonry will cause the latter to act in a way that is more akin to reinforced concrete and, particularly if the steel is pre-stressed, it will reduce the damping otherwise inherent in masonry and thereby make it more susceptible to damage, due to resonance with external dynamic forces.

A way of overcoming this might be to reinforce the masonry with a material with substantial tensile strength but with an elastic modulus closer to that of masonry. Development work on the use of cement-bound polypropylene strand has been carried out at the University of Rome under the direction of Professor G. Croci. The disadvantage of such 'soft' reinforcement is that quite large cracks may have to form before the reinforcement has any effect.

Ductile behaviour is particularly important where the provision of tensile restraint is intended to strengthen the masonry against future seismic effects.

4.5.4 Strengthening of Wall Junctions

Another form of tensile strengthening is sometimes used in refurbishment of Georgian and early Victorian terrace houses which have been built without proper bond between façade wall and cross-walls. In such houses, it is not unusual to find structural gaps of 20–30 mm at these junctions.

To stabilize the façades and ensure that the filling of those gaps remains effective, one can form rebates on the inside of the façade wall which continue in the adjacent face of the cross-wall. The rebates are half-a-brick (~100 mm) deep and one course high and extend about 750 mm from the junction along both walls. The vertical spacing is approximately 450 mm and every second rebate is cut on the opposite face of the cross-wall. Into these rebates, one then embeds reinforcing bars bent at a right angle. Alternatively one can pre-cast reinforced concrete 'elbows' and bed them into the rebates with a mortar of similar properties to the existing one (where the coursing of the cross-wall does not line up with that of the façade, one of the rebates may have to be two courses high) (Fig. 4.9). This method has the advantage of allowing friction-ally restrained slip along the joints, as opposed to the mortared-in bars, which in case of a strong enough movement may tear out large chunks of the weak

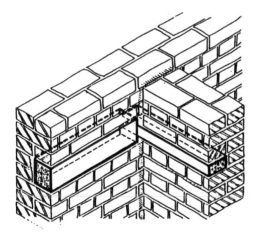

Figure 4.9 Precast concrete 'elbow ties', connecting façade to cross wall.

brickwork, bordered by diagonal cracks; this will then offer only poor resistance to further movement.

A method of strengthening such junctions, advocated in the past, consists of a stainless steel 'hairpin' bar which is poked through holes drilled in the façade and mortared into the bed-joints of the cross-wall. This is in theory attractive because it only entails the drilling of small diameter holes in the façade and raking out of bed-joints in the cross-wall to receive the legs of the hairpin. In these buildings, the cross-walls were, however, often built before the façade wall by the 'rough' gang of bricklayer's apprentices, whilst the face was built later by the master bricklayer himself. In consequence, the courses do not always line up and this, otherwise elegant, method then becomes much more difficult to carry out.

Where the appearance of an anchorage on the façade can be satisfactorily dealt with, various anchors drilled in from the façade and injected with cement grout can be used.

4.5.5 Strengthening Against Local Overstressing

When buildings are adapted to new uses, it is sometimes found that beams, whether existing or introduced as part of the adaptation, impose reactions that in turn lead to stresses in the walls, which locally exceed the limits set by building regulations or deduced from testing. One solution in that case is to spread the load from the beam bearing over a greater length of wall. This can be done by locally re-building the masonry with strong bricks, so-called engineering bricks, bedded in strong mortar. Usually such a local strengthening would be 3–6 courses deep and twice as long as it is deep. As long as the beam bearing is not near the end of the wall, the bearing stresses will disperse through the strong brickwork and reduce to acceptable levels at its underside (Fig. 4.10a).

Another method for dispersing high bearing stresses is by the insertion of a padstone. This can be a single block of very strong stone, long enough to ensure that the beam reaction, when spread over this length, does not lead to excessive stresses and deep enough to limit the bending stress in the padstone itself; this depth is usually between half and two-thirds of the length (Fig. 4.10b). Alternatively,

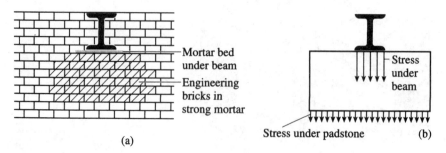

(a)   Stress under padstone   (b)

Figure 4.10 Strengthening against local overstressing under a new beam seating.

particularly if the masonry is to be plastered, the padstone can be made of reinforced concrete, but in that case provision must be made to prevent the plaster cracking over the boundary of the two dissimilar materials, for instance, by incorporating a fine metal mesh in the plaster.

Local strengthening, as described above, whether by local re-building or by insertion of padstones, clearly changes the nature, and in some cases the appearance, of the masonry. If, however, it enables the building as a whole to be retained by adaptation to a new use, it may be a price worth paying, if the alternative were to be demolition.

4.5.6 Treatment of Bulging and Splitting

Bulging is almost inherent in rubble core masonry. Unless there are fairly closely spaced bonding stones, going from one face of the wall to the other, such walls are vulnerable to frost expansion if water is allowed to enter the core because of bad details of gutters and roof or bad maintenance, etc. There have been cases when the top of the core cavity has been found empty and the bonding stones pulled in from the outside face of the wall, whilst the lower part of the wall was bulging from the excess material which has fallen into the gaps in the lower core, left after frost expansion (see Fig. 4.7). In some cases of large piers, this phenomenon is aggravated by the fact that the settling of the core materials leads to the loads having to be carried solely by the ashlar 'shell' of the piers.

These conditions require two things to be done:

1. Any cavities in which water might collect and freeze must be filled to prevent further deterioration.
2. The two skins of the wall must be tied together to enable the wall to function as one entity.

The filling of cavities can be achieved by grouting. Tying together the two skins of the wall will, however, require either the insertion and bedding of numerous bonding stones or the provision of metal 'stitching' bars, fixed at each end in the ashlar skins and bridging the core or cavity. Traditional grouting on its own is unlikely to glue the two skins together adequately; certain resin-grouting techniques are claimed to achieve adhesion to the face stones, but if the core material is weak and dusty it is doubtful if substantial connection can be achieved between the two skins of face masonry.

Anchoring the stitching bars in the outer skins of the walls can be done using expanding anchor bolts or resin anchors, each of which has advantages and disadvantages. It can alternatively be effected by using ribbed reinforcing bars (of non-ferrous metal or of stainless steel), placed in holes drilled through the wall or pier, and grouting them in at the same time as the voids in the core are being filled.

The expanding anchor bolts have the advantage that one can mechanically check that they have been tightened and therefore have a reasonable grip on the stones in which they are fixed. They do, however, need a fairly sound stone, and they must not be placed too near the edges of stones, if the expansion is not to cause cracking. They do not lend themselves to being applied to poor-quality masonry, where they may have to 'grip' in joints composed of low-strength mortar, or in badly filled joints. It has also been found that some expanding anchors relax their grip after the initial tightening; tests should therefore be carried out to check the suitability of a proposed anchor for the masonry in question.

The resin capsule anchor is fairly easy to apply and does not cause any splitting forces in the stones. In masonry with imperfectly filled joints there is, however, a tendency for the resin to run out of the hole and into any adjacent voids and thus make the anchorage of the bar ineffective. With another type of resin anchor, the anchor is placed in the 'dry' hole and the pre-mixed resin is subsequently injected; this permits a certain amount of cavity filling. For both types of resin anchors, feasibility testing, as well as pull-out testing, is essential.

Grouting-in of reinforcement bars combines the advantage of non-splitting of the resin anchor with the convenience of filling the voids at the same time. If the rubble-and-mortar core is dry, there is, however, a danger that the grout will be dewatered on its way and stiffen up in the injection hole, before it has penetrated sufficiently into the voids. Any attempt to counteract this, by flushing out the holes with water first, must however be done extremely carefully to avoid washing more fines out of the mortar and thus aggravate the condition of the core.

In Britain, the grouts most commonly used for these purposes are composed of varying proportion of lime, pulverized fuel ash and Portland cement. Pulverized fuel ash is a very fine grained, pozzolanic material; it improves the flowability of the grout, as well as being capable, in the presence of lime or Portland Cement of hardening and gaining fairly high strengths. The setting and strength gain is however much slower than for Portland Cement and this enables grouts, made with this material, to penetrate much deeper than ordinary cement grout before it stiffens and hardens.

4.5.7 Treatment of Buckling

Buckling of a wall as a whole can occur where the wall is high in proportion to its thickness, the load is applied eccentrically and there are no effective lateral restraints. 'Bonding timbers' are referred to in Sec. 3.2.3 as potential repositories of fungal infection. Where their depth is a substantial proportion of the wall thickness, they will also, when they shrink, effectively deprive the part of the wall, in which they are embedded, of its vertical structural strength. The load then becomes too eccentric to the effective cross-section and buckling results (see Fig. 4.11).

As mentioned in Sec. 4.1.5, brick cladding to buildings with reinforced concrete structures can sometimes get 'pinched' between edge beams/slab edges, due to creep-shortening of the vertical concrete structure. This can lead to buckling of the wall. Remedies against this are dealt with in Sec. 4.8.4.

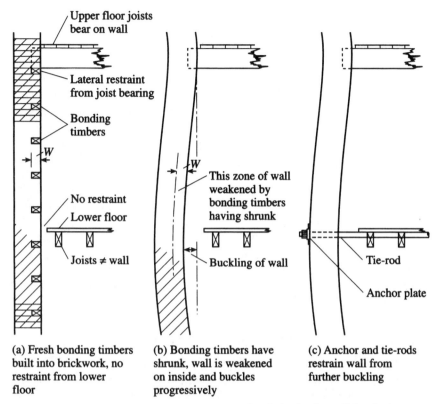

(a) Fresh bonding timbers built into brickwork, no restraint from lower floor

(b) Bonding timbers have shrunk, wall is weakened on inside and buckles progressively

(c) Anchor and tie-rods restrain wall from further buckling

Figure 4.11 Buckling of wall, due to shrinkage of 'bonding timbers' and remedial anchoring.

4.5.8 Anchoring-retained Façades

However much façade retention offends against the principles of true conservation, it has been extensively carried out and will continue to be done, when all that can reasonably be achieved by way of conservation is to preserve the streetscape.

Unless the façade is incorporated as a load-bearing structural part of the new scheme, the connections between it and the new structure must allow relative movement between the two, in the plane of the wall.

The report '*CIRIA 626 Masonry Façade Retention: Best Practice Guide*' (published by the Construction Industry Research & Development Association) contains comprehensive advice on this and other aspects of design and construction, involving façade retention.

When the interior of a building is being completely reconstructed, whilst the façade is being retained, some of the anchors, described in Sec. 4.5.6, can be used; in addition, a variation of the injected anchor, using a cementitious grout instead of resin can be considered. The anchors are first fixed to the façade, which is being restrained against wind by temporary scaffolding. They are subsequently connected to or built into the new structure being erected behind. In this way, the original façade wall is left to support its own dead load, but is being held against lateral forces by its connection to the new structure.

When used in this way the anchors, be they of expanding bolt or grouted in bar type, have to be combined with devices which allow relative movement between the old façade and the new structure (e.g. due to differential settlements) at the same time as providing adequate horizontal restraint. It should be remembered, when designing such devices, that slotted bolt holes, unless combined with low-friction washers, will not allow movement, when the bolts are done up tight. Stainless steel (or non-ferrous metal) should be used for any components that may be exposed to penetrating moisture and/or condensation. If such components will be in contact with carbon steel in such an environment, the risk of bi-metallic corrosion should be considered and, if necessary, isolation washers and/or grommets have to be provided.

The fire resistance of the connections between the retained façade and the new structure must be considered. If metal parts were to loose significant strength, or resin were to soften, the façade might collapse during a fire. Fire protection may therefore have to be provided. The CIRIA report refers to tests and gives advice on this.

4.5.9 Improving Stability by Increasing the Vertical Loading

The lateral stability of a wall or pier is proportional to its thickness and the vertical load acting on it. There may therefore be instances when strengthening of masonry walls can take the form of adding more masonry to the walls so as to increase their overall thickness. An example of this is the strengthening of masonry spires, built of certain kinds of soft stone, which sometimes may wear so thin, by weathering, that the wind stability of the spire becomes endangered. If the supporting tower can carry a slight increase in vertical load, this stability problem can be overcome by constructing a bonded masonry lining which will provide both a stabilizing dead load and additional strength to the existing shell of the spire.

An instance of this is the spire to the Holy Trinity Church, Coventry, which is an octagonal pyramid, 42 m high and 6.5 m wide at the base. The lower parts of the single skin, sandstone, masonry had been worn down by weathering from about 180 mm thickness to an average of about 120 mm, with isolated spots as thin as 80 mm. As a result of this, the margin of safety against the spire being blown over by a high wind had been reduced below the acceptable limit. A lining of brickwork was built against the inside of the sandstone shell and bonded to it with stainless steel straps and expanding anchors. The thickness of the lining was 230 mm around the lower windows, 110 mm on the solid faces of the octagon on the lower ten metres and 75 mm on the remaining 9 m height of the lining (Fig. 4.12).

Occasionally it will be found that a spire has a combination of original poor materials and workmanship, together with later unsuitable repairs, such that demolition and re-building becomes the only practical solution.

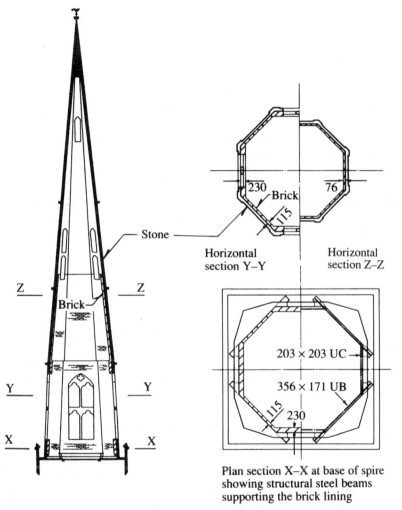

Horizontal
section Y–Y

Horizontal
section Z–Z

Plan section X–X at base of spire
showing structural steel beams
supporting the brick lining

Figure 4.12 Strengthening of stone spire by internal-bonded brick lining (*after Beckmann and Blanchard, 1980*).

4.5.10 Anti-seismic Strengthening

As mentioned in Sec. 4.3.6, a great number of buildings have survived centuries of earthquakes, despite the fact that current methods of analysis indicate that they should have failed. The survival is often accompanied by minor to moderate, repairable, damage. When deciding on which methods to use when embarking on a programme of strengthening of a substantial number of buildings of similar form and construction, it can be of great assistance to study the pattern of past earthquake damage to buildings of the type in question, *in the same region*. The damage pattern depends on the characteristics of the earthquake as well as on the type of construction of the buildings.

Another matter that should also be considered, before any strengthening scheme is drawn up, is the degree of seismic resistance to be provided. In the case of a building where the original construction is the major contribution to its cultural significance, the amount of new structure required, in order to provide full resistance to the strongest predicted earthquake, may well destroy the historic character of the building. If occupants can be safely evacuated, a lesser degree of strengthening, allowing limited damage should be permissible and, is from the conservation point of view preferable.

Many of the requirements of seismic resistance will be met, if the structure and its elements have adequate robustness and stability, as mentioned in Sec. 1.5. Generally, the first step in seismic strengthening of buildings with masonry walls is to make provisions to ensure that the plan shape remains undisturbed: corners must be prevented from behaving as hinges, T-junctions must not pull apart and straight lengths of wall must not be free to bow in plan or move laterally. This requires that the floors are made to act in their plane as stiff diaphragms or plates and that all walls are tied laterally to the floors.

The substitution of timber floors by reinforced concrete slabs is often advocated as a means of creating rigid diaphragm floors; there may, however, be other means of achieving this. It should be borne in mind that the introduction of concrete floors will add mass at high level in the building. This will lead to an increase of the inertia forces, imposed on the vertical structure during an earthquake, and is therefore undesirable. What is essential is the tying of walls to the floors. At an inspection of the damage after the 1979 earthquake in Kotor and Budva, Montenegro, buildings were observed where an entire wall parallel to the floor joists had fallen away, whilst the walls on which the floor joists were supported and which therefore had a modicum of lateral restraint were still standing.

Some of the anchoring and tying devices, described under Secs 4.5.6–4.5.8, may be useful for the purpose of providing lateral restraint to walls. It should, however, be remembered that seismic movements may temporarily *diminish* the vertical pre-load from gravity; this may reduce the holding power of expanding anchors. It should also be borne in mind that the reciprocating nature of the seismic inertia forces may cause stone(s), in which tie rods or anchors are fixed, to be pulled out by ratchet action.

Professor Duilio Benedetti of Politecnico di Milano carried out tests that demonstrated that masonry walls can withstand far greater seismic movements without serious damage if they are prevented from extending in their plane by means of tie rods with cross-heads or plates as end-anchorages. Lazar Sumanov (1999) ascribes the survival of Byzantine churches in Macedonia, through centuries, to the longitudinal timber baulks, originally built into the wall masonry. Over the years, these have however rotted. He has demonstrated, by shaking-table tests on a reduced-scale model, that replacing such timbers with pre-loaded steel tie rods increases the resistance of such masonry structures manyfold.

As well as lateral restraint, additional longitudinal tensile resistance may therefore need to be provided for the walls. Various methods of providing this are described under Sec. 4.5.3.

Where the vertical dead load on walls is insufficient to counteract the seismic effects, e.g. in the upper storeys of buildings, vertical tensile resistance has to be provided. Reinforced concrete 'columns', which in reality are just concrete casings to reinforcing bar tie rods, are very often advocated for this. If located within the thickness of the wall, they do however require vertical chases to be cut; these impair the horizontal integrity of the wall. If placed externally to the wall, such tie columns are visually obtrusive and difficult to connect to the top of the wall.

Depending on the nature of the construction of the wall, it may be possible to drill vertical holes through the height of the wall in which tie rods can be placed, terminating in rock anchors in the foundations. Drilling such deep holes does, however, require the use of water or air flush. The former may find its way through joints to the face of the masonry and/or it may be necessary to drill relief holes to allow the water to escape in order to reduce the hydrostatic pressure. As the released water carries the drilling dust in suspension, it can cause unsightly streaking. In addition, seepage can damage wall paintings. Air flush of the drilling, on the other hand, will normally produce vast clouds of dust. In the absence of wall paintings, the best course may be to grout in relief tubes to throw the dust-laden water clear of the wall and to point, and thus seal, all joints prior to using water-flush drilling.

A further development of the vertical tie bars is to tension them and thereby vertically pre-stress or pre-load the masonry. The advantages of this are that the shear resistance of the wall is enhanced by vertical pre-load and that the natural frequency of the wall is increased and hence removed from a possible resonance frequency of the earthquake motion (see Sec. 3.4.3). On the debit side, the pre-load will prevent the opening of joints and the extra damping that would otherwise result from this, and it will prevent the possible change in movement pattern that might otherwise increase the seismic resistance (see Sec. 4.3.6).

Vertical pre-stressing was rather elegantly used on the clock tower in Dubrovnik. The pre-stressing cables were placed just clear of the masonry in the re-entrant corners; they were anchored on a reinforced concrete slab at the top and in the rock under the foundations and were surrounded by fire insulation inside tubular shields. Apart from the supporting rebate for the concrete slab, the masonry is untouched and the intervention could fairly easily be reversed, if so desired.

4.5.11 Buttressing of Leaning Walls

Lateral thrust causing masonry walls and piers to lean can arise from earth pressure, thrust from arch springings or thrust from domes which have been inadequately restrained (see Sec. 4.6.1). The traditional remedy has been to provide further buttressing. No foundation will, however, carry any load without settlement so new buttresses will not have any beneficial effect until the soil under their footings has been compressed to take the load, by which time the soil under the original footings may have completed its consolidation and the movement therefore have ceased. The consequences of this can occasionally be observed in buttresses, which have been added to a wall and which are seen to be hanging from the original structure, slowly pulling away from it, rather than supporting it.

It is possible to overcome this problem by pre-loading the new buttress. The most convenient way of doing this is by using a flat-jack, a metal capsule usually circular in plan and with a cross-section like a dumb-bell (see Fig. 4.2). This is inflated by hydraulic liquid and can be used to put a pre-determined force between the existing wall and the new buttress or between the new buttress and its own foundation. In this way, it is possible to consolidate the soil under the new buttress foundation within a matter of some weeks or months, after which the gap, in which the flat-jack was inserted, can be grouted solid and the soil, having being pre-compressed, will then support the load so that the buttress will fulfil its intended function.

4.5.12 Repair of Damage caused by embedded, rusting, Ironwork

Where there are iron cramps or tie bars within 150–200 mm from the face of exposed masonry, it is futile to hope that some 'patent' resin treatment will effectively stop further rusting, by excluding moisture. It will get in through weathered pointing in the joints, and the rust that has formed will attract and hold moisture. Once the corrosion has got hold, the cramps and tie-back bars will somehow have to be removed. Long bar-cramps in square or polygonal towers and spires can sometimes be partially removed by drilling out a long core of the masonry, which includes the straight part of the iron bar, its lead sheath and some of the surrounding stone. If the ends of the bar have been turned down, local 'open surgery' is needed to remove them.

Whether the ironwork has to be replaced depends on its original intended function. Where it is clear that it was originally put in to mitigate the effects of differential settlement, which can now be considered complete, there should be no need for replacement, provided that such settlement is unlikely to resume, due to changes in subsoil and groundwater conditions. Where the iron cramps are tying back cantilevering corniches, they must be replaced. This will require temporary support of the cantilevering stonework. In church towers where there is a possibility of ringing of heavy or multiple bells, any reduction of tensile capacity is inadvisable.

If any replacement cramps and/or tie bars are judged to be necessary, they should be of stainless steel and of similar tensile capacity (but not necessarily the same section) as the iron components that they replace.

4.5.13 Remedies for rusting/missing Cavity-ties

If corrosion of the old type strip ties is so serious that it has cracked the bed-joints, the ties should be cut out and replaced with expanding, or resin capsule anchors or one of the proprietory self-anchoring pins made for the purpose. The cutting out will entail careful removal of some bricks; these should be replaced, properly bedded in, and the whole wall be re-pointed.

Where investigation of one small area has shown inadequate provision of cavity-ties, there is usually little to be gained by extending the investigation to the whole wall. Assuming deficiency throughout and installing supplementary ties, or even installing ties to 'full specification', will usually only be marginally more expensive than a full investigation.

4.6 ARCHES, VAULTS AND DOMES

4.6.1 Structural Behaviour of Arches and Vaults

Beams carry vertical loads by developing bending stresses that are essentially horizontal. Columns, pillars and walls carry vertical loads by developing largely vertical compressive stresses. Arches, vaults and domes carry vertical loads by developing stresses that follow their shape along curved paths.

In order to visualize the behaviour of these structures, it may be helpful to look at the mirror image of a simple arch: the string polygon. If a piece of string is held, slack, at two points that are at the same level and a distance s apart and weights are attached to it, it will take up a shape that depends on its length and on the magnitude and disposition of the weights. Figure 4.13a shows on the left-hand side such a string held at A and B and subjected to four weights or point loads P_1 to P_4.

The right-hand part of Fig. 4.13a shows the geometric construction that enables the tension in the various parts of the string to be established in relation to the forces P: R_p is the resultant of the forces, R_A and R_B are the reactions at A and B, respectively. The diagram is constructed by drawing the converging lines R_A, 1–2, 2–3, etc. parallel to the corresponding lengths of string on the left-hand picture.

Note that although all the loads, P, are vertical, the reactions to the tension in the string are *inclined* and therefore have a *horizontal component*: the string tries to *pull* A and B towards each other.

If one imagines the string turned upside down but the loads P still acting downwards, one gets the situation in Fig. 4.13b; the string has now become a series of struts, with compression forces that can be determined by the construction in the right-hand part. What was a string polygon has become a *thrust-line*. This concept was published as an anagram in 1675 by Robert Hooke. Following the mirror-image analogy, whilst the loads are *vertical*, the reactions are inclined and thus have a *horizontal component*; the struts form an arch which tries to thrust the springing points A and B away from each other.

The essence of this is that *arch action*, whether in an arch, a vault or a dome, *always creates a thrust*. It acts internally as a compression and externally as a pair of forces, trying to push the abutments apart.

Another important point can be deduced from the mirror-analogy of the string polygon; the shape of the line of thrust is dictated by the relative magnitudes of the loads and their disposition across the span—not by the shape of the arch itself. The amount of the horizontal thrust on the abutments depends

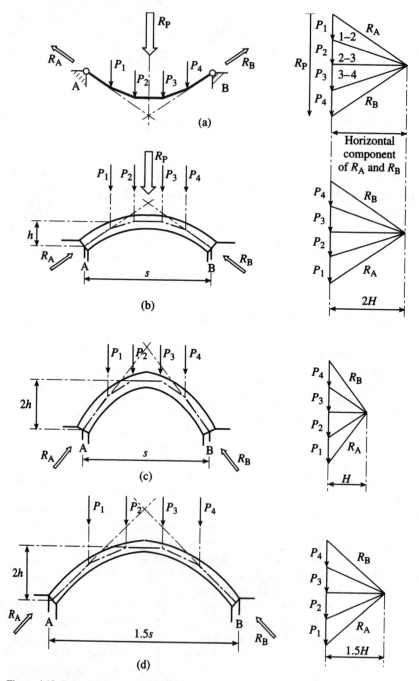

Figure 4.13 String polygon and arch thrust-lines.

on the rise of the arch. The thrust-line in Fig. 4.13c has twice the rise of that in Fig. 4.13b and the constructions show that the horizontal thrust is halved, for the same load configuration. The arch in Fig. 4.13d has the same rise as the one in Fig. 4.13c but $1\frac{1}{2}$ times the span. The *total* load is the same, but the horizontal thrust is $1\frac{1}{2}$ times that in Fig. 4.13c.

The general rule is that the horizontal thrust is *proportional* to the *total load* and to the *span* between the abutments and is *inversely proportional* to the *rise*.

The thrust of an arch or vault against its abutments may cause these to move apart. This has the consequence that the span becomes (slightly) too big for the arch. The arch will then try to compensate for this by 'opening out'; it does this by opening the joints between the voussoir stones, or bricks, on the inside (the 'intrados') at the crown and on the outside (the 'extrados') at the springings. This happens sometimes when the foundations to the abutments undergo some initial differential settlement immediately after the scaffold (the 'centring') to the arch is struck (see Fig. 4.14b).

A visually similar phenomenon can occur over a period following the de-centring, if very thick lime mortar joints have been used in combination with fairly 'thin' voussoir units: the, as yet not fully hardened, mortar is compressed, the joints get thinner and the arch or vault becomes too small for the span between the springings. The structure again reacts by opening out to bridge the span and in so doing cracks the joints on the intrados at the crown and on the extrados at the springings. The latter may not show up as clearly, being sometimes masked by the subsequent construction.

In 1987, cracks of this nature were observed during an inspection of some of the underground chambers of the Taj Mahal. Shortly after the inspection, a translation was made available of a letter written in 1652 AD (i.e. 9 years after the construction) by Aurangzeb to his father Shah Jahan (who had it built as a mausoleum for his wife): 'The building of this shrine of holy foundation is standing in exactly the same manner as it was completed in the Illuminated Presence (of Shah Jahan), except that the dome over the pious grave leaks in two or three places . . . and the seven *arched underground chambers have developed cracks . . .* ' (authors' italics). This passage demonstrated that the observed cracking had occurred only a few years after the original construction.

Another fairly common cause of cracking in arches is differential settlement of the supports to the springings. In that case, the joints will open on the intrados around the quarter point adjacent to the support going down and on the extrados around the quarter point adjacent to the 'higher' support (see Fig. 4.14c).

A similar effect can also result, if the distribution of the loads is not compatible with the shape of the arch. In Fig. 4.14d, the loads on the right-hand half of the arch are substantially heavier than on the left-hand half. It is essential for the stability of the arch that the thrust-line stays within it (see later). In order to do so, it has to come very close to the intrados at the left-hand quarter-point and has to pass very close to the extrados a bit to the right of the crown. Whenever the thrust-line moves nearer than one-third of the arch thickness towards a face of the arch, the joints will open on the *opposite* face, i.e. if the thrust-line is closer to

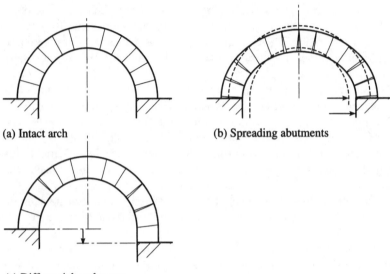

(a) Intact arch (b) Spreading abutments

(c) Differential settlement

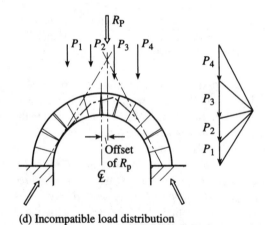

(d) Incompatible load distribution

Figure 4.14 Diagnoses of causes of arch cracking.

intrados than one-third the thickness, joints open on extrados, and vice versa (compare Fig. 4.1*d*).

The shape of the thrust-line and hence the pattern of the loading become more critical, the thinner the arch or vault is, relative to its span. The fill over the springings of vaults therefore has an important structural function, and if for any reason removal becomes necessary this should be done carefully, preserving the symmetry of the loading.

All these aspects have been dealt with in great depth by Professor Jacques Heyman in his books *The Masonry Arch* (1982) and *The Stone Skeleton* (1995), in which he quotes Coulomb to the effect that, if a thrust-line can be constructed that nowhere is closer to the extrados or the intrados than one-tenth of the thickness of the arch (or vault) the arch will be structurally stable. The edge distance of one-tenth of the thickness was originally stated for Gothic stone masonry arches, where the strength of the stone is hardly ever critical; for heavily loaded arches, or arches built of soft stone or bricks, the critical edge distance may have to be increased to, say, one-seventh or one-sixth, to avoid compression failure of the materials.

All the above assumes that the masonry spandrel panels, and/or the vaults, or any other element, carried by the arch, have little or no stiffness, nor strength, in the plane of the arch and simply acts as a load. If, however, the arch merely forms the top of an opening in a large expanse of solid masonry wall, constructed of large blocks of stone with thin joints of strong mortar, a different mechanism, first identified by Sir Christopher Wren, can develop.

In this, each part of the wall, either side of the centreline of the opening, is considered as a rigid body with a corbel (Fig. 4.15a). As long as vertical lines from the centres of gravity of those rigid bodies (including the corbels) stay within the masonry at the bottom of the opening the structure is stable and no thrusts need to be considered.

This mechanism involves some, largely horizontal, tensile forces in the virtual corbels over the arch and the reveals of the opening. In masonry, of the type described above, this would not be a problem, as the shear strength under the pre-load from the masonry above would be adequate to transfer the horizontal force at a butt joint in one course to the stones in the adjacent ones.

In reality, of course, the 'corbels' meet over the opening and can 'lean' on one another. This will generate a thrust, but the effective 'arch height' will be much greater than the visible one, and the thrust correspondingly smaller (Fig. 4.15b).

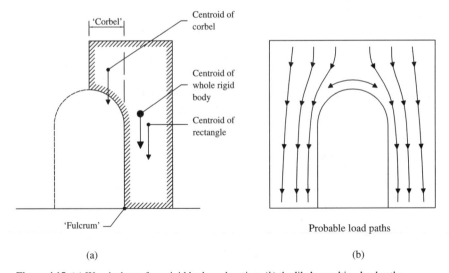

Figure 4.15 (a) Wren's thrust-free, rigid body arch action; (b) the likely resulting load paths.

4.6.2 Structural Behaviour of Domes

The structural behaviour of real domes is very complex, involving forces acting in all three dimensions and bending stresses as well as compression and tension. It is, however, possible to make two simplifying assumptions. Each of them explains a part of the overall behaviour and together they account for most of the structural strength.

The first assumption is that the dome is a 'shell', with a thickness that is very small relative to its other dimensions. If the basic principles of equilibrium are applied to a small element of such a shell together with the condition that the shell has no bending strength (due to its small thickness), a set of differential equations is obtained. These equations look very off-putting to a person with only elementary mathematics, but if the geometry of the shell is known, they can be solved and will show how the shell, as a whole, can carry its load.

When such calculations are carried out, it will however be found that whilst there is compression along the whole of the meridians (the curves following the intersections between the dome surface and vertical planes through its apex), the circumferential stresses (along the circles formed by the intersections of horizontal planes with the dome) change from compression to tension when the slope from horizontal of the surface of the dome reaches a certain value that depends on the shape of the dome and the distribution of the load (see Fig. 4.16a). For a part-spherical dome with uniform load per unit area of its surface (such as self-weight), the angle is 51.8°.

If a spherical dome has to act as a membrane shell, it therefore has to have a certain tensile strength in the circumferential direction unless it is rather shallow (as stated in Sec. 4.1, masonry is generally weak in tension).

It is however possible to shape a dome, carrying only its self-weight, so that no tension arises. Robert Hooke (cf. Sec. 1.2.1), who assisted Wren at St Paul's Cathedral, arrived (by means of experiments with weighted string polygons?) at the ideal shape which he called a 'cubico-parabolic conoid'. The inner dome of St Paul's spans a diameter of some 32 m and is only two bricks (45 cm) thick. It shows no structural cracks, despite the explosion of a Second World War bomb in the North Transept (see Sec. 3.3.4) and it has no iron ties or cramps in it. A drawing from the eighteenth century (copied from Wren's original?) shows the actual shape as consisting of three spherical 'bands' with different radii, joining tangentially and 'flattening' towards the top. The absence of structural cracking indicates that the actual shape of the dome is very close to the ideal theoretical shape, as calculated by Professor Jacques Heyman.

If membrane criteria are not satisfied, there is, however, another way in which a dome can carry its load. This is demonstrated as follows.

The second, simplifying, assumption abandons the concept of the thin shell and represents the dome by a series of arch ribs. Each arch rib consists of two opposed 'orange segments' formed by cutting the dome (with its real thickness) by two vertical planes intersecting at the apex (see Fig. 4.16b). If it is possible to construct a thrust-line for the forces acting on one such arch, such that it does not get closer to either intrados or extrados of the arch than the value appropriate to

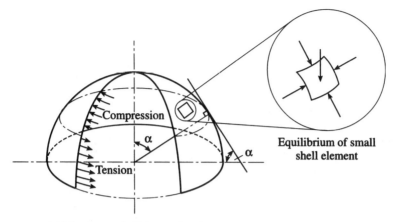

(a) Membrane-shell dome, showing
ring compression and tension

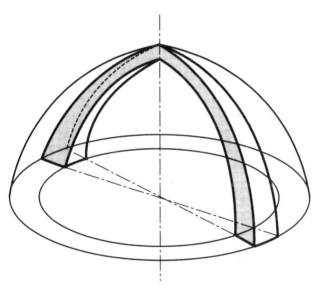

(b) 'Orange-segment' arch—Behaviour of dome

Figure 4.16 Membrane and arch rib action of domes.

the material (one-tenth of the thickness for stone and, say, one-sixth for brick), then each arch is stable and strong enough. If each arch is stable and strong enough, then clearly the dome is stable and strong enough, even though it may show cracks along the meridians.

In reality, dome behaviour will usually be a mixture of the two simplified mechanisms, postulated above: where the slope is less than the one at which the ring force changes from compression to tension, membrane action is likely to dominate (thus overcoming the objection to the orange-segment model, that it produces infinite stresses at the apex). Lower down the orange-segment behaviour will take over, unless the masonry has been reinforced with metal tension rings (and even these will stretch before developing their force).

Radial cracks in the lower part of a dome are therefore not necessarily a symptom of drastic structural shortcoming, however unsightly they may appear on the decorations inside, but if they continue to widen, it is a sign of inadequate horizontal restraint at the base of the dome. This is because, like arches and vaults, domes require horizontal restraint at their base; this is irrespective of their structural action being predominantly 'orange-segment arch' or 'thin shell membrane'.

Like arches and vaults, domes react to uneven settlement by cracking, but the crack patterns are likely to be more complex.

4.6.3 Strengthening and Structural Repair Techniques for Arches, Vaults and Domes

As the preceding paragraphs have indicated, structural distress in arched structures will usually manifest itself by cracking or opening of joints. Before any structural intervention is considered, it is however essential to ascertain whether the cracking is progressing or whether it is static and, if so, if the cause is unlikely to recur, in which case only 'cosmetic' repairs are required.

The continuing, increasing, leaning of walls and piers, subject to thrust from arches or vaults, may have been caused when the original builders misjudged the foundation conditions or got the geometry of the structure wrong. Alternatively it may be due to later generations having interfered with the structure or with the surrounding ground. The result is usually a slow spreading of the arches or vaults with correspondingly increasing tilt of the external walls or piers.

The problems of buttressing have been discussed under Sec. 4.5.11. In the case of spreading arches and vaults, an alternative to buttressing is the provision of ties, which prevent the springings of the arches from moving apart. Physical ties at springing level are usually architecturally undesirable and it is only on rare occasions that one can transfer the restraining force to the springings from a tie located at an invisible or visually unimportant level.

An example of this was however provided by the Koran school attached to a mosque in Rezayieh in north-western Iran. The author was given to understand by the local director of the National Organization for Preservation of Ancient Monuments, that the springings of the arches, supporting the domed roof, had originally had connecting ties, in the form of slender tree trunks. At some stage, the ties had been removed for aesthetic reasons, and in consequence the arches had spread so that the external walls were now leaning out so far that their stability was in jeopardy.

The remedial scheme, being carried out under the aegis of the National Organization in 1976, consisted of steel beams, placed within the thickness of the flat roof between the domes. The beams were connected to stanchions placed within the thickness of the external walls. The thrust from the arches would be transferred to the steel stanchions at about two-thirds of their height and resisted by bending in the stanchions, the top ends of which were held together by the tie beam (see Fig. 4.17).

Professor Schultze of Aachen University has provided another example in which tie rods, inserted to prevent the spreading of the abutments of a Gothic vault, had their mid-points raised above the crown of the vault by means of a steel roof truss, so as to make them invisible from below the vault.

There has been a number of instances when arches and vaults have been 'strengthened' by being suspended by grouted-in hanger bars from a concrete structure built above (and in many cases right on top of) the original masonry structure. The attraction of such an arrangement is that the concrete relieving structure can be designed and built to conform today's codes of practice, thus circumventing the ambiguities in the original construction, e.g. if sufficient buttressing is not available, the relieving structure can be made to act as a beam.

Such interventions preserve the appearance of the underside of the arch or vault, but change the function from structural to merely decorative. They should therefore be considered as a last resort, particularly when they involve casting or spraying the concrete directly on to the masonry, as in that case the intervention becomes irreversible. If provision of such an overlay structure, after careful consideration, is found to be the only solution, a separation membrane e.g. polyethylene sheeting should be used.

The shape of horizontal sections through domes, whether polygonal or circular, can make it relatively easy to provide additional circumferential strength

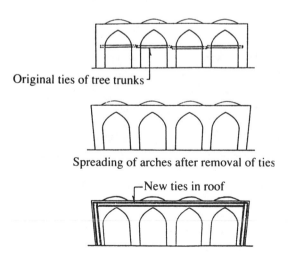

Original ties of tree trunks

Spreading of arches after removal of ties

New ties in roof

Figure 4.17 Causes of spreading arches of Koran school in Rezayie and remedial works.

and/or horizontal base restraint; removal of the roof cladding and a small thickness of the masonry will enable the insertion of a tension ring. As a single steel rod or cable may cause too high bearing stresses and hence tend to 'cut' into the masonry (especially if this is of soft brick), a number of smaller bars or cables should be used together with spreaders along meridians. Such hardware must be adequately protected against ingress of rainwater, unless stainless steel or non-ferrous alloys are used (see Secs 4.1.4 and 4.5.12). If space permits, reinforced concrete ring beams may be used, preferably with a separation membrane to facilitate a future renewal or removal, so that reversibility of the intervention is made easier.

No attempt should be made to close up cracks by tensioning the ring ties. For one thing, dislodged particles of mortar, etc. will have wedged themselves into the cracks so as to resist closure and, secondly, where cracking is caused by insufficient buttressing of the base ring, the dome will have modified its original shape and have spread so as to 'follow' the receding buttresses. Closing the cracks, would then 'shrink' the dome and cause separation cracks at the buttresses. A moderate amount of pre-tensioning may, however, be beneficial by enabling the ring ties to restrain further movement without significant extension, as such extension would be accompanied by further cracking.

4.7 LATERALLY LOADED MASONRY STRUCTURES

This heading covers as disparate structures as Gothic window tracery and retaining walls. What is common to them is that the predominant external loading on them is horizontal and that they therefore have to be treated differently from load-bearing walls and piers. They do however each have their specific problems.

4.7.1 Window Tracery

Gothic window tracery is a very special form of masonry construction. The resistance against wind loads of such windows has little to do with bending of the tracery ribs and must be created by arching within the depth (at right angle with the plane of the window) of the ribs with the window surround acting as abutment. It sometimes happens that such abutments yield to the extent that the arching action is put in jeopardy. For modest-sized windows, it may be adequate just to fill the joints which are opening and/or those in which the mortar has been eroded by weathering.

An example, on a large scale, is the great East window at York Minster, which has the overall dimensions of a lawn tennis court. Below the springing of the main arch, the tracery is supported by a system of double mullions. These are connected at the level of the springings of the window arch by a stone walkway approximately 750 mm wide and at half height with stone transomes at right angles to the window. The window was found to have a

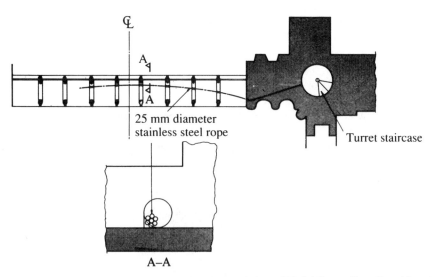

25 mm diameter
stainless steel rope

Turret staircase

A–A

Figure 4.18 Horizontal restraint to tracery of east window of York Minster (*from Dowrick and Beckmann, 1971*).

substantial horizontal bow, outwards. The resistance to external wind pressure did not give cause for concern because the bowed walkway and the tracery itself would arch between the turrets either side of the window. There was, however, great concern that suction caused by a strong westerly wind gust might suck out the tracery, together with the early fifteenth-century stained glass.

The adequacy of the structural mechanism of wind resistance of the window in this condition was very difficult to justify. One solution might have been to take down the glass and the tracery and re-build the latter with such strengthening as would guarantee the resistance against wind. This course of action was not favoured, because removal of stained glass, fixed with mortar to the tracery, almost always causes some damage to the glass.

The chosen solution was a system of horizontally draped, catenary (i.e. in the shape of a string polygon), stainless steel cables to which the ribs of the tracery were connected by means of small stainless steel turnbuckles, such as used in yacht rigging. Any tendency to outward movement of the tracery under wind suction will pull on the turnbuckles and hence tighten the horizontal catenary cables, which will then restrain any further movement. The visual effect of this arrangement has been found to be negligible (see Fig. 4.18).

4.7.2 Retaining Walls

Any wall that separates two different levels of soil is in effect a retaining wall, regardless of whether it is freestanding and creates a step in ground level or whether it 'spans' between a cellar floor and a ground floor. In the latter case, it carries the vertical load from the weight of the masonry above the ground floor in addition to its own weight, whilst the freestanding wall has its own weight (including a parapet, if any) as its only vertical load.

Common to the two wall types is the condition that the horizontal loading from the pressure of the retained earth (and possibly groundwater) is of crucial importance, and the most important difference between them is the horizontal support given by the ground floor to the wall of the cellar or undercroft.

The total horizontal force per unit length of a retaining wall is generally proportional to the square of the difference in level of the soil on the two sides of the wall and proportional to the density of the soil. The latter depends in turn on the groundwater level. The earth pressure also depends on the general slope of the ground on the 'high' side; the more it slopes upwards, away from the wall, the higher pressure and, finally, any so-called 'surcharge', such as vehicles on a road near the top of the wall, will increase the total horizontal loading, albeit intermittently.

Any change of *any* of the above-mentioned parameters will therefore affect the horizontal forces acting on the retaining wall and will hence affect its stability. Excavations, other than small exploratory pits, at the low side of a retaining wall are particularly to be avoided.

Modern reinforced concrete retaining walls often have their stability improved by ground anchors: steel cables or strands are grouted into holes drilled in the soil, sloping slightly downwards from the wall, and anchored by the friction between grout and soil. Ground anchors deliver their restraint as concentrated forces where they are anchored to the wall. Their use to strengthen historical structures would therefore seem to be limited to large civil engineering retaining walls (e.g. in railway cuttings and quay sides) that are thick enough to accommodate (reinforced) concrete spreader blocks or beams. There may also be some concern over their durability, as the grout cover to the strands may be difficult to maintain in the near-horizontal holes.

Lesser retaining walls, associated with or forming parts of buildings, can in some instances have their stability improved by treatment of the soil conditions on the 'high' side. Foremost among such measures is improvement of the drainage; this will lower the groundwater and thus reduce the density of the soil and hence the pressure on the wall. An additional benefit of better drainage is a reduction of any dampness on the inside of the cellar walls.

It should finally be remembered that, in the case of cellar walls, internal cross-walls with a spacing less than twice the height from floor to ceiling contribute significantly to the strength of the external cellar wall. This is because brick masonry has better bending capacity horizontally than vertically. If an intended new use requires the removal of such cross-walls, this must not be proposed until a thorough investigation of all relevant parameters has confirmed that the external retaining cellar wall will be structurally adequate without the cross-walls.

4.8 MASONRY CLADDING TO IRON- AND STEEL-FRAMED BUILDINGS

Many city buildings from the end of the nineteenth and the first half of the twentieth century appear at first glance to be of traditional masonry construction. Closer investigation will however show that the structure is a frame with columns ('stanchions') and beams of wrought iron or steel.

4.8.1 Common Arrangements of Cladding

In the earlier buildings with this type of structure, the façade columns had brickwork built around them and between the columns; brickwork spandrel walls were supported on the beams. The brickwork may, in turn, have a bonded cladding of stone or, in the later buildings, the stone cladding is built directly against the steel stanchions, supported on the beams and sometimes tied back to the frame with metal cramps (Fig. 4.19). From the late 1930s, it became increasingly common to encase the steelwork in concrete in order to provide fire resistance and corrosion protection.

Terra cotta was sometimes used for these buildings in a similar way as stone; as a cladding bonded to a brick backing and on some decorative features like domes, the hollow terra cotta blocks were directly supported by and tied to the members of a metal skeleton forming the structure of the 'dome'.

4.8.2 Problems and their Diagnosis

When no concrete encasement was provided, the steelwork would usually have had some paint protection, but this would often be little more than a coat of 'red oxide' applied at the works and touched up on site. In the main, reliance was placed on the masonry cladding and the roof to protect the steel from the elements and thus prevent corrosion.

Inadequate gutter details and erosion of masonry mortar joints, aggravated by poor maintenance, will however over the years allow water to get at the steel. Corrosion will start where the paint film, if any, has been damaged during the cladding construction and will then spread from there. Rust can propagate under a seemingly intact paint film; see Sec. 6.6. The severity of the corrosion is not always related to inadequacy of the painting. Water, getting in at high level, may run down the stanchions without causing much rusting, but where the vertical flow is impeded by cleats and beams, etc., the water collects and causes severe rusting.

The corrosion rarely causes significant loss of strength of the frame. It can, however, have serious consequences for the masonry cladding for the following reasons. Rust, once formed, encourages the corrosion of the underlying metal; it also occupies a volume that is between 6 and 10 times as large as the volume of iron from which it is created. As a result, the rust will form a crust that will increase in thickness with time and exert an outward pressure on the masonry cladding. This pressure leads to cracking of the masonry, usually along lines parallel to and within the width

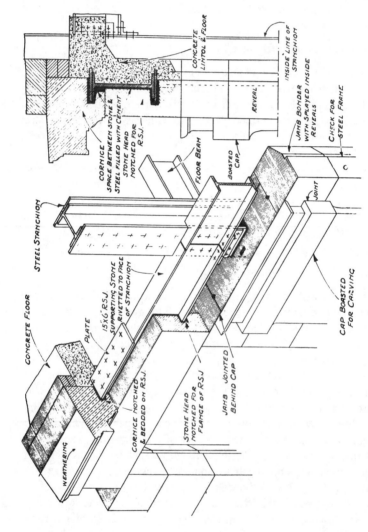

Figure 4.19 Details of masonry cladding to steel-framed building *(from Warland, 1953)*.

CORNICE
SPACE BETWEEN STONE &
STEEL FILLED WITH CEMENT
STONE HEAD
NOTCHED FOR
R.S.J.

CONCRETE
LINTOL & FLOOR

REVEAL

INSIDE LINE OF STANCHION

JAMB BONDER
WITH SPLAYED INSIDE
REVEALS

CHECK FOR
STEEL FRAME

STEEL STANCHION

FLOOR BEAM

BOASTED
CAP

JOINT

CONCRETE FLOOR

PLATE

15"×6" R.S.J.
SUPPORTING STONE
RIVETTED TO FACE
OF STANCHION

CAP BOASTED
FOR CARVING

CORNICE NOTCHED
& BEDDED ON R.S.J.

STONE HEAD
NOTCHED FOR
FLANGE OF R.S.J.

JAMB JOINTED CAP
BEHIND CAP

WEATHERING

of the steel or iron members. In some case, entire stones or brick courses are displaced, but usually the cracking disrupts the masonry units themselves, particularly when, as in the case of stone, the units are shaped with internal sharp re-entrant corners to fit round the flanges of the steel sections.

The diagnosis is simple in principle; once it is established that the building has an iron or steel frame, cracks on the façade along the lines of the metal frame members behind the face indicate serious corrosion underneath. The extent of the trouble is, however, far more difficult to ascertain, as absence of cracks does *not* prove absence of rust.

The reason for this is that due to the inaccuracies in the relative alignment of the frame relative to the cladding, the cladding will in some areas have been built hard up against the frame, whereas in other places there may be a gap, 10–20 mm wide, between the back of the cladding and the frame member.

In the first case, the rust will initially have filled any porosity and voids in the mortar bedding of the cladding and the pressure will then have built up until cracking occurred. In the second case, the rust will have been able to grow into the gap without exerting any pressure on the cladding, which will therefore, on inspection, show no cracking. If, however, at the time of inspection, the rust has just filled the gap, it may only be a few years before the pressure builds up enough for cracking to start.

These two situations can exist on the same façade within a metre or two from each other. The extent of the corrosion attack on the frame, and hence the amount of remedial work necessary, can therefore *not* be assessed from an inspection of the face only, but requires removal of the cladding.

It is quite common to find cracking, indicative of corrosion, on only 10–20 per cent of the facades of a building. If then the building owner insists on a guarantee that no further work will be required for the next 25 years, it means dismantling the whole of the cladding. Such dismantling inevitably causes damage. Replacement material, to match the existing, may be difficult to obtain in the quantities required, and the cost of such an operation is usually out of proportion to the benefit (see also Sec. 6.6.3).

A less drastic and far more economical approach is to remove the cladding where it is cracked and from adjacent parts of the frame, as far as rust is observed, to treat the rusted metal as described below and reinstate the cladding. Financial provision should then be made to cover the possibility that isolated areas, where rust was not discovered first time, may have to be dealt with in 5 or 10 years time.

4.8.3 Repair Techniques

Normal repair and/or replacement techniques will be used for the masonry with special attention being paid to achieving weatherproof joints. For the exercise to be successful, it is however essential that the corroded metal surfaces are properly treated. *All* rusted surfaces must be exposed, not just those facing the cladding; this may entail extensive removal of brickwork surrounding the frame members. All rusted surfaces must then be given radical remedial treatment against corrosion, as described in Sec. 6.6.5. The treated surfaces must be protected with an abrasion resistant paint or by heavy-duty self-adhesive tape so as to prevent breaching of the paint system during the reinstatement of the masonry cladding.

4.9 STONE STAIRS AND LANDINGS

The structural adequacy of the Georgian/Victorian 'cantilevered' stone stairs has always been a challenge to structural engineers. The bending strength of the individual stair treads would often not seem to be nearly enough to give reliable safety, and even if it were, the embedment in the wall would appear to be grossly overstressed.

4.9.1 Structural Behaviour

It is generally accepted that, in the majority of cases, the individual stair treads do not act as cantilevers in bending. For a straight stair flight with plain overlap joints, each tread receives the vertical reaction ΣP_{n-1} from the one above, along its upper overlap joint and it transmits this reaction plus its own load P_n to the tread below along its lower overlap joint. The couple M_n, resulting from the horizontal translation of these vertical forces, is transmitted in torsion by the tread to its embedment in the wall (see Fig. 4.20a). At the next tread down the load transfer is repeated and so on, until the bottom tread which rests on the ground. In such a flight, each tread has to resist the load from all the treads above it.

If the treads have rebated joints, these can transmit both vertical *and* horizontal forces. The horizontal forces on the upper and lower joints of each tread create a couple that partly counteracts the moment of the vertical forces and in this way greatly reduces the torsion on the lower treads (see Fig. 4.20b). A more detailed account of this action is given by Sam Price (1996) in *Architectural Research Quarterly*. The effect of the interlocking rebated joints does, however, depend on perfect fit, and hence the workmanship of the joints and in many cases the behaviour will be somewhere between the two, outlined above.

At the junction with a landing at the top of such a flight, a mutual support action can then develop through the in-plane compressive forces in landing and flight combining and creating a vertical reaction (see Fig. 4.20d). The in-plane forces are in turn transmitted to the wall in shear. At the junction between the landing and the rising flight the situation is however not so happy. The in-plane forces, required to create the reaction to support the flight, would have to be *tensile*, which the joint cannot transmit.

The plan in Fig. 4.20c shows a two-part landing (this would be a fairly common arrangement to limit the size of stone slabs to be handled during construction). In this arrangement, the landing slab ABFE is supported on the walls along ABD *and* along the junction with the lower flight AC. It is loaded by the landing slab CDHG along the half-joint between CD and EF (see Fig. 4.20e). The landing

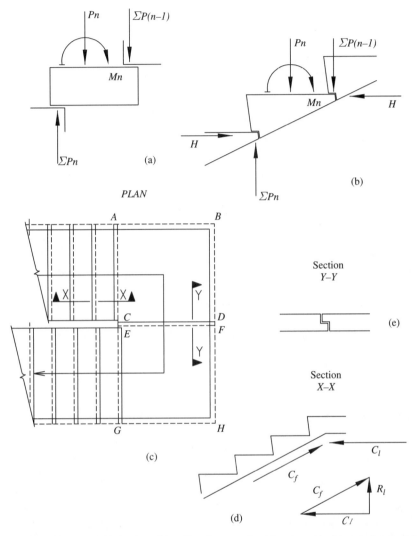

Figure 4.20 Structural behaviour of 'cantilever' stone stairs; (*a*) action with plain, overlapping joints; (*b*) action with rebated joints; (*c*) plan; (*d*) mutual support at junction of flight and randing; (*e*) section of half-joint in landing.

slab CDHG is supported on the walls along FHG *and* on the half-joint; it is loaded by the rising flight along EG. This and similar arrangements give rise to significant bending moments in the landing slabs, but nowhere near what would be the case, if they were truly cantilevering. A further factor, reducing the bending, is that the deflections of the various components, however small, will tend to make the resultants of the in-plane forces deviate from the centreline of the treads towards the supporting walls.

Sometimes there are situations where these actions cannot take place, and where considerable bending has to be resisted by the landing slab. It is also quite common to find an iron or timber beam, that has been inserted under the landing edge at some time in the past!

As the load-carrying mechanism of these stairs is often very complex, past performance is often the best guide to their adequacy, as long as they are not over-loaded, damaged or interfered with during refurbishment work. Even a wrought iron balustrade may contribute to the load-carrying capacity by distributing load from one tread to its neighbours.

4.9.2 Problems and Remedies

Occasionally it is found that a tread in an otherwise intact flight of stairs is cracked. This is usually caused by a heavy load having been allowed to drop on to the tread. Renewal, even if a replacement tread of matching stone could be made, would be a very tricky operation. It would require an elaborate support scaffold and would involve difficult cutting out of the embedded end and grout-ing in of the replacement. A much less drastic alternative is to core drill a hole along the length of the tread (which will need temporary support during this oper-ation) and grout in a steel tube, thus restoring the integrity of the cracked tread. Where the crack is some distance away from the wall, it may be possible to make undercut grooves in the top of the tread, bond in reinforcing bars with resin and make good, cosmetically, with a stonedust-based mortar.

4.10 FIRE PROBLEMS

4.10.1 Effects on Masonry Materials of Fire, Smoke and Water

The components and materials of masonry are incombustible. They are, however, affected by the temperature, to which they are raised by the flames impinging on the masonry. This temperature depends on the temperature of the flames, the time the masonry is exposed to the flames and the mass of wall or pier behind the exposed surface.

The calcium carbonate in limestone and in mortars begins to give off carbon dioxide at about 500 °C and at about 900 °C it will be almost completely 'cal-cined', i.e. transformed to calcium oxide (quicklime) and will have lost most of its cohesive and adhesive properties.

Sandstone may contain grains with chemically bound crystalline water, which when heated above 100 °C will boil and cause spalling of the surface.

Bricks are unlikely to be adversely affected by temperatures below those at which they were fired. This is usually 1000 °C or more for modern bricks, but can be considerably less for old bricks, fired in wood-fuelled kilns. When heated above the original firing temperature, the original porous, sintered, texture may soften and the bricks distort.

Smoke as such does not affect the mechanical properties of masonry. The soot deposited by smoke from burning of certain materials, such as some fuels and a range of synthetic products, may however contain chemicals, such as sulphates and chlorides, which can attack limestone and mortars.

Water from firefighting can have two effects. The first is that of producing thermal shock on surfaces heated by the fire; this can lead to spalling of the surface of the units, be they brick or stone. The second, which will be most serious in rubble-cored masonry, is that of saturating the interior of the masonry, so that even after a moderate fire, a long period of drying out will be required after reinstatement of the rest of the fabric.

4.10.2 Appraisal of Structural Masonry After a Fire

A serious fire will almost invariably cause spalling of stone surfaces, exposed to the flames, and the remaining stone may well have been weakened to some depth beyond the spalling.

In the case of limestone, a pink discolouration will indicate significant loss of strength (cf. Sec. 7.5.2) and will indicate a need to disregard the strength of the affected depth or even require its removal.

Sandstones are not so easily assessed, and the use of a 'Schmidt-Hammer' (see Sec. 7.2.4) to compare the surface hardness with that of an area, not affected by fire, may help to determine the depth to which the strength loss has penetrated. This could be done by removing, say 25–30 mm, preparing and testing a small area, removing a further 25–30 mm and so on, until a hardness, comparable to that of the unaffected masonry were reached.

Late nineteenth- and twentieth-century bricks, that have not disintegrated or cracked, are unlikely to have lost a significant proportion of their original strength, but earlier, lightly fired, bricks that have first been heated and then doused with water could be suspect.

Lime mortar that has been calcined will carbonize when exposed to the carbon dioxide of the atmosphere and should, in theory, regain its strength. There is, however, no data available to confirm this, nor how long the process may take. Cement mortars will behave as concrete (cf. Sec. 7.5.1).

4.10.3 Reinstatement after Fire Damage

Masonry, to replace damaged material, cannot be applied to the unaffected remainder of a wall or pier in arbitrary thicknesses. One has to consider a minimum structural replacement thickness of half a brick or a depth of stone, equivalent to the thickness of the ashlar.

There is also the problem of bonding the replacement to the remaining sound masonry. Attempting to do this by forming pockets to receive headers or bonding stones is likely to cause more harm than good. The best that can be hoped for is to build the replacement leaf with a mortar-filled vertical joint against the surviving masonry and using grouted-in metal ties to bond the two skins together (cf. Sec. 4.5.9).

When only a relatively small thickness has to be replaced, one might be inclined to use sprayed concrete (see Sec. 7.4.1). The deformation characteristics of the new concrete are, however, likely to be different from those of the parent masonry. This is because the concrete will shrink, while the masonry, which has been dessicated by the fire, will take up moisture from the atmosphere and expand over a period of months. Before a concrete replacement is decided upon, these potential problems should therefore be carefully investigated.

Where only a relatively small proportion of the total wall thickness has been affected, so that its load-carrying capacity is still adequate, and where the affected face is hidden from view, the best solution may be to hack off any lose or crumbling stone and not attempt to replace the lost thickness. This was done at York Minster, after the fire in 1984, on the triangle of the Central Tower masonry between the roof and the vault of the South Transept. In this case, a wire mesh was stapled to the sound masonry to catch any small fragments that might detach themselves due to the slow re-hydration and carbonation of the partly calcined layer left in place.

FIVE

TIMBER

Tithe Barn, Bradford-on-Avon: Originally constructed in the fourteenth century, the roof spans some ten metres. The curved members give additional support to the lower 'collar beam', which may have supported a storage loft in the past. They transfer some of the loads to the 'springings' of the 'arches', half-way down the walls, in a similar way to the action of a 'hammer-beam' construction. The inclined braces, between the principal rafters and the purlins, provide lateral stability. The timbers, built into the walls are vulnerable to decay, if the roof covering is damaged.

When man emerged from the caves, timber would have been his first building material. Being generally light and easy to work, it has remained the most popular one where it is freely available and it has the great virtue of coming from a source that is, in principle, renewable. Being a natural material, it is however variable in its properties and its performance, whether in a piece of antique furniture or in an old structure, depended to a large degree on the way it was selected and fashioned by the craftsman, who had learned the tricks of his trade through a long apprenticeship.

With the decline of craft traditions in the building industry, this knowledge was partially forgotten. Timber of good quality also became harder to obtain and it was sometimes used in unsuitable ways or in unsuitable conditions. The resulting bad experiences, combined with engineers' perception of the seemingly vastly superior properties of steel and concrete, led to timber being relegated to domestic floor and roof structures and little, if any, design of timber structures was taught to engineering students. The 'knock-on' effect of this was that many timber structures were condemned and replaced with concrete or steel as part of refurbishment work in older buildings. Many timber structures can, however, continue to serve their purpose for continued or new use of buildings, if their appraisal and repair/strengthening are based on proper knowledge of the making and intended behaviour of the structural elements and the response of the material to loading and environment.

It may be worth here to repeat that 'Hardwood' and 'Softwood' are traditional, if simplistic, English terms, describing timber from angiosperm, i.e. broad-leaved trees and gymnosperm, i.e. conifers, respectively; they do not refer directly to the mechanical properties of the wood, although they tend to be indicative in the field of timber *structures*. (Balsa is the prime exception, being an exceedingly light and soft material, ideal for model aeroplanes and Kon-Tiki rafts, if not for much else; it is a hardwood according to this nomenclature.)

5.1 MATERIAL PROPERTIES AND BEHAVIOUR

5.1.1 From Tree to Timber

It is stating the obvious, that timber comes from trees, but it is an important factor to bear in mind, because so many of the properties of the material are influenced by this.

In the growing tree, the layer immediately beneath the bark (the cambium) forms both longitudinal and transverse cells, which, apart from forming the internal structure of the wood, can store and conduct the sap. In the spring in temperate climates, when large amounts of nutrients have to be carried to the new growth, a larger proportion of cells are formed and the resulting wood, the 'early' (or 'spring') growth is more porous and lighter than that of the 'late' (or 'summer') growth. This leads to alternating 'cylindrical'(actually conical) layers being formed, which appear on a cross-section as the 'growth rings'.

In a very young tree all the wood is sapwood, but as the tree gets older, the flow slows down, the vessels near the centre gradually clog up and the cell contents are transformed into darker complex chemical substances that, in some species, increase the inherent, natural, durability of the wood. This inner, darker, part of the tree is called the heartwood.

The longitudinal arrangement of the cells is diverted where branches are formed; these, in fact, repeat the cell pattern of the trunk in miniature and appear on a longitudinal section of the trunk as knots. This interruption of the more or less parallel array of the fibres gives rise to a significant weakening of the longitudinal strength of the timber in the area surrounding the knot.

Trees were traditionally felled during the winter when they held the least amount of sap and hence water. This was in order to reduce the amount of moisture that the finished timber, particularly the sapwood, had to lose, before it achieved moisture equilibrium with its final environment. As timber, like any other material, shrinks when it loses moisture, this meant that shrinkage from the as-felled condition to the final condition was reduced. A more effective way of partly seasoning wood from broad-leaved trees was to ring-bark them and leave them over the summer. Modern, commercial, forestry practice generally relies on drying out the timber after felling.

Square timbers were formerly shaped in the forest by adzing and axing the felled trunks and this practice continued in Scandinavia into the first decades of the twentieth century. In Britain pit-sawing was introduced in the seventeenth century. Square timbers are now generally sawn, together with boards, at mills with seasoning facilities within easy reach, where the moisture content of the sawn articles is reduced.

Due to its non-isotropic internal structure, timber does not shrink by the same amount in every direction: the ratios between the amounts of shrinkage in the axial, radial and circumferential directions are, very approximately, 1 to 50 to 100. Furthermore, the spring-grown wood shrinks more than the summer-grown.

As a rough indication, the approximate percentages of shrinkage from 'green' to 'air-dry' condition are:

	Longitudinal	Radial	Tangential
Oak	0.03–0.4	1.1–7.5	2.5–10.6
Beech	0.2–0.3	2.3–6.0	5.0–10.7
Baltic softwoods	0.01–0.2	0.6–3.8	2.0–7.3

(The terms 'green' and 'air-dry' are explained in Sec. 5.1.2.)
This shrinkage movement is irreversible. The subsequent moisture movement, when in use, can be reversible and is about half of the initial shrinkage.

The greater shrinkage in the circumferential direction relative to the radial can cause radial fissures or 'shakes'. In a log or a square timber these run longitudinally, widening towards the surface (Fig. 5.1*a*). The proportionally greater circumferential shrinkage can also cause 'cupping' (transverse curling) of tangentially sawn

(a)   (b)

Figure 5.1 Effects of shrinkage: (*a*) fissures; (*b*) 'cupping' of boards (*from Suenson, Vol. 2, 1922*).

boards (Fig. 5.1*b*). Radially or 'quarter'-sawn planks (rare nowadays, but may be found in old buildings) do not suffer from this.

5.1.2 Mechanical Properties of Components

A considerable amount of testing has been carried out on small, defect-free, so-called 'clear' samples of wood. These test results are however only a starting point, when assessing the carrying capacity of a structural component. This is because the strength of any piece of structural timber is significantly affected by its natural imperfections, such as knots, fissures, etc.

Nevertheless, certain basic facts emerge from these tests:

1. Timber, subjected to direct tension, behaves elastically right up to failure. This takes place by the fibres tearing suddenly. Direct tensile tests are difficult to carry out, as the specimens are prone to break or shear in the jaws of the testing machine. Modern 'tensile' tests are therefore usually bending tests that measure a 'modulus of rupture'. This is about 125–135 per cent of the direct tensile strength.
2. Young's Modulus, measured in bending tests is, in the air-dry condition, about 4–5 per cent of that of steel.
3. The compressive strength parallel to the grain is somewhat less than the tensile strength; generally in the order of 70–80 per cent. This compressive strength is thus about 50–65 per cent of the modulus of rupture.
4. The compressive strength at right angle to the grain is drastically lower than that parallel to the grain. For hardwoods the 'across-grain' compressive strength is generally in the order of 25–30 per cent of that parallel to the grain; for softwoods it can be as low as 15–20 per cent. When compressed parallel to the grain, the timber fails by the fibres buckling and folding into each other and the specimen shortens by a few per cent. When compressed at right angle to the grain, the timber squashes and shortens substantially before it fails by cracking parallel to the compressive force; for softwoods, a shortening of about 30 per cent prior to failure has been observed (Fig. 5.2).
5. The tensile strength at right angle to the grain is even lower than the across-grain compressive strength: 10–15 per cent of the modulus of rupture, or 13–19 per cent of the direct tensile strength.
6. The shear strength parallel to the grain is about 65–75 per cent of the compressive strength at right angle to the grain, i.e. about 8–12 per cent of the modulus of rupture, or 10–15 per cent of the tensile strength parallel to the

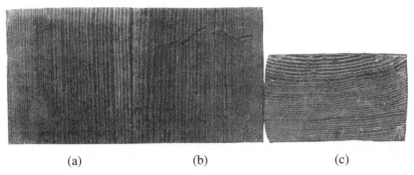

(a)	(b)	(c)

Figure 5.2 Three cubes of softwood: (*a*) unstressed; (*b*) crushed parallel to the grain, at 53.4 N/mm^2; (*c*) compressed at right angle to the grain to 28 N/mm^2 (*from Suenson, Vol. 2, 1922*).

grain. (The low strengths at right angle to the grain and in shear have been, and still are the major limitations on timber as a structural material.)

7. When the timber is 'green', all strength values are significantly lower than for the 'air-dry' condition: generally in the order of 75–90 per cent for tension and 65–75 per cent for compression.

'Green' and 'air-dry' are terms, used to describe the moisture content of the timber when recently felled, and when it has reached moisture equilibrium with the air. The moisture content is defined as the weight of the water in a given volume of timber divided by the weight of that volume, when oven dry:

$$M_c = \frac{W_1 - W_0}{W_0}$$

(This definition leads to the, at first glance confusing, result of 100 per cent moisture content of a piece of wood of which half the weight is water; similarly, a log, just felled in spring, may have a moisture content of 300 per cent.) For appraisal of existing timber structures, the moisture content can be assessed by means of 'moisture-meters'. These are instruments, that measure the electrical conductivity between two prongs, pressed into the face of the timber.

Some of the mean values of the results of many tests on clear specimens, published by the Building Research Establishment ('The Strength Properties of Timber' by Lavers and Moore, 1983), are given in Table 5.1.

This series of test results, does not include data for compression at right angle to the grain, probably due to the difficulty of defining the failure criterion; code design stresses for this mode of loading are determined on the basis of acceptable deformations.

It should be borne in mind that strength values, such as those in Table 5.1 above, are the results of tests, in which the specimen is loaded relatively rapidly. The strength of timber is however particularly dependent on the duration of loading; the strength under sustained load is only about two-thirds of that under a load applied for, say, 15–30 seconds.

Table 5.1 Results of strength tests (in N/mm²) (the percentages are moisture contents)

	Scots Pine		Douglas Fir*		European Oak		Beech	
	50%	12%	50%	12%	50%	12%	50%	12%
Modulus of rupture	46	89	54	93	59	97	65	118
Compression, parallel to grain	22	47	26	52	28	52	28	56
Shear, parallel to grain	6	13	7	10	9	14	9	16
Modulus of elasticity	7300	10 000	10 400	12 700	8300	10 100	9800	12 600

* These mechanical properties apply only to Canadian-grown Douglas Fir; UK-grown has properties nearer to Scots Pine.

For the purposes of assessment of the carrying capacity of an existing structure, it is worth noting that tests, carried out at the University of Karlsruhe, have shown no significant difference in strength between clear samples of old and modern softwoods of the same species. However, for the same species, slower growth produces closer growth rings and hence greater strength, and in trees of similar diameter, a higher proportion of the wood will be heartwood. Old timber members may have been sourced from natural forests in areas where climate and soil led to slow growth; they may therefore have better strength than modern timber of the same species. (In some areas, fertilizer is nowadays used to promote quick growth of re-planted trees, and even the rotting stumps of felled trees add nutrients to the soil. The practice of spacing re-planted trees so as to admit light and air, further maximizes the rate of growth.) Furthermore, when having to provide structurally critical components, such as long-span beams, the traditional carpenter would look for, and select, the pieces from the stock-pile, that had the closest growth rings, the straightest grain and the minimum of imperfections. In modern parlance, this was 'visual grading' (see Sec. 5.2.1) and that can result in old main timbers being stronger and stiffer than their modern equivalents, when judged merely on species and appearance. There have been cases when a conventional 'code-of-practice' assessment indicated severe overstress in beams, which were carrying their loads without any excessive deflection or other signs of structural deficiency.

5.1.3 Durability

Being an organic material, timber is susceptible to attack from insects and their larvae and from microorganisms.

Insects attack living trees and stored timber as well as timber in completed structures, regardless of whether it is damp or dry. They tend to have a preference for sapwood. In many tropical and subtropical countries the greatest danger is from termites. In termite societies it is the adult workers who gnaw the wood in the first instance. There are however a great many varieties of termites, each with their particular habits, and they would require several books to describe. Preventative measures and/or eradication is extremely difficult and a subject for the expert.

In temperate climates the greater insect damage is caused by the larvae, hatched from eggs laid in the timber by the adult insect. The larvae feed on the wood, tunneling through it until they emerge, pupate, mate and lay more eggs in the timber. They generally prefer softwoods and sapwood, but one species in particular, the 'Death-Watch Beetle', finds oak sapwood very appetizing, but it will attack the heartwood, once it finds that there is little more nourishment and breeding space to be had in the sapwood.

The activity of beetle larvae increases with temperature up to a certain level. Mediaeval church roofs survived for centuries when little or no heating was provided, but succumbed fairly quickly to the death-watch beetle, once heating was installed.

Fungal decay occurs in several forms, each caused by its particular organism. Common to them all is that they require moisture to thrive; British Codes of Practice state that at moisture contents less than 20 per cent, timber is practically immune to fungal attack. In fact, the fungi causing 'Wet Rot' die, if the timber is dried to a moisture content of less than 20 per cent.

The fungus causing 'Dry Rot', *Serpula lachrymans*, is the main cause of biological damage to timber structures in Britain. It is capable of 'hibernating' when the host timber is dried out and as it is also able to infiltrate adjacent masonry, it is generally very difficult to eradicate completely. Exposure to temperatures above 40 °C for 15–30 min will kill it, but this temperature has to reach the full depth of the affected wood and this is difficult to achieve in-situ in a building.

Common to insect- and fungal attack is that they destroy the substances that give the timber its structural strength and that if no action is taken, they progress and spread to hitherto unaffected wood and to new timber. Any project for continued or new use of a building with structural timber, even if this is only floorboards, must therefore include a full and careful examination of the timbers for signs of biological attack and, if any is found, however minuscule its extent, an assessment should be made of its significance.

5.2 APPRAISAL OF TIMBER STRUCTURES, THEIR CHARACTERISTICS AND COMMON PROBLEMS

Unless the actual loads, carried by the structure during an earlier use of the building, have been ascertained and found to have been greater than those that will be imposed by the intended new use, any appraisal must include an assessment of the strength of the structure.

5.2.1 Common Problems of Assessment of Strength

The strength of a structural member of timber depends on the species of tree, it was cut from, and the inevitable natural imperfections, present in the particular piece of wood that makes up the member. Some species of wood have a characteristic grain, which makes them fairly easy to recognize by a superficial, visual, examination (as long as the surface is not obscured by paint).

The location and age of the building can in many cases provide another fairly reliable indication of the species, because timber was in the past either cut locally or imported from only one or two sources. For instance, although the first ships, carrying Norwegian Fir timber, are said to have landed at Grimsby in 1230, English mediaeval and Tudor timber frames were usually made from local oak or elm. From the seventeenth century onwards, softwoods were imported in quantity, mainly from Scandinavia and the Baltic coasts.

The later the date of construction, the wider however becomes the area from which the timber may have come and this makes identification by age and location increasingly uncertain. Furthermore, in the case of softwoods, some species with significantly different strength properties are difficult to distinguish from each other by superficial examination. Identification of the species may therefore require expert examination, with a magnifying lens, of the grain of a freshly cut cross-section.

Identification of the species enables the strength properties of a clear piece (a piece free from any imperfections) to be assessed by reference to test results, such as those, published by the British Building Research Establishment (see Sec. 5.1.2).

Any structural member is however likely to have at least one or two natural imperfections, which will reduce its strength below that of the clear sample. The amount of the reduction will depend on the nature and extent of the imperfection. The structural significance of the reduction depends on the level of stress at the location of the imperfection. For example, the *amount* of reduction of the bending strength of a beam, due to a knot, depends on its size and whether it is situated in the bottom, tensile, face or half-way up the side. The *structural significance* of the reduction depends on whether the knot is situated where the stresses are high, or where they are low.

The effects of the imperfections are allowed for in codes of practice by stipulation of reductions of design stresses. These reductions depend on the severity of the imperfections. The amount and severity of imperfections were, and are, the basis for the commercial classifications of timber, used by the various national producing and exporting organizations, e.g. 'Select No. 1', 'Fifths', etc. These quality classes are however only descriptive of the appearance of the timber, and whilst that may give a rough guide to its structural quality, it does not properly quantify the effect of the imperfections on the strength.

For timber for new construction it is possible to assess the weakening mechanically: the timbers are passed through a machine, in which they are subjected along their entire length to a local bending moment, imposed by loaded rollers; the deflections are measured and if the deflection of a piece is greater than a certain value, set for the expected structural 'grade', the piece is given the lower 'grading' and the permissible stresses that may be used for its design have to be reduced correspondingly. The piece will be colour coded by the machine along its length to indicate its grade.

This method is obviously impractical for existing structures, and the older alternative method of visual grading has to be used. In this, the visual imperfections (or 'defects' as they are called in British codes of practice) are measured and checked against limits, tabulated for each grade in the code. Specialist assistance may be helpful for this.

The current British Standard Code of Practice, BS 5628, relies for visual grading on the rules in BS 4978. For assessment of the effect of knots, these rules require the position of the pith (the centre line of the growth rings) to be determined. This is rarely possible in an existing structure. BS 5628 is therefore of little help in appraisal of older structures. The predecessor of BS 5628: CP 112, Part 2 (1971) did however contain rules for visual grading and gave tables for design stresses, corresponding to several grades, for softwoods as well as both British-grown and tropical hardwoods.

The grading rules and permissible stresses under (unfactored) working load, are given in Tables 3 and 4 of CP 112 and tables 59–62 in Appendix A to this code. They list 'basic' stresses for clear specimens and give permissible design stresses for grades, labelled '75', '65', '50' and '40', corresponding roughly to the percentages of the 'basic' stresses. These stresses are given for 'green' and for 'dry' timber. *For 'dry' stresses to be used in design, the moisture content has to be less than 18 per cent.*

For each grade the maximum permitted 'defects' (natural imperfections) are quoted under the following headings:

Fissures (Table 59): The maximum depth of any fissure (measured with a feeler gauge or pen-knife), as a fraction of the thickness of the member, is stated for each grade.

Slope of grain (Table 60): The maximum slope of the grain (measured on any face), relative to the axis of the member, is stated for each grade.

Wane (Table 61): These are the corners missing from the full rectangular section for part of the length of a piece of timber, due to the tapering of the tree towards its top; it shows as rounded chamfers, untouched by saw or adze. The maximum size, as a fraction of the width of the face, on which it occurs, is stated for each grade.

Knots (Table 62): For a given width of surface, the maximum size of knot, permitted for each grade, is stated; this depends on the position of the knot within the width (e.g. arris, margin or middle half) and also depends on whether the member is a beam, a compression member or a tension member.

The rules limiting these 'defects' apply to hardwoods and softwoods alike. In addition to these, CP 112 contains a further grade limitation for softwoods only:

Rate of growth: The *minimum* number of growth rings per 25 mm, measured radially over 75 mm, is stated for each grade.

The 'defects', common to hardwoods and softwoods, can be fairly easily measured in an existing structure, if the timbers are accessible and the faces have not been covered up or obliterated under too many coats of paint.

The number of growth rings is however intended to be measured on a cut cross-section. This can be difficult on existing structures, where ends of members may not be accessible for inspection. Fortunately, a rate of growth, that has been so rapid that it results in a significant reduction of the permissible stress, is less likely to be found in older structures, as the builders of these were able to select the better, slow-grown, timbers for their purposes.

For beams with notches for floor joists, it may be possible to lift a few joists and inspect the grain on the notch faces; lapped joints in roof trusses may also allow inspection. Where this is not possible and it is considered essential to establish the rate of growth, a small (10 mm) core can usually be drilled from a face, where it will cause no significant damage.

Most national codes have some rules for grading of timber, which determine the design stresses to be used for new structures. Whilst CP 112 has been superseded for new construction, it remains, however, one of the most useful for the assessment of existing structures. The information in Tables 3 and 4 and the grading rules of CP 112 have therefore been reproduced in the appendix at the end of this chapter as an aid to appraisal.

It should be remembered that the strength properties of timber, and hence the permissible stresses, depend on the duration of the load. BS 5268 and CP 112 quote ratios of 1:1.25:1.5, for permanent, short-term and transient loads, respectively. The permissible stresses, in these codes of practice, refer to permanent loads.

Assuming that the species has been identified, the moisture content ascertained and the defects have been measured, one can arrive at permissible working stresses, which can be used for calculations to confirm the strength of the member. In such an appraisal, the locations of the defects are known; it is therefore possible to allocate different grades, and hence different permissible stresses, to highly and less highly stressed parts of members. In this way one can make a more favourable assessment, than when designing a new structure with timber that will be graded regardless of where along the length the defects occur. For example, if a large knot is found near the end of an otherwise perfect beam, the fact that the knot would reduce the permissible bending stress from, say, 14 N/mm^2 to 9 N/mm^2, is of little consequence, as the actual bending stress at the position of the knot is likely to be well below 9 N/mm^2, if it is, say, 13.5 N/mm^2 at midspan. When an initial calculation has indicated a moderate shortfall of carrying capacity, this approach should be tried, as it may show that strengthening will not be necessary.

When it comes to roof trusses and similar frames, the carrying capacity will however depend, not only on the strength of the members, but also on the strengths of the connections. These are generally more critical than the strength of the members and the strength of some of the traditional connections, found in old roof trusses, etc., are very difficult to assess by calculation. The reason for this is the complicated three-dimensional geometry of the carpentry of the connections in what would otherwise be straightforward two-dimensional trusses; this causes the paths of the forces to diverge out of the centre plane of the truss and become indeterminate.

Occasionally, structural members can be seen to have performed their function satisfactorily for many years. If then calculations, even when based on localized grading according to load-induced stress levels, cannot prove a fully acceptable factor of safety, load-testing may still demonstrate structural adequacy. Ross (2002) gives details of these procedures and quotes case studies.

5.2.2 Mediaeval and Early Renaissance Buildings

Very early timber-framed buildings had their main posts dug into the ground, but almost all surviving buildings were built up above ground with the lowest member of the frame being a sill beam, resting on a plinth of stone or brick. The uprights of the frame were tenoned into this sill beam.

Generally, the structures consisted of wall frames of vertical and horizontal members, usually with some diagonal bracing members, surmounted by roof structures of various configurations (Fig. 5.3). The wall frame members formed panels, filled in with 'wattle', a weave of slender branches, which supported 'daub', a clay mixture applied to both sides of the wattle to form the wall surface. In later, and/or more important buildings, the panels would be infilled with brickwork and this was also used for repair/refurbishment.

In some parts of England, 'cruck' frames were favoured; in these, the main uprights were made from curved tree trunks, split along the plane of the curve to produce two symmetrical 'blades', leaning against each other and meeting at the top ends to locate the ridge of the roof. The wall and roof members were supported by various brackets, or 'spurs', cantilevering out from the cruck blades (Figs 5.4 and 5.5). A common feature of wall frames and cruck frames was that the loads were intended to be carried by the frames and not by the material in the walls.

Many important ecclesiastical and secular buildings were, however, built with load-bearing masonry walls and these were covered with roofs, supported on

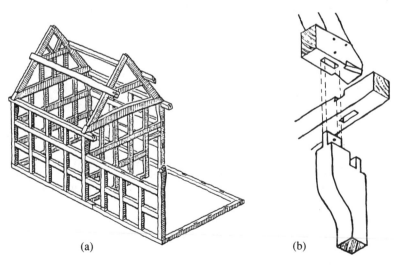

(a)   (b)

Figure 5.3 (*a*) Wall frame and roof trusses during assembly (*from Harris, 1979*); (*b*) Detail of frame connection (*after Brunskill, 1985*).

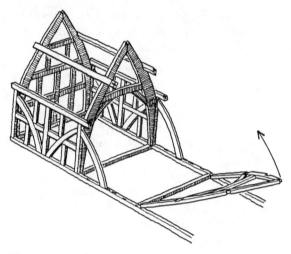

Figure 5.4 Cruck frame in cause of erection (*from Harris, 1979*).

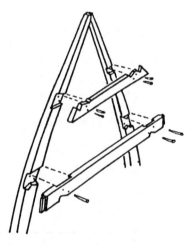

Figure 5.5 Connection details of cruck frame (*after Charles and Charles, 1990*).

substantial timber structures. Some of these were of elaborate 'hammer-beam' construction. (Structurally, this is a form of arch, supported at a level below the eaves, where the vertical pre-load, from the weight of the masonry above, is adequate to contain the horizontal thrust.) Others were little more than large beams off which principal rafters and purlins were propped.

Suspended floors of moderate dimensions would have joists spanning from wall to wall. These were commonly placed with their larger cross-sectional dimension horizontal; this enabled the use of trees that were not perfectly straight-grown without excessive straight-cutting and hence with a minimum of waste. Larger span floors had heavy main beams, spanning from wall to wall, with the joists spanning between them and supported in notches, cut in the beams.

A feature, common to all these structures, was the exclusive use of carpentered joints with no iron components. The difficulty of making an all-timber connection capable of resisting substantial tensile forces, was well understood. The configurations, generally used, avoided any joints transferring major tensions, with the exception of the joint between principal roof rafter and tie beam, where a substantial proportion of the rafter thrust could be transmitted to the tie beam by shear and friction.

Other joints might have had to resist lesser tensile forces, due to wind loads. Even then, most frames and trusses had their members arranged symmetrically about the centre of the span, so that most, if not all, of the wind forces in one direction would be resisted by members on the one half, working in compression, whilst the members on the other half would resist wind forces in the other direction.

The oak pegs, sometimes referred to as 'timber nails', usually found in these joints, mainly served to hold the parts together during erection and would not in themselves have provided any significant tensile capacity to the joints, which derived their main tensile strength from dovetailing. (Research at the University of Karlsruhe found the ultimate strength of oak 'nails' in single shear, parallel to the grain, to vary between 5 and 12 kN for both 20 and 24 mm 'nail' diameter.)

The main structural problems, likely to be encountered in buildings of this age and type, are usually caused by later alterations, carried out without concern for the structural integrity and/or by biological degradation, encouraged by poor maintenance.

In the first category belong the cutting off of diagonal braces, to form new door openings, and the removal of internal roof truss members, to create attic rooms, as well as indiscriminate notching of floor joists and even main beams, to accommodate plumbing and central heating pipe-work or electric cabling. The inherent reserve of strength of these early timber structures is such that their response to this kind of butchery is often no more than some large, and in some people's eyes picturesque, deflections. Such deflections do, however, not only reduce the serviceability of the building; they can be an indication that the margin of safety is far less than acceptable. Sagging floors should therefore not just be levelled up by the use of 'firring pieces', nailed to the joists, and roofs should not just be re-tiled, without the cause and the significance of the sagging being investigated.

The second category of problems encompasses the rotting of the feet of roof rafters, due to leaking roof cladding and blocked gutters, as well as decay of tenons, connecting uprights to horizontal members, where the panel infill has allowed water to enter and lodge in upward-facing mortices. Failure to deal with such fungal attack, *and* with its cause, can lead to accelerating deterioration, particularly if modern heating and draught-proofing is being installed in a building, previously only sparsely heated by a few open fires.

Similar problems can arise, where an originally timber-framed building has been 'overclad' with a skin of brickwork, in order to re-style the front elevation to, say, Georgian appearance. The half-brick thickness of the skin may be inadequate to prevent penetration of driving rain, but at the same time it can significantly impede evaporation and thus leave the timber permanently moist. Conversely, large timbers, built into external walls, may shrink, due to the effects of modern heating. This leads to loads being shed into the thin brick facing.

In buildings with load-bearing masonry walls, which have allowed driving rain to penetrate to the embedded ends of floor joists and -beams, these may suffer from fungal attack, particularly 'dry rot', with all the consequent problems of infection of the masonry. Similarly, rot has been found where copings or 'decorative' balustrades, forming water-shedding features, have been removed from parapets.

Re-use of buildings from this period may also raise the question of fire resistance, particularly in respect of exposed ceiling joists and beams (see Sec. 5.4).

5.2.3 Seventeenth to Early Twentieth-Century Structures

The use of timber framing continued well into the nineteenth century for domestic buildings, with those in towns and cities showing increasingly elaborate decorative treatment of the frames on the elevations. Agricultural buildings like barns, similarly went on being built as timber frames, with cladding of timber boarding. The Building Regulations, introduced in 1667 after The Great Fire, did however put an end to the use of exposed timber framing in London.

The larger spans and the greater number of stories in the mill buildings of the early Industrial Revolution did also impose heavier loads than could be carried by timber framing. This, combined with the increasing scarcity of home-grown durable hardwoods, lead to a general adoption of masonry for walls, the use of timber being increasingly limited to softwood roof trusses and floor beams. By the mid-1800s, even these were replaced in the new mills by iron and brick-vaulted floors, so as to provide an incombustible structure.

Increasing amounts of softwood were being imported from the eighteenth century onwards from Scandinavia and from the Baltic region. Although the Baltic ports, in particular, could supply softwood in prodigious sizes (Baulks, 16 in. × 16 in. and 30 ft long, were still available in the 1920s) and of excellent quality, carpentered joints in these timbers were proportionally weaker than the corresponding ones in oak. This, together with the increasing availability of wrought iron, gradually changed the detailing of the joints. Iron straps and bolts were increasingly used to make joints, capable of resisting tension. The earliest bolts were secured with wedges, driven into slots through the shank. Threaded bolts and nuts became more common in the eighteenth century.

One particular advantage of the use of iron straps in roof trusses was that, by supporting the tie beams by strap hangers from the posts, their spans could be broken into two, as in king-post trusses (Fig. 5.6) or into three as in queen-post trusses (Fig. 5.7). These developments enabled the construction of the very large roof trusses, spanning across the great widths of the nineteenth century warehouses and other wide buildings (Fig. 5.8).

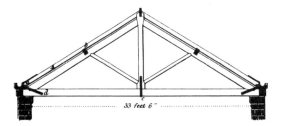

Figure 5.6 King-post truss (*from Newlands, 1857*).

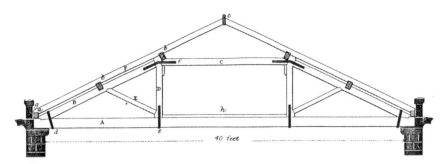

Figure 5.7 Queen-post truss (*from Newlands, 1857*).

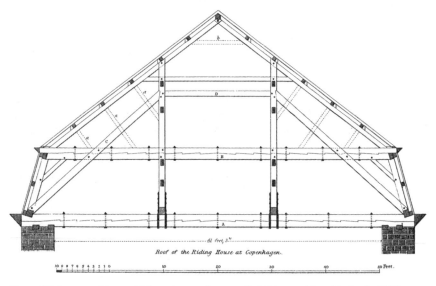

Roof of the Riding House at Copenhagen.

Figure 5.8 Large roof truss with iron straps and composite loft beams (*from Newlands, 1857*).

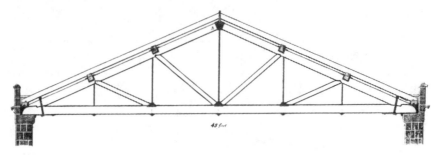

Figure 5.9 Large-span truss with iron tension verticals (*from Newlands, 1857*).

A further innovation was the use of iron tie bars in lieu of timber for tensile lattice members in the trusses (Fig. 5.9). In some countries with plentiful local timber, latticed frames with bolted joints were being used in the latter part of the nineteenth and the early part of the twentieth century to provide long-span roof structures over single-storey industrial sheds. Train sheds at railway stations were also in some instances roofed over with bolted lattice timber arches or with arches laminated by bolting boards, bent to shape, on top of each other.

Large sizes of timber are not essential for the creation of long-span roofs: 'Belfast Trusses' were essentially tied arches with close-spaced diagonal lattice members, made from timbers, no more than 3 in. (75 mm) thick, using nailed joints throughout. They were widely used from the 1860s onwards for roofs over industrial buildings, spanning up to about 30 m. In the 1940s they were used to roof over military aircraft hangars and some were built in West Africa in the 1950s.

For the majority of non-industrial buildings, timber beams continued to be the chosen floor structure. In Britain joists were saw-cut to be only 2 or 3 in. (50 or 75 mm) thick and placed with the larger cross-sectional dimension vertical. In order to stabilize the joists until the boarding had been fixed and also to spread concentrated loads on to two or more joists, herringbone bracing between the joists was introduced.

In continental Europe, square-cut floor beams at about 1 m spacing were more common and in multi-storey blocks of flats, fire separation and acoustic insulation between floors were provided by a layer of clay or similar 'pugging' on boarding, spanning between the beams and supported at half height between the floorboards and the ceiling.

Where it was desired to have floors of longer spans than could be achieved with a single beam of the available sizes of timber, resort was made to a composite beam, consisting of two softwood beams, one on top of the other, with the mating faces being cut in a 'saw-tooth' shape to provide shear connection (see Fig. 5.8), or with 'keys' of hardwood, morticed halfway into each beam at the interface. Vertical bolts were commonly used to clamp the two beams together.

Attempts were also made to stiffen large softwood beams by making them in two halves, side by side, and inserting inclined 'struts' of hardwood into chases, cut into the meeting faces of the two half-beams. The inclined struts would either meet near the top of the beam at midspan, or they would extend to the third points of the span where they would thrust against a horizontal hardwood strut, let in near the top of this 'trussed beam'. An example of this was found in the Mansion House in London.

This form of construction is unlikely to have been effective, and similar attempts at strengthening by iron tie rods, anchored at the ends of the beam and deflected to its underside by iron cross-pieces at the third points of the span, would only have increased the ultimate strength slightly and its effect on the stiffness would have been marginal. (It has been suggested that the purpose was to induce an upward camber in long beams, so as to make the floors level under self weight.) Properly trussed beams (Fig. 5.10), with the tie rods propped well down below the underside of the beam, would however have shown a significant increase of strength and stiffness above that of the timber beam on its own.

Another, later, device was the beam with a 'flitch plate'. This consisted of two beams, or two halves of one beam, sliced longitudinally and bolted either side of a plate of wrought iron or steel, of slightly smaller depth. Usually, the plate did not extend to the beam bearings and the bolts would not have achieved full composite action at ultimate loads, but at working loads the effect was significant. (The stiffness of a solid timber baulk was also often improved by sawing it in half longitudinally, reversing one of the halves and bolting them together without a flitch plate, thereby re-distributing the natural imperfections along the length.)

Continued or new use of these buildings generally meets the same problems, as arise in the re-use of Mediaeval and Renaissance structures. As joint efficiency was improved, indeterminate structural members were, however, gradually eliminated and timber sizes were reduced. In consequence, these later structures may not have as much spare capacity, with which to 'shrug off' subsequent mutilations, as do their earlier forerunners.

In domestic buildings, floor joists are sometimes found to be smaller than the size that is demanded by today's building regulations and, whilst usually strong enough, they may not be stiff enough to prevent the floor having a 'springy' feel—something that may not be acceptable to the new user. Sometimes floor joists of small cross-section, available in long lengths, were arranged to be continuous over an intermediate support, thus improving the stiffness considerably.

In buildings that have been left unused and unsupervised, fungal attack is often aggravated because the roofing has become damaged or vandalized, and the lead guttering has been stripped. The result is that the timber gets soaked by rain and does not get dried by the wind, because windows and doors have been

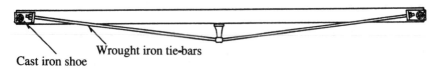

Cast iron shoe   Wrought iron tie-bars

Figure 5.10 Trussed beam.

boarded up. Where such conditions have led to 'dry rot', there is a serious risk of the masonry, supporting the timber beams and/or roof trusses, having become infected by the fungus, which can then re-infect new replacement timbers.

In domestic construction with load-bearing walls, dating from the eighteenth century onwards, the risk of fungal attack may also be greater than in earlier buildings because of the lower durability of the softwoods and because of the reduced resistance to driving rain of the smaller thickness of wall beyond the built-in end of the beam. In some houses in Bath, the top storey of the external walls was found to be a single skin of stone 120–130 mm thick, with daylight showing through the joints in places. Similar buildings have been found to have a timber frame with a self-supporting façade skin of stone. Conversely, in some substantial public buildings from this era the main floor beams have been found to be supported on pads of lead or copper in niches, allowing air circulation all around the ends of the beams.

In some instances, where load-bearing walls were too thin to allow floor joists to be supported by direct bearing, or where joists were not allowed to penetrate walls for reasons of fire resistance or weather protection, a piece of timber 50–80 mm thick would be bolted or spiked to the wall to act as a bearer for the joists. It is prudent, when appraising such a structure, to assume that the bolting, or particularly the spiking, may be suspect.

Where bolts were required to transmit the entire forces acting on the joints of, for instance a roof truss, rather than just hold the timber parts in contact, the local bearing pressure of the bolts on the timber may have caused local crushing, leading to slip in the joints. This could lead to additional deflection of the truss. A similar 'crushing slip' could occur at the shear connecting keys in composite beams, leading to increased deflection.

If such joint slip occurred early in the life of the structure, and the resulting deflections can be shown *not* to be progressing and if no load increase is contemplated, then no remedial action is required for structural reasons, although some measures may be needed to improve serviceability and/or appearance.

5.2.4 Later Twentieth-Century Structures

In the 1920s it had been realized that the major limitation on trussed timber structures was the inability to make joints that were capable of transmitting anything approaching the forces that the members could resist: carpentered joints could not transmit tension and bolts, on their own, were limited by low bearing stresses to avoid crushing of the wood.

To overcome this, a great many patterns of proprietary metal connectors were developed, with the aim of transferring forces from one member to another by carrying shear across the interface. In order to do this, the connectors had relatively large bearing surfaces, either pressed into the surface, or housed in machined recesses. The bolts in many cases only had the function of keeping the timbers from moving apart, so as to prevent the connectors to move out of their recesses.

From the late 1950s so-called 'trussed rafters' became popular for roofing over dwelling houses. These were trusses, generally made from timbers of 50 mm, or 37 mm nominal thickness, connected with punched metal plate fasteners, commonly known as 'gangnails' (which is one proprietory brand). These are thin plates of galvanized steel, punched so as to produce a large number of spikes on one side. Such plates are placed on faces of the timbers to be connected and they are then pressed together to force the spikes into the wood.

As with any other metal components, corrosion is a matter of concern with bolted joints, whether connectored or not, and similarly with gangnails, the more so as certain woods are thought to contain acidic substances. Where roofs have been kept weatherproof and the roof spaces have been sufficiently heated *and* ventilated to prevent frequent heavy condensation, the risk appears however to be minimal. (No failures of gangnailed trusses, due to corrosion of the plates, have been reported during the 40–50 years of their use.) For connections, other than gangnailed, where environmental conditions have been open to doubt, it may however be advisable as part of an appraisal to relieve a few joints of load, and temporarily partly dismantle them for examination, particularly if it is suspected that the joints may have been made with bolts without any rust protection.

Another mid-twentieth-century development was motivated by the increasing difficulty of obtaining large sizes of timber of good structural quality. The problem was overcome by gluing together boards of moderate sizes to form large members. This allowed the most highly stressed parts to be made from boards, selected to have few, if any, defects. The machinery for the accurate planing of the contact faces, and the curing regime required for durable adhesives, meant that the process was most suited to factory production.

A number of shell roofs, some of them quite spectacular, were however constructed in the 1950s and early 1960s, by means of in-situ nailing and gluing three layers of boards together, one on top of the other, with the direction of the boards in each layer being at an angle to the direction of the ones underneath (often 45 and 135°). The resorcinol- and phenol-based adhesives, that provide the strongest and most durable bond were, and are, expensive, and they require careful control of moisture content before the gluing takes place and fairly warm temperatures as well as firm clamping while the adhesive sets. The in-situ nail-gluing process did not provide these conditions. Furthermore, the gap-filling properties of these glues were not adequate for the lack of fit, to be expected. For this reason urea–formaldehyde adhesives, which were more tolerant of site conditions, as well as being less expensive, were used. Whilst moisture-resistant , they were not moisture-proof and rainwater penetration, and in some cases excessive condensation has caused a number of these structures to suffer failure of the 'gluelines', and they have had to be demolished.

The less demanding adhesives were also, as an economy measure, sometimes used for plain laminated beams, where it was assumed that the environmental conditions in the finished structure would not seriously affect their durability. As the assumed conditions may not always have pertained, appraisal of glued laminated structures should always include some check on the integrity of the gluelines.

(Failure usually shows up by delamination, leaving gaps that allow deep penetration of a feeler gauge, although they do not show up, when inspected from floor level, because intact gluelines usually are dark.)

The same period also gave rise to box beams with webs of plywood, or even hardboard. The voids between the webs were rarely well ventilated and the rooms, over which such beams were often used, were usually heated, so that if any moisture, due for instance to roof leakage, found its way into the void, ideal conditions were created for fungal attack.

Particle-boards, e.g. chipboard, have also been used for flat roof decks, but as the adhesive, used in their manufacture, is rarely of the 'water-and-boil-proof' variety, the almost universal failure of flat roof waterproofing leads to softening of the boarding.

Some, 'semi-structural' applications, such as framing for external cladding, have been carried out, using spruce. This softwood is *not* durable, unless given adequate preservative treatment. There is evidence that suppliers have not always achieved this; in many cases because the timber was not allowed to dry prior to treatment, and therefore did not take up the preservative.

5.2.5 Stability and Robustness: Interactions with Supporting Masonry

Mediaeval and Renaissance timber frames and roof trusses had to have a certain inherent stability to enable them to stand up after erection was completed, but before the infill or cladding had been applied. Wall-frames were stiffened and stabilized in their planes by diagonals, or by the rigidity of the joints between uprights and horizontals. Roof trusses did not receive the stiffening effect of the wattle-and-daub infill and therefore often had raking braces between the principal rafters and the purlins. Once completed, domestic houses would have cross-walls which would provide transverse stability to the longitudinal walls. This would be the case, even when the external walls were of masonry and the cross-walls being essentially timber framed and clad.

The earlier multi-storey, open plan, mill buildings and warehouses of the eighteenth century had timber floors. These might not have had great stiffness in their own plane (the connection of floor board to joist being just two spikes), but even so, they contributed to the stability of the external walls. (A fire in a historic dockside warehouse burnt out all five floors and the roof, leaving the six-storey high external walls unsupported. Quick, simple, external shoring of the wall was not permissible, as it would have blocked the quayside; internal installation of propping was ruled out on health and safety grounds, so the entire remains had to be demolished.) However, in several warehouses it has been observed that the inferior in-plane stiffness of the floors had allowed the external walls to gradually drift out of plumb, although the tying action of the cross beams made both walls lean the same way.

Removal of large areas of floor in such buildings, whether permanently or temporarily, as part of an adaptation, should therefore only be carried out after careful assessment of the effect on stability and, possibly, installation of a substitute structural member.

Eighteenth- and nineteenth-century multi-storey town houses often have several timber partitions. Some of these are load-bearing and have a substantial triangulated timber frame, spanning between, and bracing, the flank walls of the house; they make a major contribution to the stability of the whole structure. Ordinary partitions, even if only consisting of 'rectangular' framing and lath and plaster, also help to stabilize walls. Floor beams and joists, over a large front 'reception room', that runs the whole width between flank walls, will tie the front wall to the more stable, cellular, 'back-room' structure. (A good lath-and-plaster ceiling, if directly attached to floor, or ceiling, joists, will provide a measure of in-plane stiffness to the floor structure and hence help stability.)

The tendency, in the twentieth century, towards lighter roof structures has at times been accompanied by a reluctance to provide diagonal bracing in the plane of the roof. This may be of little consequence for 'everyday loading', but shows up whenever high winds, of the velocities that are statistically bound to occur from time to time, hit a part of the British Isles: Gables are sucked out and the roof trusses all keel over.

The features that provide stability will also sometimes help robustness. For example: Diagonal bracing in the plane of the roof may enable the roof to survive the accidental removal of the support of one principal rafter, by spanning across the gap (see also Sec. 5.3.5).

5.3 STRUCTURAL STRENGTHENING AND REPAIR TECHNIQUES

Before deciding on a scheme to strengthen and/or repair a timber structure which, on the basis of a first inspection and conventional calculations, appears to be slightly deficient, one should consider the relative merits of a more thorough inspection, with local 'regrading', enabling higher permissible stresses to be used in calculations, or even a load-test. It should be borne in mind that complete removal of floor boards (and/or ceilings) to allow full inspection may (temporarily) reduce stability. It requires evacuation of the floor (and/or the floor below), but may anyway be inherent in a proposed strengthening/repair. Load tests are expensive and disruptive to occupants, but may in some cases be less expensive than the proposed strengthening scheme. Economics apart, there may be compelling historical/'archaeological' reasons to keep as much as possible of the original structure intact.

When structural strengthening/repair is necessary, the susceptibility of timber to 'infectious' biological attack and its significant moisture movement make it essential that the structural work is not undertaken as an isolated operation, but is designed as part of a general renovation that includes the weather- and damp-proofing of the building as a whole. The methods and materials, used, must be chosen with due consideration of the condition of the entire fabric before, during and after completion of the works. *For instance: a building may have been left unoccupied and hence unheated for many years with a defective roof; as a result, the timber will most likely be very damp and absence of visible fungal attack may only mean that the temperature has been too low for it to develop. Once such a building is weatherproofed and heated, the dormant fungi will start to grow at a frightening rate unless eradicated beforehand.* The movements resulting from accelerated drying out of such a building may not be as spectacular, but they can result in damage, particularly at joints between old and replacement timber, unless the latter has been conditioned so as to follow the movement of the old.

5.3.1 Suitability and Compatibility of Materials and Methods

It is not enough to design a strengthening scheme merely to overcome a structural weakness; the design must take into account the general condition of the fabric and to what extent the structural problem is caused by moisture problems and/or biological degradation.

Where new and old timber have to share the load, consideration must be given not only to the relative stiffnesses in the final environmental condition, but also to the relative moisture movements of the two timbers after installation. It is unlikely (except perhaps for a glued joint) that neglect of this aspect would lead to structural collapse, but it could create serviceability problems.

Consideration must also be given to possible chemical interaction between repair components and the timber and/or any treatment of it: the tannic acid of fresh oak may attack carbon steel; similarly some fungicidal treatments are based on metal salts which may corrode steel. Where new metal roofing, e.g. lead or copper, is to be laid directly on new timber boarding, which has been given fungicidal or flame-retardant treatment, the possibility of chemical interaction in the presence of condensation or minor leakage must be considered and manufacturers' assurances should not always be uncritically accepted.

5.3.2 Replacement of Timber and Strengthening with additional Timber

When a member is badly rotted or has been significantly weakened by past notching by plumbers and/or electricians, the structure can be restored to its previous integrity and function by replacement of the damaged member by a new member of timber, preferably of the identical species.

If it is only a part of a long member that is affected, it may be possible to cut out and replace only that part. In that case, the jointing will usually require some overlap of old and new. Where the damage is due to fungal attack, it is often considered advisable to cut out well beyond any visible attack, in order to avoid re-infection.

This type of repair will restore the original design and behaviour of the structure, but it entails some loss of the original timber. This is in contrast to the approach of some preservationists, who would advocate the installation of a relieving structure (often of steel!) in order to preserve the maximum amount of the original timber, rot and all.

In the case of replacement of a whole member, there are two main problems: relieving the member of load during the operation and, where the member is part of a frame or a roof truss, to disconnect it from the rest of the structure and connect the replacement. (The latter problem may also present itself in the case of floor beams, housed in individual narrow niches in both supporting walls.) When only a part of a member has to be replaced, an additional problem arises: how to transmit the member forces through the joints at the ends of the replacement piece.

Compressive forces appear at first glance to be easy to transmit, but in reality it is not so simple: if two end surfaces of timber are pressed against each other, there is a tendency for the harder fibres in one to penetrate into the softer parts of the other, like the bristles of two brushes pushed together. Direct end bearing may therefore not mobilize the full strength of the parent timber, although it is probably enough in most cases.

The strength of a compression member is, furthermore limited, not only by its direct compressive strength, but also by its stiffness, i.e. its resistance to buckling. Simply butting the ends of the replacement piece would be almost tantamount to putting hinges at those points: not the best way to achieve stiffness!

The problem was, according to F. Charles, solved in the thirteenth century by the 'scissor-scarf' joint (Fig. 5.11a), a very elegant device, but one demanding perfect workmanship, and even then prone to cause splitting (see Charles and Charles, 1990). In practice, a simple scarf joint, or 'half-joint' will usually be the preferred solution to the problem of piecing in a new length of timber in a compression member (Fig. 5.11b). It may be possible to shape the ends of the scarfs, so as to make them 'self-locking' and thus in theory avoid the use of bolts (Fig. 5.11c), but in practice it is difficult to avoid splitting, due to wedge action of the splayed ends of the scarf(s). The difficulty of achieving simultaneous bearing at both cross-cut ends of the scarf may, if necessary, be overcome by 'buttering' the surfaces with a synthetic resin putty, prior to assembly.

For members subject to moderate tensile forces, the 'tabled scarf joint' (Fig. 5.12a) may allow the splicing in of a new length of timber without the use of bolts. Generally, however, a bolted 'fishplate-joint', using splice plates of timber or of steel, will be preferred where the tensile force is substantial, or where the appearance can be accepted. For beams and other members subject to significant bending, a longer version of the 'fishplate' joint, with the splice plates in the plane of bending, will generally be adopted (Fig. 5.12b).

If the end of a floor beam, built into a masonry wall, has suffered from 'dry rot' and there are difficulties in changing the conditions at the beam support in the wall, so as to prevent a repeated attack, the splice plates and the new length of timber can be replaced by a pair of steel channels, so as to isolate the remaining sound part of the existing beam from the damp wall (Fig. 5.13). This, in effect, is a borderline case between replacement and strengthening by metal devices.

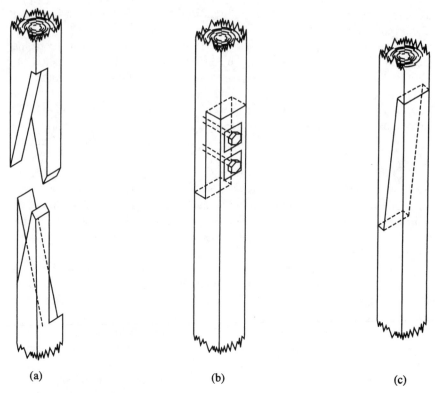

Figure 5.11 Splices for replacement compression timbers: (*a*) 'scissor-scarf'; (*b*) simple, bolted scarf; (*c*) 'self-locking scarf'.

Figure 5.12 Splices for replacement timbers in tension or bending: (*a*) 'tabled scarf joint'; (*b*) bolted 'fishplate' joint.

Occasionally existing timber joists and/or beams, whilst generally sound, are not strong enough or stiff enough to be acceptable for a proposed new use of the building. It may then be possible to remedy the deficiency by creating composite action with new, additional, timber components.

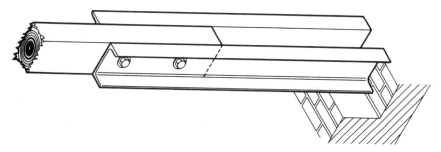

Figure 5.13 Replacement of rotten beam end with steel channels, to prevent re-infection of timber from fungal spores in the wall.

In the case of joists, it may be possible to lift the floor boards and place ply-wood sheets over the joists, to act as a compression flange. The plywood must then be fixed to the joists with screws at close centres, capable of transmitting the horizontal shear forces. Transverse joints between sheets must not occur within the middle three quarters of the joist span, unless the plywood is lap-jointed, so as to be able to transmit the entire compressive flange force.

When the joists are notched over the top of a beam, it may be possible to increase its depth by glueing and screwing boards on top of it, between the joist ends (which can be cut back, if necessary, to make room for the strengthening, provided adequate bearing of the joists on the beam is left). It may be advantageous to make the additional boards of a timber with a higher Young's modulus than the parent beam, in order to get increased effect within the depth available (Fig. 5.14). When such strengthening by 'lamination' is carried out, the moisture content of the new material should be equal to, or less than, that of the existing timber, to avoid subsequent additional deflections, due to differential moisture movement.

5.3.3 Strengthening by Metal Straps, Ties or Plates

The techniques adopted in the seventeenth century onwards for new timber trusses, i.e. the provision of metal straps to resist tensile forces at connections, can equally well be used to strengthen and/or repair existing structures.

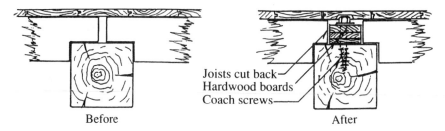

Joists cut back
Hardwood boards
Coach screws

Before   After

Figure 5.14 Strengthening of floor beam by 'top lamination'.

St Mary's Church in Sandwich was rebuilt in the seventeenth century after an earthquake had destroyed most of the fabric. The roof trusses over the rather wide nave were made of softwood, with wrought-iron straps reinforcing those connections that were subject to tensile forces. (These straps were connected to the timbers by the 'forelock' bolts then in use; these were tightened by a wedge being driven into a slot in the shank of the bolt and then bent over.) The configuration of the trusses was however structurally unsound and led to large bending moments in the tie beams, due to the inability of the walls to prevent the feet of the rafters from spreading. In one or two of the trusses the tie beam had, in fact, failed in bending at some time in the past and nineteenth-century strengthening straps with threaded bolts were in evidence (Fig. 5.15). Even so, there were gross distortions of the tie beams and an adequate margin of safety of the roof could not be demonstrated.

The remedy, chosen with a view to causing as little visual impact as possible, was to install ties of 16 mm diameter steel rods, connecting the rafter

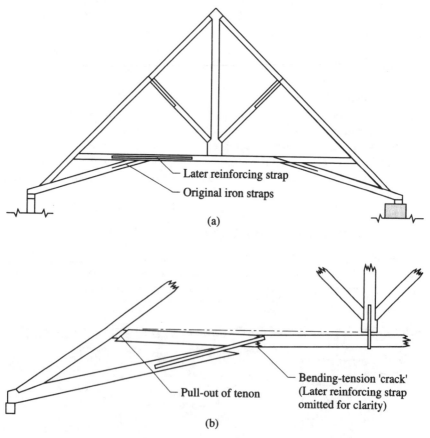

(a)

(b)

Figure 5.15 (a) Diagram of 'badly designed' roof truss; (b) Bending failure of tie beam (*after Stocker and Bridge, 1989*).

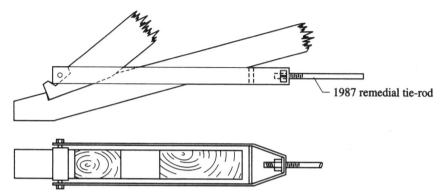

1987 remedial tie-rod

Figure 5.16 Strengthening tie rod to truss in Fig. 5.15, with 'stirr-up' connection to rafter feet (*after Stocker and Bridge, 1989*).

feet (Fig. 5.16). This enabled the original structure to be preserved, including the iron strengthening straps and only the members with serious beetle attack had to be renewed.

Eighteenth- and nineteenth-century iron technology was being used in other ways for strengthening of existing timber structures; one example is flitching: Where the strengthened beam will not be seen, it is often possible, after relieving the beam of its imposed load, to bolt a 'flitch' plate on either side (In the case of ceilings that may not be disturbed, this may require secondary beams or joists to be temporarily removed.)

Another impediment to external flitching occurs when the secondaries are supported on the beam in notches that go down too far to allow plates of adequate depth to be applied, without interfering with the existing notch supports. In such a situation it may be possible to split the beam in two, by a vertical saw cut down the middle, and insert the flitch plate in the saw cut (after moving the halves a small distance apart) and bolt it through. This process will be easier to carry out in a satisfactory way, if the beam can temporarily be taken out of the building and the work done in a workshop.

Where the soffit of the beam is exposed to view, or temporary removal is not practicable a fairly recent technique is to create a deep vertical slot in the centre of the beam by making multiple vertical saw-cuts, of a depth such as to leave a few centimetres of original timber on the soffit. A steel plate, of a thickness such as to be a snug fit in the slot and of a depth to finish flush with the top of the beam, is placed in the slot and bolted through (Fig. 5.17). The plate may run the full length of the beam, so as to provide reinforcement over the supports, or, if only bending strengthening is required, it may stop short, leaving the beam ends solid.

Successful application of this idea has been reported, but it requires extremely accurate setting-out of the bolt-holes in the beam to ensure that they line up with the holes in the steel plates, and that the bolts are a tight fit. It is

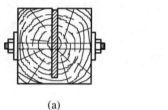

(a)   (b)

Figure 5.17 (*a*) Cross-section of beam with strengthening flitch plate, placed in sawn slot; (*b*) Cross-section of beam with bolted-on top plate.

extremely difficult to achieve this. One possible method would be to drill through the timber first, insert the steel plate and mark the centres of the holes by 'drilling' from one side of the beam through the pre-drilled holes. The steel plate could then be lifted out and drilled through 'off site' before being re-inserted. An easier option would be to use a tee-section, coach-screwed to the top of the beam, but any long-existing sag of the beam would mitigate against it.

One could envisage a similar operation, in which the steel plate was bonded to the timber with an epoxy adhesive, instead of being bolted through. This would get over the difficulties of bolt-hole alignment and the continuous bond over the entire interfaces between steel and timber should ensure full composite action. It is however not certain that one and the same adhesive would provide perfect bond to both timber and steel, when applied under site conditions.

Another question about this proposal, that has not so far been fully answered, is what the effect would be of the moisture movements of the timber, relative to the resin and/or the steel. If the operation were carried out in a building, that had been kept reasonably heated and dry during its previous use and which were not being opened to the weather during the works, then the subsequent humidity fluctuations of the air surrounding the beam would be relatively small. The moisture content of the timber would therefore not vary much and there would be little moisture movement.

If, however, the building was semi-derelict, with broken windows and leaking roof, then the moisture content of the timber, when the work was started, would be high. A significant amount of drying shrinkage would then take place after the building had been weatherproofed and was being heated in the course of its new use. In the case of a bolted flitch plate, the timber can slide against the steel and thus shrink against an only slight frictional restraint. If the plate has been glued in, the restraint against shrinkage at the interface will, however, be practically complete and the relative movement has to be accommodated by distortion of the cell structure of the wood adjacent to the interface. There are, at present, no published data about this potential problem.

Where the situation allows it, the easiest way of dealing with the problem may be to coach-screw a plate to the top of the beam. In this way, the steel does not contribute to the carrying capacity in a very efficient way, but it involves much less intervention than the alternatives and can be the only safe way to proceed if there is a historically important and delicate plaster ceiling, in contact with, and only tenuously connected to, the underside of the beam.

5.3.4 Resin-based Reinforcement Devices

The following variant of the 'glued-in' flitch-plate, described above, has been used in a number of instances, so far apparently with success: A central vertical slot is formed, as described above. A number of steel reinforcement bars are placed in the slot, bedded in an epoxy resin, which completely fills the slot and bonds the bars to the timber (Fig. 5.18).

This method was used in the Hyde Park Barracks in Sydney to restore the carrying capacity of the floor beams, in which the heartwood had been largely consumed by termites. In this case, however, there was little need for repeated deep saw-cuts; once the top of the beam had been opened, it was mainly a case of 'carving' out the affected timber back to sound wood, before placing reinforcement in the bottom of the cavity and then filling it with a resin compound to 'replace' the termite-eaten timber.

The same reservations about the differential moisture movement between timber and resin do however apply equally to this type of strengthening as they do to the glued-in flitch plate.

Where, in temperate but damp climates, an end of a beam built into an external masonry wall has rotted, the following operation has been carried out, so far successfully: The beam is propped and the rotten end is cut off. Rods of glass-fibre reinforced resin (or of stainless steel) are placed in holes drilled at an angle through the sound timber, so as to enable the rods to extend into the space that was occupied by the removed, decayed, beam end. The rods are bonded to the timber with a suitable resin adhesive and the 'amputated' beam end is re-cast in a 'resin-mortar' around the reinforcement rods (Fig. 5.19).

Test results are available, which demonstrate the adequacy of the adhesive jointing of reinforcement bars to resin and resin to timber. Load tests have also

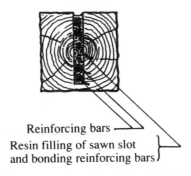

Reinforcing bars
Resin filling of sawn slot
and bonding reinforcing bars

Figure 5.18 Cross-section of beam, strengthened by resin-bonded reinforcing bars.

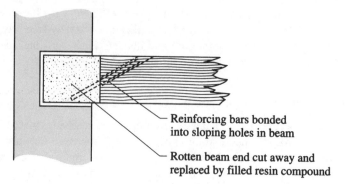

Reinforcing bars bonded
into sloping holes in beam

Rotten beam end cut away and
replaced by filled resin compound

Figure 5.19 Rotten beam end replaced by reinforced resin cast.

been performed and show that this procedure results in full restoration of the load-carrying capacity. The method has, however, not been in use for long enough to ascertain the life expectancy of the bond between resin mortar and timber under the various environmental conditions that may be encountered. What may give rise to uncertainty, is, as for the 'glued-in' flitch plate mentioned above, the moisture movements of the timber, relative to the non-wood material. Where conditions remain fairly constant from before the operation, through the repair process and in the refurbished building during its new use, serious problems would appear to be unlikely, but there is no proof as yet.

With all resin repairs it must be remembered that resin compounds have their own special properties, some of which are quite different from those of wood. Advice on any potential problems, that might arise, should therefore, in each case, be sought from an expert, who is not connected with the manufacturers of resin nor with promoters of any of these repair processes. A *Guide to best Practice* has been published by The Timber Research and Development Association.

5.3.5 Improving Stability and Robustness—Anti-seismic Bracing

The measures that will improve the stability of timber structures will, in principle improve their seismic resistance. For this latter purpose the hardware may however have to be somewhat heavier.

The beneficial effect of diagonal members on stability and robustness has been mentioned in Sec. 5.2.5. Where such members are missing, it may be possible to add them. If they are badly connected to the main members, the connections can perhaps be strengthened. Crossing steel ties can be a convenient way of overcoming the shortfall, if their appearance can be tolerated, and as long as fire considerations do not dictate otherwise (see Sec. 5.4.1).

Timber structures, being light, are not subjected to large inertia forces arising from their own mass, and they have an inherent flexibility that enables them to absorb a fair amount of movement without damage. In many cases, however, they support heavy cladding, e.g. masonry wall infill or tiled or slated roofing, which transmit their inertia forces to the frame and/or the roof structure and which are likely to be damaged by quite small movements. Also, timber floors may provide the only stabilizing restraint to masonry walls.

For pre-1950s' structures it is almost impossible to provide the level of seismic resistance, required by codes of practice for new construction, without drastically altering the character of the building and destroying the historic constructional details; it is, however, usually possible to enhance the resistance significantly by fairly simple and inexpensive means. For example, if, in a timber frame with light-weight cladding, the frame connections are inadequate for structural safety or do not have sufficient stiffness to prevent damage to the cladding, it may be possible to strengthen and stiffen the frame in its own plane by introducing sloping bracing members. The simplest way of achieving this may be by using crossing diagonal ties of steel flats, bolted or screwed to the timber members. Earthquake accelerations may cause forces in roof trusses and frames in the opposite direction to those due to gravity and hence cause tension in what was intended to be joints working in compression. In that case, the addition of some bolted steel straps may be all that is needed to provide the necessary tensile capacity of the connections.

As mentioned in Sec. 4.5.10, Professor Benedetti's tests demonstrate that masonry walls can withstand greater seismic movements without serious damage if tensile restraints prevent them from extending in their plane.

If, in a timber frame, the joints can be strengthened against tension, again by the addition of bolted steel straps, this will enable the frame members to act as restraining ties and significantly improve the resistance of the infill walls. At the same time, some measures may have to be taken to improve the resistance of the infill against lateral movement out of the frame.

In many traditional roof structures the only resistance against horizontal movement, parallel to the ridge, is provided by the connections between purlins and principal rafters; this may well be inadequate to prevent all the trusses from simultaneously keeling over in the same direction and it will often not offer adequate restraint to masonry gables. This deficiency can be overcome by the provision of crossing diagonal flat steel ties, running at about 45° from eaves to ridge and bolted to the principal rafters. At the same time, anchor straps should be embedded into the masonry and connected to the main timbers, so as to provide restraint to the gable triangle.

It has been found that one of the causes of serious earthquake damage to traditional masonry buildings is movements which distort the horizontal plan shape, making rectangular wall patterns become lozenge shaped. This distortion can be largely prevented by connecting the walls to the floors and making the floors act as rigid horizontal diaphragms. The wall-to-floor connections can be made by straps of flat steel, embedded in the masonry joints and screwed to the floor joists. The horizontal stiffening of the floors can be done in two ways: either by means of crossing diagonal flat steel straps, bolted or screwed to the floor joists, or by screwing plywood sheets to the floor joists, as described in Sec. 5.3.2.

5.4 FIRE PROBLEMS AND PREVENTION

A common 'knee-jerk' reaction to proposals for continued or new use of timber-framed buildings is that they constitute an excessive fire risk. There is no getting away from the fact that timber is a combustible material, but that does not mean that a timber structure cannot remain standing after a fire, that lasts one or even two hours. A closer look at the behaviour of structural timber in fire will often dispel the initial alarm and show that the level of risk can be reduced to an acceptable level. It should also be remembered that the fire tests, like those of BS 476, do not simulate real fires, but only create standard conditions.

5.4.1 Basic Behaviour of Timber Structures in Fire

All timber is combustible. The timbers of the structure will however rarely constitute a major fire risk, as they are usually remote from any likely source of accidental fire (except in the case of faulty electrical wiring clipped directly to structural members). All timbers are not equally inflammable; some softwoods, particularly the more porous ones (and those with a high resin content), ignite more easily and encourage a more rapid spread of flame than do some hardwoods, like oak.

Tar and similar substances, used in the past as fungicidal treatment, will encourage the spread of flame and will also increase the 'fire load' (the amount of material available for sustaining a fire). Solvent- and oil-based new treatments will do likewise, at least temporarily until the carrier fluid has evaporated.

All timber, exposed to sufficient heat, 'pyrolyses' to produce flames and charcoal, which then, if the heat is maintained, eventually burns to carbon dioxide. The rate, at which the initial charring progresses inwards from the exposed surface, depends on the rate of heat gain. For practical purposes, the British code of practice BS 5268—Part 4 quotes 40 mm per hour for most softwoods (50 mm per hour for Western Red Cedar) and 30 mm per hour for hardwoods, such as oak, with a density greater than 650 kg/m^3. For free-standing columns and fully exposed tension members, the code recommends 25 per cent higher values.

The strength of the uncharred timber, when it has cooled, can for practical purposes be taken as unaffected by the fire. This is due to the high insulating value of the charred wood, as well as to the inherent nature of the strength of timber. (A thin layer of the wood just inside the char will lose some strength during the fire, but will recover on cooling down.) This behaviour is in contrast to that of iron and steel; these conduct the heat quite rapidly, weakening from 200 °C upwards with their strength reducing to half of its initial value at 600 °C.

The retained strength of the uncharred timber means that if a fire, however intense, is put out before too much of the cross-sections of the timber members have been charred, the timber structure will still be able to carry its load, albeit with a reduced safety margin. An exposed steel structure, by way of contrast, will often distort and collapse in a similar fire, before it can be put out.

For this reason, the Achilles' heel of many seventeenth century and later timber structures is often to be found in the metal components, used in the connections, particularly the bolts. Due to the high conductivity of the metal, the bolts char the timber surrounding them and in this way seriously weaken the connections before the strength of the members has been significantly reduced. Other weak points are connections where, due to twisting or warping of the timber, the members are no longer in close contact with each other; at such connections, the fire will 'search out' the gaps and char the 'internal' surfaces of such joints.

It follows from the above that, provided 'thin' elements of the structure, such as floor boards, are protected against fire, the heavy structural members will often remain capable of carrying their load, and thus prevent collapse of the building, until the fire has been put out. This will not necessarily obviate the need for rebuilding after the fire, but it will make the fire fighting less hazardous and hence more likely to prevent total loss.

The main points about fire safety, to be observed when refurbishing a timber structured building, are therefore:

1. Adequate means of escape; the occupants of the building must be able to get out, before they are overcome by smoke. (The majority of fire fatalities die from asphyxiation, not from burns.)
2. Protection of thin structural members to reduce charring sufficiently to allow them to fulfil their structural function for the required period of fire resistance.
3. Protection of heavy structural members and/or connections of such, which have metal components, e.g. flitch beams and metal strap connections in roof trusses, that will conduct the heat from a fire so as to cause internal charring of the timber. Similarly, joints with gaps, allowing access for the heat to parts of thinner cross-section than the main members, e.g. tenons or scarfed ends.

These provisions apply equally to any strengthening devices, that may be required for reasons of decay of the original timbers, or for other structural reasons. Metallic strengthening components may have to be shielded if the whole structure is not being protected; the possibility of resin repairs losing their strength earlier than the timbers that they were intended to strengthen, must also be considered. BS 5268 implies that the resorcinol/phenol/urea–formaldehyde adhesives, used in most factory-made laminated timber components, char at the same rate as the timber, but it leaves a question mark over other glues, including epoxies, which may loose strength at lower temperatures.

5.4.2 Protective Measures

There are a multitude of measures which can detect or stop a fire developing and/or spreading, such as smoke detectors, sprinklers, compartmentation, etc. These are very effective, but they will not be discussed here, because they come under the heading of fire engineering, a specialist discipline in its own right and one that should be involved in all major conservation projects.

The protective measures, to be described here, are those that would be attached to, or be installed immediately adjacent to, structural elements.

There are on the market a number of laquers, paints and pastes, which offer protection by being *intumescent*, that is, when exposed to flame, they swell up and form a bubbly, insulating, crust, which protects its substrate from the heat. Their effect is, however, very limited, when compared with the natural char, and their main usefulness is in slowing down the spread of flame. There are other flame-retarding preparations, but they all fail to prevent ignition once the timber is engulfed in flames from other sources.

Where the structure does not form part of a historical interior, which has to be preserved, the provision of a fireproof ceiling, of plasterboard or similar, will be the most effective way of protecting the structure from fire from below. (Gypsum-plaster and plasterboard are particularly effective for fire protection, due to the high content of chemically bound water in the set plaster; this is released on exposure to the heat of flames.) Where it is imperative to preserve a ceiling of timber, and also necessary to prevent spread of fire upwards, the only solution may be to build a floor of fireproof construction over it.

When repair or strengthening involves splicing on new timber with bolted connections, it must be ensured that the mating faces of the connected timbers are in tight contact throughout, if necessary, by re-tightening after shrinkage has taken place. Bolt heads and nuts should be countersunk and covered with boards of fire resisting insulation material, to prevent heat from a fire being conducted to the interior of the timber members. Sometimes, the simplest way of insulating the bolts is to dowel on 'coverplates' of timber, thick enough to provide the necessary insulation for the required period, before being completely charred themselves.

5.5 PRECAUTIONS AGAINST BIOLOGICAL DEGRADATION

The many references to insect and fungal attack in the earlier parts of this chapter are prompted by the fact that biological degradation is one of the worst obstacles to conservation of timber structures. Time and time again one comes across otherwise perfectly preserved buildings, where a closer examination of the roof reveals rafter feet so badly rotted that nothing short of complete replacement can remedy the situation.

Any appraisal of a traditional building for continued or new use must therefore include a thorough inspection for signs of present or past rot or beetle infestation.

This can be difficult: it may involve removing carpets and lifting floorboards in order to inspect joist ends embedded in external walls; it may require removal of small areas of roofing or cutting holes in ceilings in order to inspect rafter feet and wall plates in inaccessible roof spaces, but it is essential, if one wants to avoid nasty surprises later.

Particular care is required with buildings which have in the past had little or no heating: An infestation that has progressed slowly or has lain dormant, due to the unfriendly environment in the past, can 'wake up' with a vengeance with the introduction of central heating.

Any attack, that is discovered, must be dealt with by means, that are appropriate to both the building and the nature of the attack: It was once thought that fumigation once a year was a suitable, and possibly the only practical way, of dealing with death-watch beetle in the roof structure of Westminster Hall, which could be evacuated until the fumes had cleared. It has, however, recently been found that it does not deal with the larvae deep inside the wood, whereas it kills their predators, such as spiders; it has therefore been discontinued, also for reasons of health and safety. It was, of course, never acceptable for a building used for continuous human occupation.

Insect control and eradication are generally matters for the specialist. Advice should, if at all possible, be obtained from an expert who is *not* connected with firms selling insecticidal preparations.

One precaution, that may have structural implications, is 'defrassing': removal of affected sapwood in order to evict the larvae in it and allow subsequent insecticidal treatment to penetrate into the heart where the next generation of eggs would be deposited. This is clearly only practicable where the proportion of sapwood is modest, so that the structural weakening of the members and the connections does not become excessive.

The traditional remedy against wet rot is to dry it out; this will kill it and if there is enough unaffected cross-section left for structural purposes, no more needs to be done, as long as moisture is prevented from reappearing, e.g. by regular maintenance of roofing and gutters.

Dry rot was traditionally dealt with by 'amputation' at about 1 m beyond the limit of any visible attack and replacement with sound timber. Where the affected timber had been in contact with masonry, new timber had to be be encased in copper sheaths or in bituminous membranes to prevent re-infection. An alternative is to replace the removed length of timber by a steel extension (see Fig. 5.13).

Where the attack is confined to the end 10–20 mm of joists, embedded in a wall, it may be possible to get away with cutting the joists just inside the face of the wall and supporting the cut ends on metal joist hangers, mortared into the, cleaned-out, pockets in the wall. This does however require the work to be carried out extremely carefully to avoid fungal substances, capable of re-growth, being transferred to the hitherto sound timber.

The, at one time recommended, practice of saturating the masonry with insecticidal solution is now deprecated, partly because of the possible effects on the occupants of poisonous residues in the masonry, and partly because the treatment tended to leave large amounts of moisture in the masonry, making it difficult to dry out, and thus mitigated against providing an environment that did not encourage fungal growth. Removal of fungal matter on the surface of the masonry should, however, be carried out, as the *serpula lachrymans* has the ability to transport moisture from its source, e.g. a damp part of the wall, to otherwise dry timber.

The overriding consideration, in preventing fungal attack on timber, is however to keep it dry. Ventilation of the environment of the timber is therefore of paramount importance.

APPENDIX TO CHAPTER 5

Working Stresses and Grading Rules from British Standard Code of Practice CP 112: Part 2 (1971)

Whilst this code of practice has been superseded for new construction by BS 5628, the following excerpts have been reproduced by kind permission of British Standards Institute as an aid to assessment of existing timber structures (for which BS 5628 is not suitable—see Sec. 5.2.1, pp. 143).

CP 112: Part 2: 1971

Table 3. Green stresses and Moduli of Elasticity

Standard name	Bending and tension parallel to the grain					Compression parallel to the grain				
	Basic	75 Grade	65 Grade	50 Grade	40 Grade	Basic	75 Grade	65 Grade	50 Grade	40 Grade
	N/mm²	N/mm²	N/mm²	N/mm²	N/mm²	N/mm²	N/mm²	N/mm²	N/mm²	N/mm²
SOFTWOODS										
a. Imported										
Douglas fir	15·2	11·4	9·7	7·6	5·9	11·0	8·3	6·9	5·5	4·5
Western hemlock (unmixed)	13·1	9·7	8·3	6·6	5·2	10·3	7·6	6·6	5·2	4·1
Western hemlock (commercial)	11·7	8·6	7·6	5·9	4·5	9·0	6·6	5·9	4·5	3·4
Parana pine	11·7	8·6	7·6	5·9	4·5	10·3	7·6	6·6	5·2	4·1
Pitch pine	15·2	11·4	9·7	7·6	5·9	11·0	8·3	6·9	5·5	4·5
Redwood	11·7	8·6	7·6	5·9	4·5	8·3	6·2	5·2	4·1	3·1
Whitewood	11·7	8·6	7·6	5·9	4·5	8·3	6·2	5·2	4·1	3·1
Canadian spruce	11·0	8·3	6·9	5·5	4·5	8·3	6·2	5·2	4·1	3·1
Western red cedar	9·0	6·6	5·9	4·5	3·4	6·2	4·5	3·8	3·1	2·4
b. Home-grown										
Douglas fir	14·5	10·7	9·3	7·2	5·9	10·3	7·6	6·6	5·2	4·1
Larch	13·8	10·3	9·0	6·9	5·5	9·7	7·2	6·2	4·8	3·8
Scots pine	11·0	8·3	6·9	5·5	4·5	8·3	6·2	5·2	4·1	3·1
European spruce	8·3	6·2	5·2	4·1	3·1	6·2	4·5	3·8	3·1	2·4
Sitka spruce	7·6	5·5	4·8	3·8	3·1	5·5	4·1	3·4	2·8	2·1
HARDWOODS										
a. Imported										
Abura	13·8	10·3	9·0	6·9	5·5	10·3	7·6	6·6	5·2	4·1
African mahogany	12·4	9·3	7·9	6·2	4·8	9·7	7·2	6·2	4·8	3·8
Afrormosia	22·1	16·5	14·1	11·0	8·6	15·9	11·7	10·3	7·9	6·2
Greenheart	37·9	28·3	24·1	19·0	15·2	27·6	20·7	17·9	13·8	11·0
Gurjun/keruing	17·2	12·8	11·0	8·6	6·9	13·8	10·3	9·0	6·9	5·5
Iroko	20·7	15·5	13·4	10·3	8·3	15·2	11·4	9·7	7·6	5·9
Jarrah	19·3	14·5	12·4	9·7	7·6	15·9	11·7	10·3	7·9	6·2
Karri	22·1	16·5	14·1	11·0	8·6	16·5	12·4	10·7	8·3	6·6
Opepe	25·5	19·3	16·5	12·8	10·3	22·1	16·5	14·1	11·0	8·6
Red meranti/red seraya	12·4	9·3	7·9	6·2	4·8	9·7	7·2	6·2	4·8	3·8
Sapele	19·3	14·5	12·4	9·7	7·6	15·9	11·7	10·3	7·9	6·2
Teak	22·1	16·5	14·1	11·0	8·6	16·5	12·4	10·7	8·3	6·6
b. Home-grown										
European ash	17·2	12·8	11·0	8·6	6·9	11·0	8·3	6·9	5·5	4·5
European beech	17·2	12·8	11·0	8·6	6·9	11·0	8·3	6·9	5·5	4·5
European oak	15·9	11·7	10·3	7·9	6·2	11·0	8·3	6·9	5·5	4·5

NOTE. These stresses apply to timber

Compression perpendicular to the grain			Shear parallel to the grain					Modulus of elasticity for all grades	
Basic	75/65 Grade	50/40 Grades	Basic	75 Grade	65 Grade	50 Grade	40 Grade	Mean	Minimum
N/mm²	N/mm²	N/mm²	N/mm²	N/mm²	N/mm²	N/mm²	N/mm²	N/mm²	N/mm²
1·79	1·59	1·31	1·72	1·24	1·10	0·83	0·69	10 300	5 900
1·38	1·17	1·03	1·52	1·10	0·97	0·76	0·62	9 000	5 500
1·38	1·17	1·03	1·38	1·03	0·90	0·69	0·55	8 600	5 200
1·52	1·31	1·10	1·52	1·10	0·97	0·76	0·62	8 300	4 500
1·79	1·59	1·31	1·72	1·24	1·10	0·83	0·69	10 300	5 900
1·52	1·31	1·10	1·38	1·03	0·90	0·69	0·55	7 600	4 100
1·38	1·17	1·03	1·38	1·03	0·90	0·69	0·55	6 900	4 100
1·38	1·17	1·03	1·38	1·03	0·90	0·69	0·55	8 300	5 200
1·03	0·90	0·76	1·24	0·90	0·76	0·62	0·48	6 200	3 800
1·72	1·52	1·31	1·38	1·03	0·90	0·69	0·55	9 000	4 500
1·79	1·59	1·31	1·52	1·10	0·97	0·76	0·62	9 000	4 500
1·72	1·52	1·31	1·38	1·03	0·90	0·69	0·55	5 300	4 800
1·10	0·97	0·83	1·10	0·83	0·69	0·55	0·41	5 900	3 100
1·10	0·97	0·83	1·10	0·83	0·69	0·55	0·41	6 600	3 400
2·34	2·07	1·72	2·07	1·52	1·31	1·03	0·83	8 300	4 500
2·07	1·79	1·52	1·72	1·24	1·10	0·83	0·69	7 900	4 100
4·14	3·59	3·10	2·62	1·93	1·65	1·31	1·03	10 300	6 900
6·20	5·38	4·62	4·83	3·59	3·10	2·41	1·93	17 200	12 400
3·10	2·76	2·34	2·34	1·72	1·52	1·17	0·90	12 400	8 300
4·14	3·59	3·10	2·34	1·72	1·52	1·17	0·90	9 000	5 900
4·14	3·59	3·10	2·34	1·72	1·52	1·17	0·90	10 300	6 900
4·83	4·14	3·59	2·48	1·86	1·59	1·24	0·97	13 800	8 300
5·52	4·83	4·14	3·10	2·28	2·00	1·52	1·24	12 400	7 600
1·79	1·59	1·31	1·52	1·10	0·97	0·76	0·62	7 600	4 100
4·14	3·59	3·10	2·34	1·72	1·52	1·17	0·90	9 700	6 200
4·14	3·59	3·10	2·34	1·72	1·52	1·17	0·90	11 000	6 900
3·10	2·76	2·34	2·76	2·07	1·79	1·38	1·10	10 000	6 600
3·10	2·76	2·34	2·76	2·07	1·79	1·38	1·10	10 000	6 600
3·10	2·76	2·34	2·48	1·86	1·59	1·24	0·97	8 600	4 500

having a moisture content exceeding 18 %.

CP 112: Part 2: 1971

Table 4. Dry stresses and Moduli of Elasticity

Standard name	Bending and tension parallel to the grain					Compression parallel to the grain				
	Basic	75 Grade	65 Grade	50 Grade	40 Grade	Basic	75 Grade	65 Grade	50 Grade	40 Grade
	N/mm^2	N/mm^2	N/mm^2	N/mm^2	N/mm^2	N/mm^2	N/mm^2	N/mm^2	N/mm^2	N/mm^2
SOFTWOODS										
a. Imported										
Douglas fir	18·6	13·1	11·0	8·6	6·6	14·5	10·3	8·6	6·6	5·2
Western hemlock (unmixed)	15·9	11·4	9·3	7·6	5·9	12·4	9·3	7·9	6·2	4·8
Western hemlock (commercial)	14·5	10·0	8·6	6·6	5·2	11·0	8·3	6·9	5·2	4·1
Parana pine	14·5	10·0	8·6	6·6	5·2	12·4	9·3	7·9	6·2	4·8
Pitch pine	18·6	13·1	11·0	8·6	6·6	14·5	10·3	8·6	6·6	5·2
Redwood	14·5	10·0	8·6	6·6	5·2	11·0	7·9	6·6	4·8	3·8
Whitewood	14·5	10·0	8·6	6·6	5·2	11·0	7·9	6·6	4·8	3·8
Canadian spruce	13·8	9·7	7·9	6·2	5·2	11·0	7·9	6·6	4·8	3·8
Western red cedar	11·0	7·6	6·6	5·2	3·8	9·0	5·9	4·8	3·4	2·8
b. Home-grown										
Douglas fir	17·9	12·4	10·7	8·3	6·6	13·8	10·0	8·3	6·2	4·8
Larch	17·2	12·1	10·3	7·9	6·2	13·1	9·3	7·6	5·5	4·5
Scots pine	15·2	9·7	7·9	6·2	5·2	11·7	7·9	6·6	4·8	3·8
European spruce	11·0	7·2	5·9	4·8	3·4	9·0	5·9	4·8	3·4	2·8
Sitka spruce	10·3	6·6	5·5	4·5	3·4	8·3	5·2	4·1	3·1	2·4
HARDWOODS										
a. Imported										
Abura	16·5	12·1	10·3	7·9	6·2	13·8	10·0	8·3	6·2	4·8
African mahogany	15·2	10·7	9·0	7·2	5·5	13·1	9·3	7·6	5·5	4·5
Afrormosia	26·2	19·3	15·9	12·4	9·7	22·1	15·2	12·4	9·3	7·6
Greenheart	41·4	31·0	26·9	20·7	16·5	30·3	22·8	19·7	15·2	12·1
Gurjun/keruing	22·8	14·8	12·4	9·7	7·9	19·3	13·1	11·0	8·3	6·6
Iroko	23·4	17·6	15·2	11·7	9·3	19·3	14·5	12·1	9·0	7·2
Jarrah	23·4	16·9	14·1	11·0	8·6	20·7	15·2	12·4	9·3	7·6
Karri	26·2	19·3	15·9	12·4	9·7	22·1	15·9	13·1	9·7	7·9
Opepe	29·0	22·4	18·6	14·5	11·7	24·8	18·6	15·9	12·4	9·7
Red meranti/red seraya	15·2	10·7	9·0	7·2	5·5	13·1	9·3	7·6	5·5	4·5
Sapele	23·4	16·9	14·1	11·0	8·6	20·7	15·2	12·4	9·3	7·6
Teak	26·2	19·3	15·9	12·4	9·7	22·1	15·9	13·1	9·7	7·9
b. Home-grown										
European ash	22·8	14·8	12·4	9·7	7·9	15·2	10·3	8·6	6·6	5·2
European beech	22·8	14·8	12·4	9·7	7·9	15·2	10·3	8·6	6·6	5·2
European oak	20·7	13·8	11·7	9·0	7·2	15·2	10·3	8·6	6·6	5·2

NOTE. These stresses apply to timber having

Compression perpendicular to the grain			Shear parallel to the grain					Modulus of elasticity for all grades	
Basic	75/65 Grades	50/40 Grades	Basic	75 Grade	65 Grade	50 Grade	40 Grade	Mean	Minimum
N/mm²	N/mm²	N/mm²	N/mm²	N/mm²	N/mm²	N/mm²	N/mm²	N/mm²	N/mm²
2·62	2·34	1·93	1·93	1·34	1·21	0·90	0·76	11 700	6 600
2·07	1·72	1·52	1·65	1·21	1·07	0·83	0·66	10 000	5 900
2·07	1·72	1·52	1·52	1·14	0·97	0·76	0·62	9 300	5 500
2·21	1·93	1·65	1·65	1·21	1·07	0·83	0·60	9 000	4 800
2·62	2·34	1·93	1·93	1·34	1·21	0·90	0·76	11 700	6 600
2·21	1·93	1·65	1·52	1·14	0·97	0·76	0·62	8 300	4 500
2·07	1·72	1·52	1·52	1·14	0·97	0·76	0·62	8 300	4 500
2·07	1·72	1·52	1·52	1·14	0·97	0·76	0·62	9 000	5 500
1·52	1·31	1·10	1·38	0·97	0·83	0·69	0·55	6 900	4 100
2·48	2·21	1·93	1·52	1·14	0·97	0·76	0·62	10 000	4 800
2·62	2·34	1·93	1·72	1·21	1·07	0·83	0·66	9 700	4 800
2·48	2·21	1·93	1·52	1·14	0·97	0·76	0·62	9 700	5 500
1·65	1·38	1·24	1·24	0·90	0·76	0·62	0·45	6 900	3 800
1·65	1·38	1·24	1·24	0·90	0·76	0·62	0·45	7 200	3 800
3·45	3·10	2·48	2·41	1·65	1·45	1·14	0·90	9 300	4 800
3·10	2·62	2·21	1·93	1·34	1·21	0·90	0·76	8 600	4 500
6·21	5·17	4·48	2·76	2·07	1·79	1·38	1·10	12 100	7 900
9·31	7·93	6·90	5·52	3·93	3·38	2·62	2·14	18 600	13 400
4·48	3·79	3·45	2·62	1·86	1·65	1·28	0·97	13 800	9 300
6·21	5·17	4·48	2·62	1·86	1·65	1·28	0·97	10 300	6 900
6·21	5·17	4·48	2·62	1·86	1·65	1·28	0·97	12 100	7 900
7·24	6·21	5·17	2·76	2·07	1·72	1·34	1·10	15 500	9 700
8·27	7·24	6·21	3·72	2·48	2·21	1·65	1·34	13 800	9 300
2·62	2·34	1·93	1·72	1·21	1·07	0·83	0·66	8 300	4 500
6·21	5·17	4·48	2·76	1·86	1·65	1·28	0·97	11 000	6 900
6·21	5·17	4·48	2·62	1·86	1·65	1·28	0·97	12 400	7 900
4·48	3·79	3·45	3·10	2·28	2·00	1·52	1·24	11 400	7 200
4·48	3·79	3·45	3·10	2·28	2·00	1·52	1·24	11 400	7 200
4·48	3·79	3·45	3·10	2·07	1·72	1·34	1·10	9 700	5 200

moisture content not exceeding 18 %.

CP 112: Part 2: 1971

Table 58. Minimum Number of Growth Rings per 25 mm for Softwood Grades

Softwood Grade	Minimum number of growth rings per 25 mm
75	8
65	6
50	4
40	4

There are no limits for hardwood grades or for laminating grades.

A.8 Fissures

A.8.1 Measurement. The size of fissures at the end of the member should be taken as the distance between lines enclosing the fissures and parallel to a pair of opposite surfaces (see Fig. 12). The sum of all fissures measured at either end of the member should be taken for the purpose of assessing the magnitude of the defect.

The size of fissures occurring on the surface of the member should be measured by means of a feeler gauge, not exceeding 0.15 mm thick, and the maximum depth of fissures occurring at any cross-section should be taken for assessing the magnitude of the defect.

A.8.2 Grading limits. To qualify for a particular grade, a hardwood or softwood member should not have fissures whose size, expressed as a fraction of the thickness of the member, exceeds the maximum value given in Table 59.

Size of fissure is $A + B$

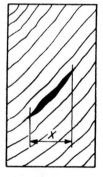

Size of fissure is X

Fig. 12. Fissures.

CP 112: Part 2: 1971

Table 59. Maximum Size of Fissures

Grade (hardwoods and softwoods)	Maximum size of fissure expressed as a fraction of the thickness of the member
75	0.3
65	0.4
50	0.5
40	0.6

Outside the middle half of the depth of the end cross-section, and at a distance from the end equal to three times the depth of the piece, and on compression members, the depth of fissures may be $1\frac{1}{2}$ times the amount permitted in Table 59.

For laminating grades there is no limit to the depth of fissures, but any having an angle of less than 45° with the wide face are not permissible.

A.9 Slope of grain

A.9.1 Measurement. The slope of grain should be measured over a distance sufficiently great to determine the general slope, disregarding slight local deviations.

The method by which slope of grain is measured should be either:

(1) by taking a line parallel to the surface checks; or

(2) by the use of a grain detector (see A.17).

A.9.2 Grading limits. To qualify for a particular grade, a hardwood, softwood or laminated member should not have a slope of grain steeper than the value given in Table 60.

Table 60. Maximum Slope of Grain

Grade (hardwood, softwood and laminated members)	Maximum slope of grain
75	1 in 14
65	1 in 11
50	1 in 8
40	1 in 6
LA	1 in 18
LB	1 in 14
LC	1 in 8

CP 112: Part 2: 1971

A.10 Spiral grain. Where spiral grain occurs, the slope of grain should be determined by measuring the worst slopes of grain on the faces and on the edges and taking the square root of the sum of the squares of the slopes. For example, if these slopes are 1 in 18 and 1 in 12, the combined slope is

$$\sqrt{\left(\frac{1}{18}\right)^2 + \left(\frac{1}{12}\right)^2} = \frac{1}{10} \text{ or a slope of 1 in 10}$$

NOTE: This procedure is intended to apply where it is deemed necessary to measure the slope of grain: in most cases, it will be sufficient for the grader to use his judgment in selecting timber containing spiral grain to an undesirable extent.

A.11 Wane

A.11.1 Measurement. The amount of wane on any surface should be the sum of the wane at the two arrises, and should be expressed as a fraction of the width of the surface on which it occurs (see Fig. 13).

A.11.2 Grading limits. To qualify for a particular grade, a hardwood or softwood member should not have wane whose total width, expressed as a fraction of the width of the surface on which it occurs, exceeds the amount given in Table 61.

For laminating grades, no wane should be permitted.

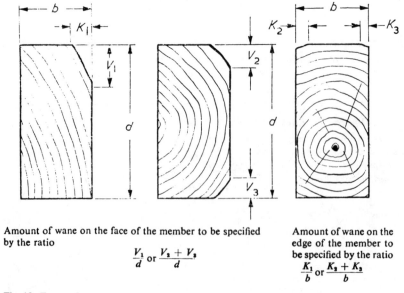

Amount of wane on the face of the member to be specified by the ratio

$$\frac{V_1}{d} \text{ or } \frac{V_2 + V_3}{d}$$

Amount of wane on the edge of the member to be specified by the ratio

$$\frac{K_1}{b} \text{ or } \frac{K_2 + K_3}{b}$$

Fig. 13. Extent of wane.

CP 112: Part 2: 1971

Table 61. Maximum amount of wane

Grade (hardwoods and softwoods)	Maximum amount of wane, expressed as a fraction of the width of the surface on which it occurs
75	0.1
65	0.1
50	0.2
40	0.2

A.12 Wormholes

Scattered pin holes and small occasional wormholes are permissible in all grades. All pieces, however, showing active infestation should be rejected or subjected to preservative treatment in accordance with CP 98*.

A.13 Resin

Resin pockets should be measured and accepted on the same basis as fissures. Substantial exudations of pitch or resin should be excluded from the faces of Grades LA, LB and LC.

A.14 Sap wood

Sap wood, whether bright or blue-stained, is not a structural defect and is acceptable.

A.15 Other defects

All pieces showing fungal decay, brittleheart or other abnormal defects affecting strength should be excluded from all grades.

A.16 Knots

A.16.1 Measurement

A.16.1.1 *Splay knots.* A splay knot should be measured only on the edge of the piece and its size taken as the width between the arris on which it occurs and a line touching the knot parallel with the arris (see Fig. 7).

A.16.1.2 *Arris knots.* The size of an arris knot depends on whether the knot is on the heart side of the piece or on the side furthest from the pith.

Where it emerges on the heart side of the piece, the knot should be measured as a splay knot.

Where it emerges on the side further from the pith, the size should be taken as the width of the knot on the edge between the arris on which it occurs and a line touching the knot parallel with the arris, plus one third of its depth on the face (see Fig. 8).

CP 98, 'Preservative treatment for constructional timber'.

CP 112: Part 2: 1971

A.16.1.3 *Edge knots.* The size of an edge knot should be taken as the width between lines containing and touching the knot and parallel with the arrises of the piece (see Fig. 9). Splay and arris knots should be assessed as edge knots.

A.16.1.4 *Margin knots.* Where a margin knot breaks through and shows on an edge, its size should be taken on the face of the material as the width between the arris and a line parallel with the arris touching the far side of the knot. Where the knot does not break through the arris, the measurement should be taken in the same way as for edge knots (see A.16.1.3 and Fig. 9).

A.16.1.5 *Face knots.* The size of a face knot should be taken as the average of its largest and smallest diameters (see Fig. 9).

A.16.1.6 *General.* Knots should be measured between lines enclosing the knot and parallel to the edges of the wide faces. If two or more knots are in line, i.e. partially or completely enclosed by the same parallel lines and separated lengthwise by less than 200 mm, the effective width of the knots should be the distance between two parallel lines which enclose the knots.

Where two or more knots occur in the same cross-section, the sum of their maximum sizes, determined as in the previous paragraph, should be taken when determining the grade.

Two or more knots of maximum size should be permitted if they are separated in a lengthwise direction, by a distance, measured centre to centre of the knots, of at least 300 mm.

In the assessment of knots, pin knots, i.e. knots with a diameter not exceeding 3 mm, may be disregarded.

A.16.2 Grading limits. To qualify for a particular grade, a hardwood or softwood member should not have knots whose sizes exceed the maximum values given in Table 62.

To qualify for a particular grade, a laminated member should not have knots whose size exceed the maximum values given in Table 63.

Table 62. Maximum permissible size for knots in sawn and precision timber

All dimensions in mm. Column headings show the structural grade (75, 65, 50, 40 Grade).

Width of surface	Beams: Edge, arris and splay knots 75	65	50	40	Margin knots 75	65	50	40	Face knots, in centre half of depth 75	65	50	40	Knots on any surface of compression member 75	65	50	40	Knots on any surface of tension member 75	65	50	40
16	4	6	8	11				6	4	7	9	10	4	7	9	10				6
19	5	7	9	13			4	7	5	8	10	11	5	8	10	11			4	7
22	6	8	11	14			5	8	6	9	11	13	6	9	11	13			5	8
25	7	9	13	16			6	9	6	10	13	15	6	10	13	15			6	9
32	8	12	16	20		4	8	11	8	12	16	18	8	12	16	18		4	8	11
36	9	13	18	22		5	9	12	9	13	17	20	9	13	17	20		5	9	12
38	10	14	19	23		6	10	13	10	14	18	21	10	14	18	21		6	10	13
40	10	15	20	24		6	10	14	10	15	19	22	10	15	19	22		6	10	14
44	11	16	22	26	4	8	12	15	11	16	21	24	11	16	21	24	4	8	12	15
50	13	18	25	29	5	9	14	17	13	19	23	27	13	19	23	27	5	9	14	17
63	16	23	31	37	7	12	18	21	16	23	29	33	16	23	29	33	7	12	18	21
75	19	28	37	43	9	16	21	25	19	27	34	39	19	27	34	39	9	16	21	25
100	25	37	50	56	13	20	29	33	25	35	45	51	25	35	45	51	13	20	29	33
125	31	47	63	69	17	25	37	41	32	43	56	64	32	43	56	64	17	25	37	41
150	37	56	75	83	20	30	44	48	38	51	66	74	38	51	66	74	20	30	44	48
175	41	61	82	88	24	34	52	56	44	59	75	84	44	59	75	84	24	34	52	56
200	44	66	87	93	28	39	59	64	50	66	85	94	50	66	85	94	28	39	59	64
225	47	70	92	97	32	44	67	72	55	72	92	101	55	72	92	101	32	44	67	72
250	51	75	97	102	35	48	75	79	60	78	99	108	60	78	99	108	35	48	75	79
300	54	79	107	112	40	59	88	92	69	91	114	122	69	91	114	122	40	59	88	92

NOTE 1. Two or more knots of maximum size are not permitted in the same 300 mm length.

NOTE 2. For members subjected to simple bending on a single span, e.g. floor joists, the knot size quoted may be increased outside the middle third of the span. These may increase proportionally towards the ends to sizes 25 % greater than the quoted values.

NOTE 3. For widths of surface other than these listed, e.g. for processed timber, the maximum size of permissible knots may be obtained by interpolation, rounding up to the nearest millimetre.

CP 112: Part 2: 1971

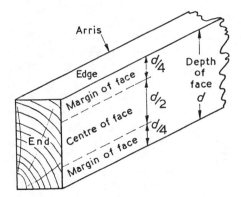

Fig. 6. Edge, face and arris.

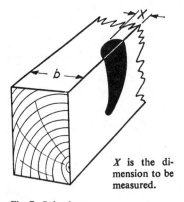

X is the dimension to be measured.

Fig. 7. Splay knot.

X and Z are the dimensions to be measured.

The size of the knot is taken as $X + \dfrac{Z}{3}$

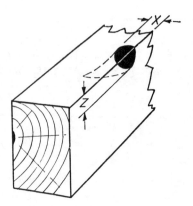

Fig. 8. Arris knot.

CP 112: Part 2: 1971

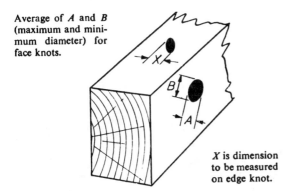

Average of *A* and *B* (maximum and minimum diameter) for face knots.

X is dimension to be measured on edge knot.

Fig. 9. Edge and face knot.

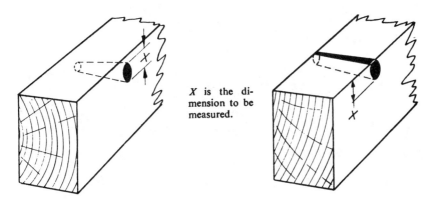

X is the dimension to be measured.

Fig. 10. Margin knots.

SIX

IRON AND STEEL

Sheerness Boatstore. Column-Beam Junction: The first rigid frame building in the world. Its columns and secondary beams have integral end plates, indicating that they are of cast iron. The main beam is built up from plates and angles, riveted together. This indicates either wrought iron or steel; however, the date of construction, 1858, pre-dates the acceptance of steel for structures, pointing firmly to wrought iron. (*M.N. Bussell phot.*)

Our way of life today and our so-called civilization owe as much to people like Abraham Darby and Samuel Crompton as to John Keats and William Wordsworth. Similarly, there are many buildings from the late eighteenth century onwards which, by virtue of the technology of their construction and their architecture, are as worthy of conservation as the sometime abodes of the literati of the period.

Some of these buildings have in fact been preserved as museum exhibits, as for example at Ironbridge, where Abraham Darby for the first time smelted iron with coke, but there are many more buildings with iron or steel structures, which merit conservation, but which cannot be kept as museum pieces. For these buildings to be kept, they must continue to be used, or a new use must be found for them.

Assessment of the adequacy of such structures for continued or changed use requires knowledge of the significantly different properties of cast iron, wrought iron and steel.

6.1 MANUFACTURING PROCESSES AND MATERIALS' PROPERTIES

6.1.1 Ancient Methods v. Coke Smelting

In ancient times and up to the end of the middle ages, iron was produced from the ore by repeated heating and hammering to physically remove impurities. The furnaces, then in use, were not capable of producing temperatures high enough to melt the iron to separate it from the impurities, and the amount that could be heated at one time was limited. What therefore emerged were small ingots with a pasty consistency, to be shaped by forging, and 'wrought' into artifacts.

The development, in the fifteenth century, of the blast furnace, made it possible to produce iron in liquid state, capable of being cast in moulds. These early blast furnaces were fuelled with charcoal. This restricted their size and thus their output (if they were too high, the charcoal would crush and choke the fire; if they were too wide, the draft would be insufficient to produce the necessary temperature).

In 1709 Abraham Darby succeeded in smelting iron, using coke as fuel. This process, in use up to this day, enabled a quantum leap in output to be made.

Cast iron, whether from a charcoal-fired furnace or smelted with coke, does however have profoundly different properties from forged or wrought iron and the differences are significant for its performance in structures.

6.1.2 Cast Iron: Manufacture, Composition and Properties

In the blast furnace (see Fig. 6.1), the iron ore reacts with carbon monoxide to form an alloy of iron and 2–5 per cent carbon. This carbon content lowers the melting point of the iron from about 1500 °C to about 1200 °C and gives the molten iron a low viscosity so that it can easily be cast in intricate shapes.

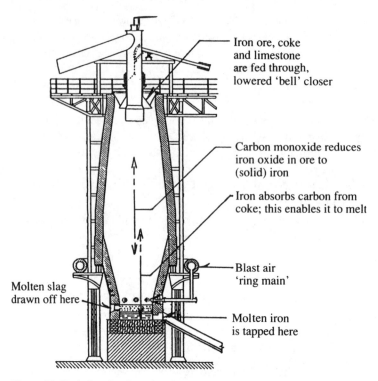

Iron ore, coke and limestone are fed through, lowered 'bell' closer

Carbon monoxide reduces iron oxide in ore to (solid) iron

Iron absorbs carbon from coke; this enables it to melt

Blast air 'ring main'

Molten slag drawn off here

Molten iron is tapped here

Figure 6.1 Vertical section through early twentieth-century blast furnace (*after Suenson, Vol.1, 1920*).

The 'raw' iron from the earlier blast furnaces contained quite a few impurities and was usually cast into bars, known as 'pigs', which were allowed to solidify before being purified by remelting for the production of the desired castings. (Nowadays the raw iron is usually transported in liquid state to the refining furnace.)

The carbon content of the molten iron was fairly high: between 2.5 and 4.5 per cent. This was beneficial during the casting process, because it made the iron flow easily. It did, however, give the finished iron less than perfect mechanical properties for structural purposes, as during the cooling of the iron, the carbon separated out into angular graphite crystals. These acted as internal stress raisers, which made the iron brittle. Whilst its compressive strength was good, 600–750 N/mm^2, its tensile strength was low and very variable: 75 and 160 N/mm^2 being the lowest and the highest results of a series of 51 tests, performed by Eaton Hodgkinson in the 1830s. Whilst the variability has been reduced since then, and the minimum strength improved, the basic characteristics remain the same for today's ordinary 'grey' cast iron, as opposed to the specially treated, malleable, 'spherical graphite' cast irons, used for special applications.

The stress–strain diagram for cast iron is curved almost from the origin (see Fig. 1.15) and Young's modulus (as far as it can be defined under these circumstances) varies between 84 000 and 90 000 N/mm^2 for compression and between 66 000 and 94 000 N/mm^2 for tension. (The values given above for the mechanical properties are metric equivalents of figures quoted by Twelve-trees in 1900; R.J.M. Sutherland quotes more recent, slightly higher values in his paper *Appraisal of Cast and Wrought Iron (as Materials)* given to the Symposium on Seaside Piers 12 November 1991.)

The moulds for iron castings were, and still are in the majority of cases, made of sand with a small proportion of clay to give it cohesion. It was tamped round wooden formers, or 'patterns', placed in two-part boxes, which would be taken apart to allow removal of the patterns. Funnels and vents would then be formed in the sand to allow the entry of the molten iron and the escape of air and the box would be reassembled ready for casting (see Fig. 6.2).

The texture of the mould sand was reproduced 'in negative' on the castings as a gritty surface finish and this is often an aid to identification, as are the generous radii to re-entrant corners (see Sec. 6.3.1).

The shrinkage of the iron after it has solidified is substantial: 1.4 per cent linearly. This would be of little consequence for the strength, if the castings were free to cool and shrink uniformly. However, thin sections cool more quickly than thicker ones and the mould material restrains the shrinkage. This can lead to 'cast-in' tensile stresses, which will reduce the resistance to external tensile forces of the finished casting. Another consequence of the shrinkage is that, if one wants to make replacements for broken castings, and tolerances are not generous, the pattern for the mould cannot be made simply by taking an impression of an existing intact specimen; it has to be enlarged to compensate for the shrinkage.

Cast iron had the advantage of being easily made into almost any desired shape; the disadvantages were that once cast, the shape could not be modified by forging. Furthermore, the variable, low tensile strength, sometimes exacerbated by the presence of voids and blow-holes, and the brittle mode of fracture, gave no warning of impending collapse. The risks could be reduced by careful design of the components, by rigorous quality control in the foundry and by proof-loading, but the usual practice was just to use very large factors of safety in design. These drawbacks were absent from wrought iron.

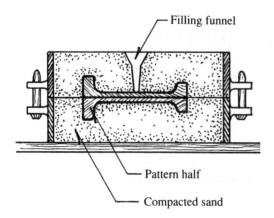

Figure 6.2 Cross-section through mould for cast-iron I-section.

6.1.3 Wrought Iron: Manufacture, Composition and Properties

The ancient methods of extracting iron from the ore at relatively low temperatures produced an alloy which, due to its relatively low carbon content, could be forged and was tough–properties that would have made it ideal for structural use. Unfortunately, the quantities in which it could be produced were quite inadequate for it to be used in structural members in general. It was only the invention of the 'puddling' process by Henry Cort in 1784, that enabled the production of wrought iron on a scale that made it a practical material for structures, in the western world.

In the puddling furnace the pig iron, with its high carbon content, was remelted under an oxidizing flame (a flame blown with more air than necessary for complete combustion). The extra oxygen burned away some of the carbon in the iron, resulting in proportions that would not become graphite crystals during cooling. This also led to the melting point of the iron rising and, as a consequence, the iron became less fluid. In order to expose the whole volume of the iron to the oxygen, it was 'stirred' with a wrought iron rod. After a while the consistency changed from a liquid to a dough, which was turned over and further 'kneaded' or 'puddled' to promote the decarburization (see Fig. 6.3).

The oxygen, however, not only burned away the carbon, it also oxidized the surface of the iron, forming a crust, or 'slag' of oxides and other impurities, some of which were entrapped in the iron. To expel some of the slag, or at least break it up and distribute it, the lump of iron 'dough' was flattened under a steam hammer (see Fig. 6.4), folded over and hammered again and this process was repeated several times, with the lump, or 'bloom', being reheated as necessary to maintain a workable consistency. Each of the first reworkings improved the properties of the material and the number of reworkings was sometimes specified as an indication of the quality required for a particular use.

After the hammering, the bloom was passed through rollers and the net result was a metal with a low carbon content, generally between 0.1 and 0.5 per cent, capable of further forming by forging or rolling (see Fig. 6.5). Because some of the oxide crusts were left in the iron, the effect of the rolling was to give the material

Figure 6.3 Iron puddling in the mid-nineteenth century (*from Derry and Williams, 1960*).

Figure 6.4 Steam hammer (*from Derry and Williams, 1960*).

a laminated texture with thin layers of very pure iron alternating with layers of slag, almost like plywood. These slag layers had little effect on the mechanical properties in the direction of rolling. They were, however, sometimes porous, so that water could penetrate and cause corrosion well into the section and not just on the surface. To make matters worse, they sometimes contained phosphorus, which accelerated the corrosion.

The mechanical properties of wrought iron are akin to those of mild steel: Tensile and compressive strengths are approximately equal, generally varying between 250 and 420 N/mm^2. These are ultimate strengths; the laboratories of the time did not always measure the yield points. The values of yield stress that were recorded (sometimes labelled 'elastic limit') generally ranged from 150 to 300 N/mm^2. Young's modulus varied between 150 000 and 220 000 N/mm^2. Wrought iron could be jointed by forge-welding: hammering together the parts, whilst at bright

Figure 6.5 Rolling mill, c. 1875 (*from Derry and Williams, 1960*).

red heat, thus expelling the oxides on the interfaces. This was a highly skilled operation, the success of which could only be demonstrated by proof-testing.

The laminar texture of wrought iron only appears to the naked eye after heavy corrosion, when the lamina separate at the edges which then take on an appearance reminiscent of 'flaky pastry'. It is, however, readily observed on a ground and polished surface under the microscope, and this is one way of distinguishing it from steel.

6.1.4 Steel: Manufacturing Processes, Composition and Properties

A common feature of all processes for making structural steel is that it is produced in the liquid state.

The earliest method was invented by Henry Bessemer in 1855. Molten pig iron was poured into a 'converter': a pear-shaped container, open at the narrow end. Compressed air was then forced through the iron from numerous injection nozzles in the base. The oxygen in the air, bubbling through the liquid iron, would burn off the carbon in an exothermic reaction (fuelled also by the silicon in the iron (see Fig. 6.6)). This raised the temperature of the 'melt' sufficiently to keep the low-carbon 'iron', or steel as it had now become, liquid enough to enable it to be poured into block, or 'ingot', moulds by tilting the container. The ingots were subsequently stripped out of their moulds and put in a 'soaking pit', where they were kept at red heat prior to be taken to the rolling 'stands'(frames with pairs of rollers).

Bessemer's invention enabled a team of workmen to produce 5 tons of malleable steel by a single 'blowing' in 20 minutes, where it would have taken one and a half days to make the same quantity of wrought iron by puddling. This seemed to open the way for real large-scale production of steel for structural purposes.

Bessemer's original process had two drawbacks: firstly, it could not remove any phosphorus from pig iron smelted from the most common British ores (which *did* contain phosphorus) and the resulting steel therefore became brittle; secondly,

Figure 6.6 Nineteenth-century Bessemer converter 'blowing' (*from Derry and Williams, 1960*).

the rapidity of the process made it difficult to control the amount of carbon that had been removed and this led to a variable product. (The moment of turning off the air-blast was determined by watching the change in the colour of the flame, emerging from the open 'neck' of the converter-vessel!)

The first drawback was largely overcome by changing the material of the lining of the converter vessel from an 'acid' silica-rich fireclay to a 'basic' one, based on Dolomite, a development made in 1879 by S. Gilchrist Thomas. The second disadvantage (and together with this, also the first) was eliminated when William Siemens introduced the 'Open-Hearth Furnace', fired by gas and pre-heated air; his process also allowed the recycling of iron and steel scrap.

The properties of early mild steel, as used in structures, were very similar to those found in today's material regardless of the method of manufacture. Allowance has however to be made for the fact that the techniques, available then, could not achieve the uniformity and level of mechanical properties taken for granted today. Twelvetrees quotes ultimate tensile strength between 400 and 495 N/mm^2, elastic limit between 280 and 310 N/mm^2 and Young's modulus 201 000 N/mm^2. (The yield-point was usually not recorded at that time.) The higher strength values of today's steel are due to more powerful rolling machinery, capable of compressing the steel at lower temperatures, as well as improved alloying compounds.

The metal, thus produced, was generally ductile at normal temperatures, but it could become brittle at low temperatures (the size of the tear in the side of the 'Titanic' has been thought to be aggravated by brittleness, due to the temperature of the sea; modern steels are less brittle in the cold). The steel was rolled in a great variety of sections, which could, if desired, be locally modified by hot forging. It could however not be forge-welded as wrought iron could, and any connections had therefore to be made by bolting or riveting.

Welding, by acetylene/oxygen flame or by electric arc, by which the parts are joined by melting together, was tentatively introduced during the 1920s. It did not become an established practice in British constructional steelwork until the 1940s, when radiographic and other methods of quality control had been developed. As late as in the 1960s, the steelwork fabricator for the Crystal Palace Sports Centre chose to rivet some of the steelwork, in order to keep his riveting shop gainfully employed.

6.2 STRUCTURAL FORMS AND USES, HISTORICAL DEVELOPMENT

6.2.1 Early Uses of Wrought Iron

Wrought iron was used in China for a suspension bridge, as early as 600 AD. In India, the pillar by Qutb Minar in Delhi is said to date from the second century AD and the great sun temple at Konarak on the east coast of Orissa, built in the thirteenth century AD, had beams of rectangular cross-section approximately 250 mm deep, 200 mm wide and 11 m long. (Examination in 1987 of a fracture surface of one of these beams showed that they had been made by forge-welding from bars about 65 by 85 mm in cross-section.)

The first European use of wrought iron in structures was generally in applications where its (comparatively) high tensile strength was of great benefit, even though the quantities available were small: connection straps for timber trusses, tie rods, tension rings for domes, etc. There are small tie rods in Norwich cathedral, dating from about 1100 AD and ties of 2 × 2 in. (50 × 50 mm) cross-section were used in Westminster Abbey in the thirteenth century. Later, Wren used wrought iron extensively in St Paul's Cathedral in London.

In the 1780s there were a number of substantial buildings in France, in which wrought iron tied arches and open web beams were used, the latter in combination with hollow tile floors, but the Revolution and the ensuing upheavals temporarily stopped further development of these techniques there. During the 1830s Chartres cathedral was however reroofed after a fire with cast iron arched rafters connected at eaves level with wrought iron ties and a few years later, a rather more complicated roof structure, combining cast and wrought iron was used at St Denis cathedral.

6.2.2 Cast Iron in 'Fire-proof' Mill Construction

In Britain a number of disastrous fires in multi-storey textile mills prompted a search for a construction that would prevent a local fire (starting, for example, by a candle dropping on accumulated cotton waste on the floor) from rapidly engulfing the whole building and causing the entire internal timber structure to collapse and cause appalling casualties. The first of these mills was constructed with tile floors carried on brick barrel vaults ('jack arches'), spanning between timber beams which were protected against fire by a substantial covering of lath and plaster. The beams were supported on the external walls and on cast iron columns, generally at the third-points of the width of the building. The first example of these buildings was constructed at Derby in 1793 by William Strutt. This was a six-storey structure, about 60 m long and 10 m wide. More buildings of this type followed and in 1796–97 Charles Bage designed and supervised the construction of a mill in Shrewsbury with an internal structure entirely of cast iron with jack-arches (see Fig. 6.7). Tie rods were provided between the beams to counteract the thrust of the vaults, but in some of the mills they were at a level where they were not very effective.

It should be noted in passing that Abraham Darby—the grandson of the inventor of coke smelting—had erected his cast iron bridge at Coalbrookdale in 1779.

Simultaneous with this development of the multi-storey mill buildings, cast iron was increasingly used for items like columns, supporting galleries in churches, intermediate supports for roofs over single-storey dock sheds, etc. Highly decorated open-web arches were used in the 1840s for the roof of the Bibliothèque Ste.Geneviève and the dome over the Reading Room of the British Museum, dating from the 1850s, has cast iron ribs. About the same time some buildings in America were constructed with facade panels and frames of cast iron.

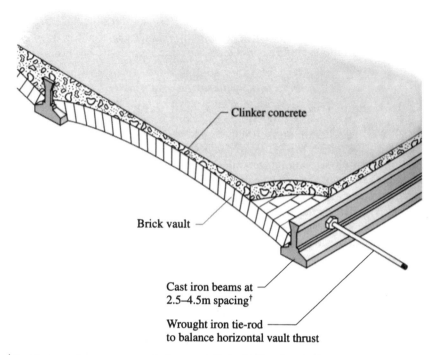

Clinker concrete

Brick vault

Cast iron beams at
2.5–4.5m spacing†

Wrought iron tie-rod
to balance horizontal vault thrust

†Cast iron was in common use for beams until the 1850s, after that it was gradually
superseded by wrought iron and, from the 1890s, by steel

Figure 6.7 Jack-arch floor, supported on cast-iron beams.

The variable strength of cast iron and the brittle nature of its fracture meant
however that very large safety factors had to be used in the design of beams and
this sometimes resulted in undesirably large sections for longer spans. When the
Palace of Westminster was rebuilt after the fire in 1834, it was a requirement that
the structure should be incombustible. This led to the use of cast iron beams in
the floors and cast- and wrought-iron roof trusses. The use of iron in the roofs
included the roof coverings of large cast iron 'tiles'. As with other very special
buildings, all components were proof-tested before being incorporated into the
structure; this justified the adoption of lower factors of safety.

The size of section that could be rolled in wrought iron was limited by the
weight of the lump or 'bloom' that could be manhandled between the steam hammer
and the reheating furnace, and by the power of the rolling machinery available then,
but with Henry Fielder's 1847 patent for building up sections from plates and angles,
riveted together, the way was open for the use of large wrought iron beams. (Parallel developments by Zores and others took place in France about the same time.)

The Crystal Palace of 1851 can be said to mark the culmination of the use
of cast iron for all members of a building structure (columns *and* beams), as well

as being an early and spectacular example of prefabricated construction. For structural members stressed mainly in compression, cast iron did however continue to be widely used right up to the 1930s. (In mechanical engineering, cast iron has been developed for use as an entirely reliable material for, even very large, cylinder blocks, pump casings, etc. The building industry has been slow to adopt this technology.)

6.2.3 Wrought Iron/Cast Iron Building Structures 1850–1900

In industrial buildings wrought iron took over from cast iron as the material for the beams (Fig. 6.8a), whilst cast iron was retained for the columns; these were generally of hollow round section, although cruciform sections were also popular. In the floor constructions, the brick jack arches were gradually supplanted by concrete vaults, as often as not using clinker from coke-fired boilers as aggregate and, in the earlier applications, lime as the cementing agent.

The next step was to space the main (primary) beams, with their columns, further apart and bridge the spans between them with small I-sections, so-called filler joists (Fig. 6.8b), which were then embedded in a concrete slab, again using clinker aggregate.

Towards the end of this period riveted I or double-I-section stanchions were sometimes used in lieu of round cast iron columns, mainly for cellular plan buildings, where they were more easily integrated with the subdividing walls.

So far, however, the external walls provided the support for the ends of the beams. This meant that they had to be designed according to existing rules-of-thumb for load-bearing walls; it also meant that construction had to alternate between the comparatively quick frame erection and the rather slow bricklaying. This was the situation in Britain and in Europe, generally, whilst in North America a development took place, which will be described in Sec. 6.2.5.

6.2.4 Roof Trusses and Large-span Structures

Roof trusses were a fertile ground for invention and experimentation. There are examples of roof structures made from castings, bolted together, roof trusses with compression members of cast iron, sometimes mixed with timber and tension members of wrought iron, as well as tied arches of various kinds. (Some of these trusses may have a bottom cord, raised for part of its length above the level of the supports. This means that they partly act as arches and may therefore exert a horizontal thrust on the supporting walls.)

Being the age of the great railway expansion, there arose the need (or wish) for roofs to span over multiple tracks and platforms. Wrought iron, and later steel, proved eminently suitable for such structures, a great many of which were built, and of which a fair number still survive.

Another field for large-span structures was exhibition halls. Just like the Great Exhibition in London, national and international exhibitions were held in a number of European countries and just as in 1851, the latest technology was used

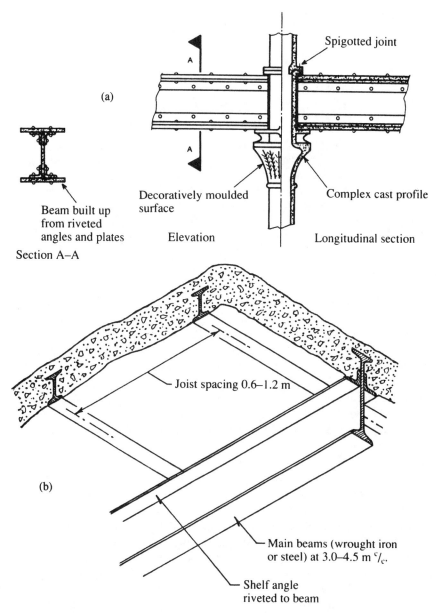

(a)

Spigotted joint

Beam built up
from riveted
angles and plates

Section A–A

Decoratively moulded
surface

Elevation

Complex cast profile

Longitudinal section

Joist spacing 0.6–1.2 m

(b)

Main beams (wrought iron
or steel) at 3.0–4.5 m $^c/_c$.

Shelf angle
riveted to beam

Figure 6.8 Typical iron-framed construction: (*a*) Cast iron column with wrought iron beams (*after Blanchard et al., 1982*); (*b*) Filler-joist floor.

in the second half of the nineteenth century to boost the host-nation's prestige. Some other public buildings were also provided with large roof spans, a notable example being the wrought iron lattice arches of the Bibliothèque Nationale reading room in Paris, dating from the 1860s. For many of these, tied arches of some kind, usually latticed, formed the primary structure, but ribbed domes, in which half-arches radiated from a compression ring under a lantern at the top, as in the British Museum Reading Room, were also popular.

6.2.5 Steel-framed Buildings

The use of iron or steel frames for the internal structure had enabled incombustible construction to be achieved and large open spaces to be created. However, the continued use of load-bearing masonry for the external walls, particularly when archaic rule-of-thumb design requirements had to be used, meant that for multi-storey buildings the external walls had to be very thick in order to carry the loads from the floors. Such walls were slow to construct and, due to their weight, required large foundations, all of which added significantly to the building costs and limited the height of building that could, for practical purposes, be constructed.

This limitation was not too onerous whilst the only vertical access for people was by stairs, but the development of the electric lift in the 1880s, combined with the increasing scarcity of land on the island of Manhattan in New York and in other American cities led to commercial pressures to build higher. The solution was found in the steel-framed building, in which all the loads were carried by the steel beams and columns (or 'stanchions', as they are sometimes called), and the external walls became merely a cladding with the purpose of keeping out the weather and providing fire separation. This not only allowed the walls to be made thinner and lighter, but by allowing the steel frame to be erected in one operation ahead of all the other trades, it speeded up the entire construction. (On the other hand, the heavy masonry walls had provided the lateral stability for the whole structure; in framed buildings with light façade walls, this had generally to be achieved by bracing or by rigid joints, capable of transmitting bending moments between beams and columns.)

The first 'skyscraper'—a 13-storey building—went up in New York in 1889, but in Europe, and in Britain in particular, Building Control Legislation did not recognize the structural function of the external stanchions and even when these external stanchions were provided, building control officers insisted on external walls conforming to the rules-of-thumb for thickness of load-bearing walls. Thus, whilst the Ritz Hotel in London, built 1900–1901, had a complete steel frame, erected ahead of everything else, its external walls were as thick as if they had been load-bearing.

This state of affairs persisted in Britain until the passing of the London Building Act 1909, which for the first time permitted fully framed construction to be designed and utilized as such and, in addition, stated permissible working stresses for cast iron, wrought iron and mild steel. (Even so, fire regulations in London mitigated against buildings higher than 25 m, which required special permits, imposing special provisions and hence extra cost.)

For many steel-framed buildings, constructed during the first half of the twentieth century, all the structural work was let as one contract to the steelwork fabricator. The foundations therefore took the form of steel-beam grillages, embedded in mass concrete.

6.2.6 Stability and Robustness

As long as iron and/or steel structures were essentially used as 'in-fills' in 'traditional' masonry buildings, stability was, and is, provided by the masonry walls, together with the floors, acting as horizontal diaphragms.

Structural robustness of old textile mills and similar buildings is however another matter: The slenderness of cast iron columns, together with the brittleness of the material, makes them vulnerable to accidental impact, etc. Most of these structures did have some devices, tying together the beams at the columns, but these are not always strong enough to allow the beams to bridge over a broken column by 'catenary' action. Introducing a new use, involving moving stock by fork-lift trucks, into such buildings may therefore not be a good idea, unless the robustness has been ascertained, and if necessary improved. The Institution of Civil Engineers' design and practice guide *Structural Appraisal of Iron-framed Textile Mills* discusses this problem in detail.

Twentieth-century steel frames had to stand up on their own until the walls and floors had been constructed. They were therefore stable in themselves. Stanchions would however rely on beams, framing into them in two directions, to limit their free height to one-storey height. This must be borne in mind when large floor openings are to be created as part of a refurbishment.

As for robustness of steel frames, there are photographic records of buildings having had two or three ground floor bays completely removed by bomb blast during the Second World War, with the rest of the six storeys, above, surviving, albeit with a rather drastic sag.

6.3 IDENTIFICATION OF MATERIALS AND CONFIRMATION OF PROPERTIES

Appraisal of the capacity of any structure requires knowledge of the mechanical properties of its material. As described in Sec. 6.1, the properties of cast iron, wrought iron and steel are significantly different. Identification of the true nature of the metal is therefore essential.

The characteristics of the metals and the methods used to form them, significantly influenced the shapes of the structural members and this often makes identification 'by inspection' possible (see Fig. 6.9). Knowledge of the age of the structure is also useful, as the three metals have flourished, in turn, as structural materials during the nineteenth and twentieth centuries (see Table 6.1).

6.3.1 Initial Identification by Shape and Surface Texture

The simplest way to identify cast iron is to consider whether the geometry of the structural component could have been produced by rolling. Anything with integral web stiffeners could not have been rolled. 'Fish-belly'-shaped webs and/or flanges are similarly unlikely to have been rolled (there is one exception: a few very early

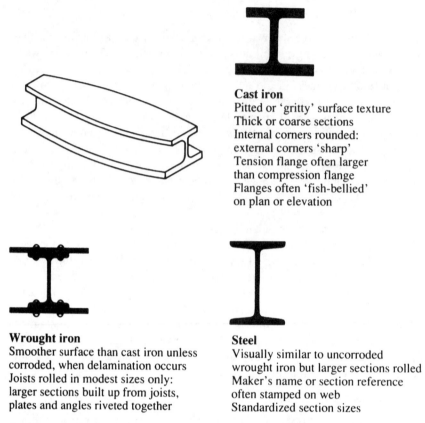

Cast iron
Pitted or 'gritty' surface texture
Thick or coarse sections
Internal corners rounded:
external corners 'sharp'
Tension flange often larger
than compression flange
Flanges often 'fish-bellied'
on plan or elevation

Wrought iron
Smoother surface than cast iron unless
corroded, when delamination occurs
Joists rolled in modest sizes only:
larger sections built up from joists,
plates and angles riveted together

Steel
Visually similar to uncorroded
wrought iron but larger sections rolled
Maker's name or section reference
often stamped on web
Standardized section sizes

Figure 6.9 Characteristics of ferrous metals, as used in beams (*after Blanchard et al., 1982*).

railway rails were of tee-section, with a web of varying depth, produced by special grooved rollers). *Anything that could not have been rolled, must have been cast.*

The sand moulds, used for forming cast iron usually left a 'gritty' surface on the metal (the exception being high grade ornamental castings). This may be masked by numerous coats of paint and removal of these on a small area (by wire-brushing, or similar) may be helpful to the identification. To limit the effects of cooling shrinkage in the mould (which could produce locked-in tensile stresses) generously rounded roots were provided at re-entrant arrises.

The basic shapes of cast iron are also indicative: Columns were usually of circular cross-section, more often than not hollow, but earlier structures often had cruciform column sections. Base-plates and heads were cast integrally with the shaft and were often ornamented in various ways; the heads in particular often displaying leafy 'Corinthian' and similar capitals, serving as bearing for the beams.

Beams were shaped in a way that allowed for the relatively low tensile strength of cast iron, with more material in the bottom (tensile) flange than in the

Table 6.1 Key dates in the chronology of structural iron and steel in Britain
(*after Blanchard et al., 1982*)

	Cast iron	Wrought iron	Steel
1792	First beams		
1794	First columns		
c. 1800	First I-beams		
1840		First built-up beams (plates and angles)	
1860	Decline in use for beams		
1877			Board of Trade approved for use in bridge-building
1882–1889			Forth rail bridge
1885			First rolled joist (Dorman Long)
1909	London Building Act gives design rules and permissible stresses		
1914	Virtually extinct in new work	Virtually extinct in new work	
1937			BS 449 published

top. Early beam sections were inverted 'Tees' or triangles, providing springings for the brick arches, used for supporting the floors. Later, I-beams were used; these had larger bottom flanges, often adopting a structurally advantageous 'fish-belly' shape in plan or elevation.

Connections were usually simple bearings, sometimes with wedges or locking rings securing beams to columns or to each other. Hollow circular columns, if cast in one piece, would tend to show slight vertical seams, diametrically opposed, where the two mould halves had been joined. Sometimes columns were cast in two halves, which were then brazed together along the two longitudinal (vertical) seams; this would make them less robust.

Wrought iron was formed like steel by rolling. It therefore has a smoother surface than cast iron. If, however, it has suffered severe corrosion, it will have become delaminated at the edges. The limitations of the earliest rolling mills meant that sections had to be built up from angles and plates, riveted together. Subsequently rolled I-sections were produced, albeit initially of modest size, and these were often strengthened by additional flange-plates, riveted on.

Wrought iron, being stronger in tension than cast iron, was also used extensively for chain links for suspension bridges, tie rods in roof trusses and arched structures and for rivets and bolts. Wrought iron tie rods, rivets and bolts were often used together with other structural members of cast iron.

Sound steel is difficult to distinguish from sound wrought iron, due to the similar way and the similar forms, in which they are produced. However, steel will rust only on the surface, but will not delaminate like wrought iron. From the early part of the twentieth-century rollers for steel sections were often 'engraved' with the name of the manufacturer and the particulars of the section (e.g. Appleby–Frodingham 8 in. $\times$ 4 in./17^{lb}, meaning an I-section, 8 in. high by 4 in. wide over the flanges, weighing 17 pounds per lineal foot, manufactured by the Appleby-Frodingham Steel Company). This can assist the identification, as these British Standard section sizes were not codified until 1904, by which time wrought iron had largely been superseded by steel.

6.3.2 Age of Structure

As the chronology of the use of iron and steel is fairly well defined, it can help the identification of the metal, if the date of the original construction of the build-ing (and of any later modifications) can be established. Many older buildings have a plaque with the year of completion somewhere on the elevation; alterna-tively, there may be a foundation stone with this information. Some of the sources, mentioned in 3.1, may also help in this respect.

If some indication of the date has been obtained, the Table 6.1 can be used to at least eliminate the unlikely, or, with luck, to identify the probable materials, used in principal structural members. It should however be remembered that smaller components of wrought iron, such as straps, holding back sculptures, can be found in mediaeval buildings, and that sizable structural elements which were made up of smaller components, for example tension chains for masonry domes, were used well before the 1790s. (Wren used both chains and hanger rods of wrought iron in St Paul's Cathedral.)

Table 6.1 in its present form applies to British developments after the indus-trial revolution. It may be a guide for former colonies. For continental Europe and for the USA the chronology may be somewhat different. It is worth noting that the metals were not always produced in the country, in which they were used. Belgian steel beams have been identified in at least one Edwardian building in London and British wrought iron in a textile mill, in Lille in France, dated 1900. In the latter case, the angles and plates had imperial dimensions, but the rivet spacing was metric, indicating that the material was imported, but that fabrication had been done locally.

In some cases, shape and age combined will be enough for a reliable identi-fication, but in other situations, particularly when both wrought iron and steel are possible, some laboratory tests will be necessary.

6.3.3 Metallurgical Analysis

For identification purposes only very small samples are required: Wrought iron and steel can be identified by chemical analysis of a few grammes of drilling swarf, but a better indication of the properties will be obtained from a micro-scopic examination of the polished and etched surface of a sample measuring about 15 $\times$ 15 mm. A similar metallurgical examination should always be carried out for cast iron, as different crystal structures, with different mechanical proper-ties can occur for the same chemical composition.

Samples should be obtained by sawing or core drilling. Flame cutting can produce dangerous notches and cooling stresses in cast iron, whilst for wrought iron and steel, the sample to be taken would have to be substantially larger to allow for removal of the flame-affected zone, which would otherwise give misleading results of the analysis.

For early steel a full chemical analysis should be carried out, in order to check the contents of sulphur and phosphorus, as these can cause some loss of toughness. Similarly, if strengthening by welding is considered, a full analysis is usually required.

6.3.4 Strength Testing in the Laboratory and on Site

After an initial assessment of the structural capacity, it is often desirable to obtain more accurate information on the mechanical properties of the metal(s). As mentioned in 3.5, tests are expensive and obtaining sufficient samples, for the test results to be meaningful, can be difficult. Before any testing is commissioned, the need for the testing and the likely benefit of the tests should therefore be carefully examined.

Generally, the mechanical properties, that should be ascertained, are: (a) The ultimate tensile strength, (b) The yield point, if any, and (c) The elongation at failure. The testing of these is normally done on samples, that are 200 mm long and 50–100 mm wide, but in special cases it may be possible to test samples as small as 100 by 25 mm (as these are non-standard, the tests are likely to be more costly and the results may not be directly comparable with the values on which today's design stresses are based). The orientation of the samples is important: Steel and wrought iron have different strength properties parallel to the direction of rolling and at right angle to it and usually the governing stress is in the direction of the former.

The part of the member, where the sample is taken, is also important: The thinner parts of rolled sections, i.e. the webs, have more mechanical work put into them in the rolling process than the (thicker) flanges and this results in them having a slightly higher strength. As the flanges are usually the most highly stressed parts, test results from only web samples could be a little optimistic.

The location of the samples is even more critical for cast iron: The grain structure is finer and the strength higher, the quicker the iron cools from its molten state and the difference is far more pronounced than for steel and wrought iron. In addition, there is a greater risk of microcracks and porosities forming in the thicker sections during cooling. Samples thicker than the usual standard 25×25 mm test pieces for modern material are therefore likely to give pessimistic results, if the factor of safety already allows for porosities, etc.

Another consequence of the strength gain with rapid cooling is that cast iron members have a thin 'crust', or skin, of higher strength next to the outer surface. If, as is sometimes done, the test sample is machined, to facilitate the clamping in the testing machine, this skin, which contributes a small amount to the strength of the member, will be lost and a slightly too low value will result from the test. A final warning about sample location: even if beams and columns are both of the same type of metal, they do not necessarily come from the same source, so test results for one may not be applicable to the other.

Removal of samples should be carried out so that the sample is not damaged, nor the parent member significantly weakened by the operation. Flame-cutting is quick and easy but, as mentioned earlier, it may induce cooling stresses in cast iron and it leaves a heat-affected zone on the sample. This has to be removed for all metals. In-situ flame-cutting also tends to be dimensionally inaccurate. Hack-sawing is slow and laborious, but is to be preferred and it can be speeded up by initial 'postage stamp' drilling. The latter can also be combined with cutting with an abrasive disc; but this again tends to leave a heat-affected zone, albeit of much smaller extent than that of flame-cutting. Whatever method is used, the formation of sharp, stress-raising, re-entrant corners in the parent member should be avoided, if necessary by pre-drilling with a generous diameter. Needless to say, the samples must not be taken, where this would reduce the cross-section dangerously.

When removal of tensile test samples of cast iron in sufficient numbers or of sufficient size is impractical, it may be possible to carry out wedge penetration tests on discs 35–50 mm in diameter. This test has so far only been used by the British Cast Iron Research Association for new cast iron; hence there is no experience of this method on old structures.

Work, carried out at Karlsruhe University, has however shown a good correlation between the strength of cast iron in tension, when tested on conventional samples, and the transverse splitting strength of cylindrical samples, machined to be 23 mm long and 11.5 mm in diameter. As such samples can often be taken from hollow round columns, without significantly weakening them, this method could prove useful. Karlsruhe University has also found ultrasonic techniques useful for detecting imperfections in cast iron columns, such as core eccentricity, porosity, slag inclusions and 'cold joints' (i.e. where the surface of the iron, flowing round the core former from two sides, had started to solidify before the two 'streams' met, so that the two streams did not melt together).

It can be useful to carry out simple in-situ non-destructive tests, such as hardness tests, in addition to the laboratory tests. For a given batch of material, the hardness increases with the strength and the test can therefore be used to check the variability of the material, or to locate the weakest members, in order to reduce the number of samples to be taken for laboratory testing. (The hardness test can also be used after a fire, to check whether any weakening of the metal has resulted from exposure to the heat.) BS 1452 gives expected ranges of hardness for different grades of modern cast iron

6.4 ESTABLISHING THE STRUCTURAL CAPACITY

6.4.1 Adequacy of original Loading Requirements for continued or new Use

As discussed in Sec. 2.4.3, if a structure has carried a certain load for a considerable period in the past, without showing any signs of having been overloaded and if it has not suffered any significant deterioration (e.g. due to corrosion) nor been damaged by past alterations, it is clearly capable of carrying the same load for a further term of use.

Often, the problem is to establish what the past loading has, in reality, been. As mentioned in Sec. 2.4.3, where actual loading has been observed within living memory, or can be documented by written or photographic records, structural adequacy 'by force of habit' is easier to justify.

6.4.2 Justification by Calculations, based on published Information on Strength and/or Permissible Stresses

If cross-sectional dimensions have been measured with moderate accuracy and the approximate date of construction has been ascertained, geometrical section properties, such as cross-sectional area, moment of inertia and section modulus, can, for all but very early rolled sections, be found in the *Historical Structural Steelwork Handbook*, by The British Constructional Steelwork Association. Where this is not possible, accurate measurements and subsequent calculation of section properties has to be carried out.

When section properties and the approximate date of construction have been ascertained, it may be possible to justify a structure by calculation, using the 'blanket' values for mechanical properties, published at, or soon after, the time of construction. Twelvetrees published values of ultimate strength about 1900, together with recommended factors of safety (see Tables 6.2 and 6.3). These factors were

Table 6.2 Material properties given by Twelvetrees (*from CIRIA 'Structural Renovation of Traditional Buildings', Report 111, 1986*)

	Tons/in.^{2} (as quoted)	(N/mm^2) (approx. equivalent)
Cast iron (average values)		
Tensile strength	8	120
Compressive strength	38–50	590–780
Transverse (flexural) strength	15	230
Shearing strength	6–13	90–200
Elastic limit	1	15
Young's modulus		
Compression	5467–5879	84 500–90 500
Tension	4262–6067	66 000–93 500
Wrought iron		
Tensile strength	18–24	280–370
Compressive strength	16–20	245–310
Shearing strength	75% of tensile strength	
Elastic limit	10–13	155–200
Young's modulus	10 000–14 280	155 000–221 000
Mild steel		
Tensile strength	26–32	400–495
Shearing strength	70–75% of tensile strength	
Elastic limit	18–20	280–310
Young's modulus	13 000	201 000

Table 6.3 Safety factors, recommended by Twelvetrees
(after CIRIA, 1986)

	Dead load	Live load
Cast iron		
Beams	5–6	8–9
Columns	5–7	8–10
Wrought iron and steel		
Beams	3–4	5–6
Columns	4–5	6–7

recognized as covering the ignorance of the actual strength of the material, as well as the variation thereof, together with the uncertainties of the loadings. Whilst this may seem a tall order, they were nevertheless widely used at the time.

Simpler to use are the permissible stresses, quoted in the London Building Act 1909 and shown in Table 6.4. They have the additional advantage of being widely accepted by Building Control Authorities in Britain, even when used together with today's loading requirements. For steel, they were superseded by the permissible stresses in BS 449, 1937, given in Table 6.5. BS 449 has been revised several times for new construction, but the stresses quoted in Table 6.5 are a useful starting point for assessment of steel structures from that period.

Similar regulations or codes were in force in most industrialized countries, although the dates of their initial promulgation varied, as did the numerical values, reflecting the quality of the local materials and local design practice. It should be remembered that these permissible stresses (as opposed to ultimate strengths and yield points) incorporate both 'load factors' and 'material factors'

Table 6.4 Permissible working stresses quoted in the London Building Act 1909 *(after Blanchard et al., 1983)*

	Tons/in.^{2}	(N/mm^2)
Cast iron		
Tension	1.5	23
Compression	8.0	124
Shearing	1.5	23
Bearing	10.0	154
Wrought iron		
Tension	5.0	77
Compression	5.0	77
Shearing	4.0	62
Bearing	7.0	108
Mild steel		
Tension	7.5	116
Compression	7.5	116
Shearing	5.5	85
Bearing	11.0	170

Table 6.5 Permissible working stresses for mild steel, quoted in BS 449:1937 (*from Blanchard et al., 1983*)

	Tons/in.^{2}	(N/mm^2)
Tension	8.0	124
Compression	8.0	124
Shearing	5.0	77
Bearing	12.0	185

and should therefore be used in the calculations together with *unfactored* service loads (see Sec. 2.2.4). There is, however no *a priori* reason why structures should be assessed according to the codes of practice, in force at the time of their design, if later, valid, information is available.

In Britain, The Highways Authority has issued a Standard BD 21/01: *The Assessment of Highway Bridges and Structures*, which quotes somewhat higher values of strengths for wrought iron and steel to be used with appropriate partial safety factors. These values, which are given as a guide are:

	Yield stress	Material factor
Steel (pre-1955)	230 N/mm^2	1.05–1.30
Wrought iron	220 N/mm^2	1.20

where the material factor is to be used with customary partial factors for loading.

BD 21/01 also gives formulae and graphs, from which the following values for permissible stresses in cast iron have been derived:

In compression generally: 154 N/mm^2.

In tension the permissible stress depends on the amount of live load, relative to the permanent load (LL/DL is the ratio of live load to dead load, i.e. permanent load):

LL/DL	0.25	0.50	0.75	1.00	1.50	2.00
f_t (N/mm^2)	44.56	39.26	36.17	34.17	31.70	30.25

(Permissible shear stresses for building structures will, according to BD 21/01, be identical to the permissible tensile stresses.)

There appears to be no information explaining the higher stresses allowed by BD 21/01, compared with the London Building Act 1909. One reason may be that the Act overall was too conservative in its approach to iron and steel structures (it was after all the first official document in Britain to accept steel frames as self-supporting). Another may be that material for bridges was subject to stricter specifications and quality control than that likely to be found in buildings.

The ductility of wrought iron and the test data available could be a valid reason for treating it on the same basis as steel; BD 21/01, if anything, appears to trust its reliability more than that of early steel, against which it contains a warning.

Taking as a basis the currently stipulated loadings and the above-quoted permissible stresses, or the strengths with the appropriate partial safety factors, calculations of beams is fairly straightforward; apart from the stress computation one only has to check that the compression flanges of the beams are adequately restrained against lateral buckling by the floor structure, particularly in the case of timber floors. BS449 has reduction factors for slenderness; other similar design codes have similar provisions to deal with slenderness.

Filler-joists (see Sec. 6.2.3) may require special consideration: BS 449: Part 2: 1969 recognizes the contribution of the concrete encasement to the strength, allowing a working stress of $(165 + 0.6\,t)$ N/mm^2, where t is the thickness in mm of the concrete over the compression flange (as long as the joist is fully embedded and 25 mm $\leq$ t $\leq$ 75 mm). Alternatively, BS 449 allows such floors to be assessed as composite sections (although not all building control authorities accept this without some justification of the shear transfer at the iron/concrete interface).

The coke clinker aggregate, often used in the concrete in filler-joist floors, can however cause severe corrosion of the iron or steel, if it has been exposed to damp conditions for a prolonged period. (The reason for this is that such concrete is often very porous and the aggregate contains a high proportion of unburnt coal, and with it sulphur, which oxidizes to sulphur dioxide, which combines with moisture to form sulphurous acid, which in turn may oxidize further to sulphuric acid.) Clinker concrete filler joist floors should therefore always be checked for corrosion of the joists and beams by random opening up and, if necessary, the reduced effective section properties should be ascertained and used in the assessment calculations.

For columns, the London Building Act gave permissible stresses for varying slenderness ratios and different end-fixity conditions and there is published information, based on test results, on the ultimate compressive strength of cast iron columns as a function of their slenderness. Some of this information is shown in graphical form on Fig. 6.10.

Use of this information sometimes gives results that suggest that the columns are grossly overloaded by the dead load alone, or that the live load capacity is almost non-existent, whilst at the same time past performance of the structure shows that this is clearly not the case. In such a situation it can sometimes be advantageous to ascertain the strength of the material by testing, as described in Sec. 6.3.3 and to calculate the carrying capacity from first principles. (This approach can also be successful for cast iron beams.)

More detailed information on this and other aspects of calculation checks is given in the Steel Construction Institute's Report: *SCI—P—138: Appraisal of existing Iron and Steel Structures* (Bussell, 1997).

6.4.3 Calculations, based on Test Results

For wrought iron and steel the formulae in BS449 and similar codes can be used with confidence, as long as the end-fixities are properly assessed and the permissible working stresses (or , for partial factor codes, the design strengths) are

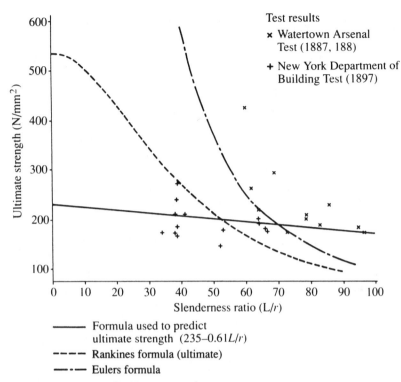

Figure 6.10 Ultimate carrying capacity of cast-iron columns as predicted by formulae and found by tests (*from Blanchard et al., 1983*).

adjusted by the ratio between the strength derived from the test results and the strength, on which the code formulae in question are based.

For cast iron it is not so simple: slender columns fail by buckling and for this failure mode the values for the modulus of elasticity and strength *in tension*, as well as in compression may be critical. This means not only that these properties have to be established by the tests (and for cast iron, such tests are much more difficult than for compression), but as the moduli are different for tension and compression, the calculations become very tricky indeed, the more so as they also have to take into account the initial eccentricity that may originate from inaccuracies of the castings.

In the case of solid cross-sections, this eccentricity, often manifesting itself as curvature, can and should be measured by conventional techniques; for hollow circular columns an extra complication arises from the possible lateral displacement of the internal, 'core' former (see Fig. 6.11). This has to be ascertained for a representative sample of all the columns by means of ultrasonic thickness measurements at 120° spacing around the circumference, at the top, at mid-height and at the base. The ultrasonics should be calibrated by measurements, using a hooked piece of wire, through holes drilled through the thickness of the metal.

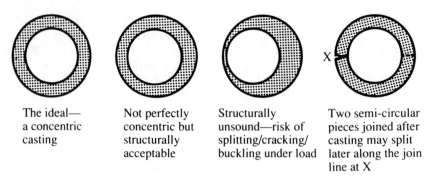

| The ideal— a concentric casting | Not perfectly concentric but structurally acceptable | Structurally unsound—risk of splitting/cracking/ buckling under load | Two semi-circular pieces joined after casting may split later along the join line at X |

Figure 6.11 Cross-sections of hollow cast iron columns, showing possible degrees of imperfection (*from Blanchard et al., 1982*).

Assessment of the carrying capacity of cast iron columns from first principles, as outlined above, not only involves tricky and laborious calculations, but also necessitates a very substantial (and hence costly) amount of testing, requiring a substantial number of samples, which it is often difficult to obtain. It can, however, be successful. The principles and problems of the calculations are described by Blanchard et al. (1983) and by Bussell in *SCI—P—138* (1997).

6.4.4 Load Tests as a Basis for Assessment

Load tests can sometimes be used to justify an existing structure, but should generally be a last resort. The reasons for this are as follows:

1. As mentioned in Sec. 3.5.3, a substantial number of bays would have to be tested, to ensure that the test results were representative of the entire floor area.
2. To establish a satisfactory margin of safety for the desired new use, one would normally apply a test load 25 per cent higher than the live load stipulated for such a use. If the structure were incapable of carrying that load, it would be damaged by, or might even collapse under the test load, whereas it might still have been safe and serviceable under a lesser load, corresponding to a different use.

It may, however, be possible to get a satisfactory answer from a single load test, if, by means of metallurgical analyses and in-situ non-destructive comparative strength tests, one can identify the floor area that contains the weakest structural members.

The above applies generally, but in particular to horizontal members. However, columns in multi-storey buildings commonly diminish in cross-section the higher up the building they are. It may therefore be necessary to test load columns at several levels. If, however, a column on one of the lower floors were to collapse under its test load, progressive collapse would result, unless heavy staging were provided in the test area throughout the total height of the building.

Even if such scaffolding were installed, it would have to leave a gap of some 50–75 mm at the level being tested, to allow for the normal deflection of the horizontal members under load. If a sudden failure of, say, a cast iron ground floor column were to occur under test loading, the sudden drop, across the gap, of the floor structure and the column above might precipitate severe damage through the height of the building. The Steel Construction Institute's guide *Appraisal of Existing Iron and Steel Structures* gives more detailed advice on load testing.

6.5 STRENGTHENING AND REPAIR TECHNIQUES

6.5.1 Preserving the Historical Evidence—Relieving Loads by Adding New Members

If, through the procedures described in Sec. 6.4, it has been established that the structure, in its existing condition, is not adequate for its intended use, the question arises, whether it can be made so by strengthening and/or repair. When deciding on the feasibility of repair and/or strengthening, and when choosing the method, any special conservation restraints, e.g. those imposed by listing, must be established. These must then be considered together with the technical aspects, discussed in the following. It should be remembered that whilst, engineering-wise, the same technique can be used both for the repair of a damaged member and for the strengthening of an undamaged, but structurally inadequate one, that may not be the case from the point of view of conservation. *An honest repair to a damaged member* may well be acceptable, even if it is non-reversible. However, the same technique applied *to strengthen an undamaged* member may not be accepted. In the meantime, propping and perhaps jacking may be necessary in order to reduce the stresses in the members, until the strengthening/repair becomes effective.

Some of the strengthening methods, outlined in the following paragraphs, will alter the appearance of the structure, or may add anachronistic elements (e.g. strengthening plates) to it. This is frowned upon by preservationists as 'falsifying the historical evidence', particularly when the strengthening is irreversible, as in the case of welded reinforcements.

If such a conflict arises, a compromise may have to be struck between either preserving the structure unaltered, and accepting a very restricted, or even impossible, further use of the building, or implementing a structural upgrading with some loss of authenticity. Such a compromise may entail leaving the original beams untouched and inserting new steel beams and/or columns between them to relieve their load. Where this approach is structurally feasible, and visually acceptable, it may be the simplest solution, as it avoids any interference with the original structural members.

6.5.2 Limitations imposed by the Properties of the Materials

Structures of mild steel, dating from the first half of the twentieth century, can usually be strengthened by welding on reinforcing plates, where necessary. Some early steels, particularly those made by the Bessemer or early Thomas process (see Sec. 6.1.4), may however not be suitable, due to their excessive content of impurities. Confirmation, that the metal is suitable for welding, must therefore be sought by a full chemical analysis (see Sec. 6.3.2).

Wrought iron does not, generally, lend itself to having anything welded on to it: the heat from the welding arc may cause the surface iron lamina to separate from the remainder, or it may melt the slag laminae, with their high silica content, which then contaminates the weld and in that way causes it to become too brittle.

Cast iron is also unsuitable for ordinary welded strengthening for the following reason: When steel is welded, the oxides that are formed due to the heat, float as a slag on top of the molten metal in the weld pool. Cast iron will however melt before the oxides and the weld pool will be contaminated with slag particles, resulting in a very poor weld strength, if any. Special procedures are available, by which cast iron can be welded, but the choice of metal for any attachment is very restricted, as is the position in which the welding has to be carried out; it will therefore only rarely be practicable for in-situ repair or strengthening. It is in any event a task for the specialist, *not* for the general steelwork erector.

6.5.3 Bolt-on Strengthening Devices

When it is structurally desirable to add strengthening plates to a section, but welding has been ruled out on any of the grounds mentioned above, bolting-on of such plates may be considered. It should, however, be borne in mind that to achieve the necessary shear transfer capacity, a large number of bolts may be required and hence a large number of holes may have to be drilled in the flange(s) of the original member. Site drilling of such holes can be a difficult and arduous task, particularly if it has to be done overhead into the bottom flange of a beam. For moderate numbers of floor beams of simple section, this method may however be a viable solution, particularly if existing composite action with a concrete slab can be utilized, so that strengthening is only required for the bottom flange.

High-strength friction-grip bolts are far more effective than ordinary 'black' bolts for connections in steelwork. In order to mobilize the friction, they do however require the contact surfaces of the parts, to be connected, to be plane, parallel and free from rust or paint. This may be difficult to achieve, when strengthening an old structure and unless one can be confident, that the correct surface condition can be ensured by grit-blasting, etc. and that the old surfaces are flat, it will be better to design the strengthening on the basis of black bolts.

Whether the plates are to be applied to a beam or a stanchion (column), they should be pre-drilled, offered up and used as templates for the holes in the original member. The bolts should then be inserted and tightened, following closely on the drilling, so as to minimize the portion of the member that has been weakened by the drilling and not yet strengthened by the bolting-on of the plate. Propping and careful jacking may, for the same reason, be necessary to reduce the stresses in the member in the temporary condition. This applies particularly, but not exclusively, to cast iron.

For sections built up from plates and angles, connected by riveting and for compound sections, consisting of I-sections with additional, riveted, flange plates, the heads of the original rivets prevent contact between the strengthening plate(s) and the original flange(s). Only rarely will it be practicable to fit packing plates (which would have to be suitably drilled or 'scalloped') through which the bolts would pass and thus provide the shear connection. The bolts would in any case be less efficient, due to the intermediate plate. The alternative of drilling out the rivets, in order to fit the strengthening plate in contact with the original flange, would leave the member seriously weakened until the new plate had been fully bolted on.

6.5.4 Welded Strengthening Devices

As mentioned in Sec. 6.5.1, mild steel members will in most cases lend themselves to having strengthening plates (or angles) welded on. This is a practical proposition even for compound and similar riveted members, as the necessary packing pieces can be adequately welded to the original flange, and to the plate, to effectuate the shear transfer. (This method has been used to strengthen some beams as part of a conversion of a building at the Trocadero in London.)

It may sometimes be sufficient to strengthen one flange and where there is a choice, it should be borne in mind that welding downhand (i.e. on top of the top flange) is significantly easier than any other position and will therefore be less costly and will more readily result in good weld quality. It may therefore sometimes be worthwhile to remove the floor screed in order to expose the top flange (see Fig. 6.12).

If the shear capacity of a riveted compound beam has to be improved, a possible solution may be to weld vertical plates, connecting the top and bottom edges of the flanges on both sides. This will overcome any inadequacy of the web-to-flange connection, as well as the weakness of the original web plate.

In the design of such welded strengthening, care should be taken to avoid great disparity between the thickness of the original flange and that of the reinforcing plate, as this can cause serious welding difficulties. The planning and programming of such work must take account of the incendiary potential of weld spatter and any combustible material must be kept well away from the welding site.

6.5.5 Adhesive-bonded Fibre Composites

The frequent difficulties of attaching strengthening plates by traditional means, i.e. bolting or welding, have prompted the adoption of adhesive-bonded plates of 'Fibre-Reinforced Plastics'. This development has proceeded in parallel with that for concrete structures, and common to both is the fact that the high cost of the materials is often compensated for by the high strength-to-weight ratio, which greatly facilitates the installation, often dispensing with the need for propping. The properties of the materials are briefly described in Sec. 7.4.4. It should be noted that, as the Young's modulus of the fibre-composites is generally less than

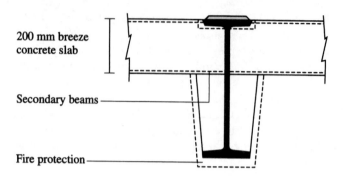

200 mm breeze
concrete slab

Secondary beams

Fire protection

New mild steel
strengthening plate

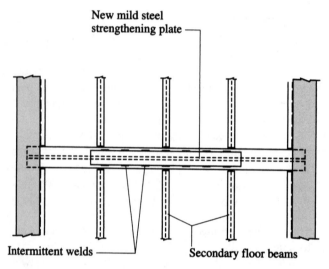

Intermittent welds

Secondary floor beams

Figure 6.12 Strengthening of steel beam by welded-on top plate.

that of steel, the strengthening strips, or 'plates', have to be significantly thicker
than those used for concrete. Aggressive environments and fire risk need special
consideration. The Institution of Civil Engineers' guide: *FRP Composites: Life
Extension and Strengthening of Metallic Structures* gives detailed information
and design guidance.

6.5.6 External Strengthening Devices

Where headroom requirements and/or architectural constraints allow it, it may be
easier and structurally more efficient to change the structural action of a member
than to strengthen it as such. For instance, trussing a beam by means of an under-
slung tie rod can increase its load-carrying capacity greatly, whilst requiring a
very small amount of additional material and only a modicum of work. As so
often, the success or otherwise of such a scheme will depend entirely on carefully
worked-out details, particularly the connections of the tie rod to the ends of the

beam. (An early railway bridge, built with trussed cast iron beams, had the tie rods connected in such a way that they *increased* the bending moments in the beams; it duly collapsed.)

If the existing open floor areas are to be subdivided for the new use, or if for other reasons the original clear spans are no longer required, it may be acceptable to insert intermediate vertical supports, so as to reduce the spans of the beams. This will require the building of foundations for the new columns and, in the case of multi-storey buildings, special detailing of the new columns to ensure safe transmission of the loads from the base of one column to the head of the one below (the beam webs will rarely be capable of transferring the column loads).

In the case of cast iron beams with unequal flanges, the fit of the new column under the beam should be adjusted, so as to ensure that the 'reverse' moment over the column is reduced and does not overstress the (smaller) top flange in tension.

6.5.7 Creating Composite Action

In jack-arch floors with 'loose' fill over the brick vaults, the beams may be strengthened by replacing the fill with concrete (using lightweight aggregate concrete will minimize or eliminate any increase of the dead load). The bond between the existing iron/steel and the new concrete must however be investigated. Rivet heads on the top flange *may* be adequate as shear connectors.

This suggests the possibility of strengthening single-section beams by providing shear connectors to the concrete slab. There are, however, great practical difficulties in doing this: Bolts, acting as shear connectors, would have to be a tight fit in the holes; this is difficult at the best of times and the difficulty would be compounded by the need to drill through two materials, that require different drill bits for successful penetration.

For beams of steel, a possible solution might be to core the holes through the concrete slab with a diameter sufficient to allow shear connector studs to be attached to the top flange of the steel beam with the standard welding gun and to subsequently fill the core holes with a non-shrinking grout. For wrought iron beams, welding would not be permissible, but the procedure could be adapted by drilling the beam flange through the core holes and insert tight-fitting bolts or, alternatively, 'pins' of tightly coiled spring steel, which are slightly tapering and can be driven into pre-drilled holes (trade name: DLR pins).

6.5.8 'Stitching' of Cracked Cast Iron

It sometimes happens that cast iron members crack due to accidental impact and/or release of cooling stresses along an imperfectly fused confluence of the molten iron in the mould. If the structural function of the member does not subject such a crack to direct, significant, tensile stresses, it may be possible to largely restore the integrity of the member by interlocking metallic 'stitching'. This entails drilling short lines of holes across the crack, with three or four holes either side; the holes are joined with a sawn slot and a 'key', usually of brass, which is a tight fit in the slot and holes, is driven in to lock the two sides of the crack together. In some cases, overlapping holes are used instead of holes and slots, but the principle is the same. (The trade names of two of these systems are 'Metalok' and 'Metal Stitching'.) The efficacy of the method can be judged from the fact that it has been used for reinstatement of engine cylinder blocks, cracked by frost.

6.5.9 Improving Stability and Robustness

Iron and steel roof structures are usually only required to carry light roof cladding, often incorporating large glazed areas. They will normally have wind-bracing incorporated in their design, and this will in most cases provide adequate stability. Ribbed domes are similarly inherently 'self-stable' (triangulated, so-called 'Schwedler' domes may however become locally unstable if some of the top 'rings' are dismantled as part of a refurbishment). Because the total load on these roofs is relatively light, a collapse of part of the structure is unlikely to seriously overload the floor structure below. Disproportionate collapse is therefore improbable, and robustness is unlikely to be a problem for such roofs.

Old mill buildings, with their elongated plan-form and their absence of cross-walls, rely for their stability on the horizontal diaphragm action of the floors. If the floors are of jack-arch construction or of concrete, this is usually not a significant problem.

For timber floors, the diaphragm action may however be suspect. If a proposed conversion, e.g. into flats, requires provision of reasonably substantial cross-walls through the height of the building, this could overcome the problem as long as these walls are properly founded. If large open floor areas have to be preserved, e.g. for office use, bolted-on horizontal diagonal bracing, combined with longitudinal ties just inside the façade walls, could be used, but this would complicate the overall fire-proofing. When timber floors have to be retained, significant diaphragm action can be provided by underlaying the floor boards with plywood sheets, securely screwed to floor joists and beams, so as to prevent lozenging of the originally rectangular beam and joist pattern.

For historical reasons, the guidance document to the British Building Regulations interprets robustness by limiting the floor area that may collapse upon removal of a structural member to 70 m^2 or 15 per cent of the total area of one storey, whichever is the less (this may be reasonable for a block of flats, but less so for an open-plan office block). This criterion will generally require all parts of the structure to be tied together. Whilst many early factory buildings have quite good tensile connections between beams and between beams and columns, the column-to-column connections do not offer much, if any, tensile resistance. Vertical tying of the columns is likely to be tricky, and it may therefore be preferable to investigate the alternative of 'bridging', i.e. the ability of the horizontal structure to 'survive' the removal of a column, by catenary spanning, albeit whilst suffering large deflections. The Institution of Civil Engineers' guide *Structural Appraisal of Iron-framed Textile Mills* gives detailed advice on these aspects.

Fully 'steel-framed' buildings from the early half of the twentieth century, such as those pre-dating the 1909 London Building Act, may rely for their stability on their masonry walls. This should be borne in mind, when a proposed adaptation entails removal of walls. If stairs can be enclosed by substantial walls and/or new lift shafts be made with concrete walls or with diagonally braced steel frames, the floors, which in such buildings are likely to be of concrete, will then give stability to the remaining walls. When assessing the robustness of structures with filler-joist floors, it should be remembered that the ends of the filler joists may only be resting on the top flange of the main beams, or held in position by a single small bolt.

6.6 CORROSION PROBLEMS AND REMEDIES

6.6.1 Corrosion Processes

Corrosion is an electrochemical process of a similar nature to that which goes on in a torch battery, but without the benefit of producing useful electrical energy. In the zinc-carbon cell of the battery the two substances, or electrodes as they are generally called, are connected by an electrolyte (the sticky stuff that oozes out, when the battery leaks) and when another connection is made by a wire, or similar, a current will flow and make the bulb light up. At the same time, some of the zinc dissolves in the electrolyte and the longer the current flows, the more zinc goes into solution. However, even before any current is made to flow, small amounts of zinc have already been dissolved locally in the electrolyte and the areas, where this happens may act as electrodes towards other areas of the zinc. This creates tiny local currents and causes slow local consumption of zinc, i.e. corrosion, eventually resulting in the familiar leaking of the battery.

In ordinary corrosion or rusting of iron and steel, small areas of the surface, in contact with water of different oxygen concentrations, may act as electrodes and small areas of impurities, such as slag, may do likewise. The result is that small areas of rust begin to form. Once this has begun to happen, the process will go on, as long as there is water present to hold iron and oxygen in solution.

The two conditions, which are necessary for rusting to take place, are that *oxygen*, i.e., air, *and water* can get to the surface of the metal. If the air is contaminated with aggressive substances, such as sulphates and/or chlorides, the rate of corrosion increases significantly.

There is another kind of accelerated corrosion which is caused by *bi-metallic action*: If two dissimilar metals are in contact in a damp environment, one of them will corrode at an accelerating rate, whilst the other will hardly be affected at all. The unaffected metal is sometimes referred to as being more 'noble' than the corroding one. Thus, if copper is allowed in contact with iron (or steel) or with zinc or aluminium, these will suffer bimetallic corrosion *if moisture is present*.

Similarly, zinc and aluminium will suffer bimetallic corrosion if in contact with steel (or iron), *if moisture is present* and ordinary steel (or iron) will corrode in contact with stainless steel, *if moisture is present*. It is for this reason, that stainless steel fixing bolts for cladding should be separated from structural members of carbon steel by insulating grommets.

A fact to be borne in mind, when confronted with apparently badly corroded steelwork, is that rust occupies between *six* and *ten* times the volume of the iron (or steel) from which it was formed.

For reasons connected with the composition of the 'casting skin', intact cast iron appears to be less susceptible to serious corrosion damage than steel (the generally thicker sections of cast iron members also suffer a proportionally smaller loss of strength for the same amount of rust). The outer surface of wrought iron, being very pure, is also corrosion resistant, but the oxide laminae make rolled sections of wrought iron liable to severe delamination in environments conducive to corrosion.

6.6.2 Common Problems of exposed Steelwork

Exposed steelwork was (and is) normally given a protective coating of paint. The paint film does however break down with time under the influence of water, air and ultraviolet light, and once even the smallest area of metal is bared, corrosion sets in. Because the rust (iron oxide) occupies a higher position in the galvanic series than iron, the corrosion can, and will, spread under the remaining intact paint film. Because the rust occupies a greater volume than the steel, the paint film then gets lifted from the metal.

On surfaces that get washed by rain and dried by wind, this will happen within a certain timespan, dictated by the resistance of the paint to general atmospheric weathering. In locations on the structure which are sheltered from the wind, and from which water cannot freely drain, moisture, laden with atmospheric pollutants such as sulphates and chlorides, will be trapped and remain for prolonged periods, during which corrosion of the areas that have lost their paint will continue, for much longer than on the free-draining, wind-dried surfaces. At any one time, corrosion damage will therefore be far more severe in such moisture traps, even if no great volumes of water are present; *moisture from condensation is enough to generate corrosion.*

The detail design of many steel structures, from the first half of the twentieth century or earlier, did unfortunately not take this into account; moreover, the details that allow moisture to lodge and remain, also make cleaning and re-painting exceedingly difficult and, sometimes, impossible. An example of such details is the common truss members of two angles riveted together, back to back, with a gap, dictated by the thickness of the gusset plates, of only 10 or 12 mm. In such structures corrosion may sometimes cause significant structural weakening.

Railway station train sheds are among structures, that were subjected to particularly severe conditions in the past, being exposed to exhaust steam, mixed with smoke containing sulphur dioxide. On the credit side, they were usually subject to a regular maintenance routine, during which re-painting would be carried out, as well as circumstances permitted, even if some of the gaps and crevices could not be properly treated.

6.6.3 Common Problems of Steelwork Encased in Masonry

In buildings with an internal iron or steel structure and external loadbearing walls, the beam ends would be supported on the walls, usually through a padstone, to a depth of between half and two-thirds of the wall thickness. They would thus have a fair thickness of masonry protecting them from the weather. With the introduction of the full steel frame, the external walls could be made thinner and often, in order to reduce the projection of the steel stanchions inside the building, only a minimum thickness of masonry was provided on the outside of the stanchions and tie beams. The protection of the metal, usually given only a token coat of paint, was then heavily dependent on the masonry. Natural stone and plain brick masonry kept the rainwater away from the steelwork by repelling some of it, but temporarily absorbing a substantial proportion of it, which it would subsequently release back into the air by evaporation. In this way it acted like a heavy overcoat, as opposed to the 'plastic mac'-action of modern curtain walling, which needs to be absolutely impermeable. Glazed brick and terracotta will, unless the pointing is perfect, suffer water penetration by capillary action through the joints, with the glazing impeding later drying out by evaporation.

Heavy corrosion may also have taken place, due to badly maintained gutters and downpipes. If downpipes are blocked, the gutters overflow and the water finds its way into the masonry. This also happens when lead-lined or asphalted gutters are not maintained. The water then finds its way to the (normally unpainted) metal and as the masonry is never impermeable enough to exclude air, corrosion will then progress, unseen, for some time.

The time when signs of trouble appear depends on how tightly the masonry has been fitted round the steelwork: As mentioned above, rust occupies a much greater volume than the parent metal. If all gaps between brick (or stone) and steel are fully filled with mortar, rusting may not start until the lime or cement has carbonated (see Sec. 7.3.1), but after that, the rust will soon fill the pores in the mortar and as more rust is formed, it will exert pressure on the masonry, which will then crack (Fig. 6.13*a*). If, however, gaps, say 10–20 mm wide, have been left between the flanges of the frame and the back of the masonry, the rust can have expanded into the gap for a long time (50 years or more) before the gap is filled, but once this has taken place, the rust will exert pressure on the masonry and crack it (Fig. 6.13*b*).

As on the same building there may be gaps in some locations and not in others and as the width of the gaps may vary between 20 mm and nothing, there is no way in which a visual inspection of the exterior can establish the existence or the extent of a corrosion problem on such a building. Another difficulty is that corrosion does not necessarily occur where the water gets in; it can run down a stanchion, causing little damage, until it finds a ledge on which it can collect, and this is where rusting occurs. Cracking may also be due to rusting of iron cramps forming part of the detailing of the masonry cladding, without necessarily indicating corrosion of the main steel structure.

If some cracking is evident, a limited opening up may indicate how far the rusting extends beyond the cracking *in that location*, but in that location only. A possible method of ascertaining an unseen problem would be to carefully drill a core hole

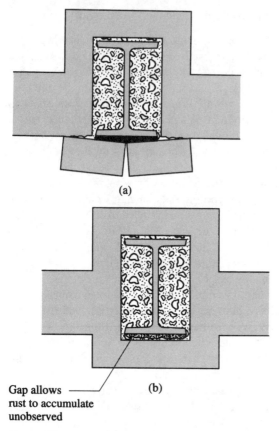

(a)

(b)

Gap allows
rust to accumulate
unobserved

Figure 6.13 Rusting steel member encased in masonry: (*a*) Gaps between steel and masonry fully mortared; (*b*) Gaps left empty between steel and masonry.

perpendicular to the face until the core barrel touches the face of the metal. By inspection of the end of the core *and* the bottom of the hole, it should be possible to ascertain the presence and thickness of any build-up of rust. When reinstating such an inspection hole, one must not create an entry for water where there was none before.

The appearance of the corrosion on built-in steelwork, exposed by opening-up, can be quite horrific. However, due to the volume expansion, mentioned in Sec. 6.6.1, a 10 mm thick layer of rust only indicates a loss of between 1.0 and 1.6 mm of steel. This is rarely structurally significant.

Absolute assurance of the absence of any problem does however, in principle, require complete removal of the masonry cladding with inevitable consequential damage and no guarantee that it will be put back as well as it was originally built (leaving aside the difficulty of procuring matching masonry material to replace that broken during the removal).

In London, after the 1940s, it became compulsory to encase steel frames in concrete and at the time of writing there have been no documented cases of corrosion damage to such encased frames. (The $1:2:4$ concrete, used for this encasement, would have had a comparatively high cement content and may therefore have had better durability and have offered better protection than might be expected from, say, 1970s concrete.) Unfortunately, this requirement was abolished when the London Building By-laws were superseded by the 1986 Building Regulations (which only dealt with health and safety and thermal insulation).

6.6.4 Common Problems of Clinker Concrete in Filler-joist Floors

As mentioned in Sec. 6.2.3, many late-Victorian and Edwardian buildings have floors of clinker-aggregate concrete with embedded iron or steel joists, spanning between main, usually riveted compound, beams, which have their upper part embedded in the concrete. The concrete produced with clinker aggregate would, as explained in Sec. 7.3.5, be porous and thus offer only poor corrosion protection to the metal, the more so as the cement content usually was low.

Furthermore, the clinker would usually contain particles of unburnt, or partially burnt, coal or coke, with substantial proportions of sulphur in them. As the concrete was porous, this sulphur would slowly oxidize to sulphur dioxide: SO_2. If moisture is present, the sulphur dioxide will form sulphurous acid: H_2SO_3, which in turn will oxidize further to sulphuric acid: H_2SO_4. This is, of course, a very strong acid and where such floors have been left open to the weather for any length of time, severe corrosion of the embedded iron- or steelwork may have resulted.

When buildings with this type of floors are to be converted, special care must therefore be taken to provide weather-screening of window openings and also a temporary roof, if removal of the original roofing is required. When such a conversion or refurbishment is being considered, an investigation should be carried out, at an early stage, to ascertain whether the floors have been subjected to damp conditions in the past and whether corrosion, severe enough to significantly affect the structural integrity and capacity, has taken place. When a conversion may involve drilling holes through, or demolishing filler-joist floors, a sample of the concrete should be analysed beforehand to ascertain if the concrete contains substances that may pose a special health risk—in one case, clinker concrete was found to contain asbestos!

6.6.5 Radical remedial Treatment of Corrosion

As mentioned in Sec. 6.6.1, continuing corrosion requires a supply of oxygen and moisture and, furthermore, the presence of rust promotes corrosion, whereas the presence of certain other metals, such as zinc may provide some protection. It follows that to prevent further corrosion one must, as far as possible, exclude air and water and, to allow for imperfections in the exclusion system, remove existing rust and, preferably, arrange for some zinc to be in contact with the iron or steel.

In principle this means the following treatment:

1. Identifying and eliminating, as far as possible, the source of the water that is causing the corrosion.
2. All surfaces must be cleaned down to bright metal.

3. A 'zinc-rich' priming paint must be applied *immediately*.

4. This priming coat must be protected with several coats of water- and air-excluding paint.

Cleaning is most effectively carried out by grit blasting, but even this cannot reach into the narrow gaps between 'back-to-back' angles and other crevices, resulting from unfortunate detailing, such as described in Sec. 6.6.2. These can only be cleaned by disassembling the whole structure. This would mean drilling out all rivets and, as riveting is no longer a 'live' trade, re-assembly would have to be done with some kind of bolt. This would destroy the authenticity and is therefore often not permissible on historical grounds. That apart, dismantling will almost certainly result in some damage, the repair of which will add to the already almost prohibitive cost. (An alternative may be to seal the gaps between the angles with a waterproof compound.)

Radical remedial treatment is therefore rarely practicable and generally it is not justified.

6.6.6 'Palliative' Treatment of Corrosion

If an exposed structure has survived 50 or a 100 years without suffering more than moderately severe corrosion and if it is not, in its future use, going to be subjected to a worse environment than in the past, cleaning with a de-scaling gun and vigorous wire-brushing will suffice as preparation for a paint system that will offer adequate protection for a good many years and which will be fairly easily maintainable (in contrast to the very special paints, referred to above).

For steelwork with 'heavy' masonry cladding, encasement in good quality concrete may be the answer. The masonry immediately in front of the steel would have to be removed and 'thinned down' to accommodate the concrete casing. The concrete would then be cast on to the cleaned steel and the masonry subsequently put back. This procedure may however entail some local damage to the masonry and consequential repair/replacement.

Galvanic protection is sometimes advocated. If properly adjusted and monitored, it can be effective, but it appears not to be something that can be installed and left unattended. Whilst it may therefore be appropriate on a public monument, it may not appeal to the average building owner.

6.7 FIRE PROBLEMS AND REMEDIES

6.7.1 Effects of Fire on Iron and Steel

Iron and steel are incombustible. This fact encouraged their use in 'fire-proof' construction in the nineteenth century. It does however not mean that buildings with iron or steel structures are inherently safe in a fire. The behaviour of the metals in a fire is governed, not by the temperature of the flames, but by the temperature attained by the metals themselves. This will depend on the length of time of exposure to the heat as well as the temperature of the fire.

When the temperature of a metal structural member is raised, it will expand. Within the normal range of ambient temperature this expansion rarely causes any trouble. In a fire, however, the temperature of the metal may be raised in the order of 500 °C and this will cause an expansion of 6 mm per metre length. If the member is not free to expand, one of two things will happen: either the member will push the surrounding, restraining, structure out of position, or if that structure is too strong to be displaced, the heated member will buckle. In both cases serious permanent structural damage is likely to result.

At the same time as the metal expands, its mechanical properties change. Up to about 200 °C the changes are not significant (for some mild steels the strength actually increases). Above this temperature both yield point and ultimate strength of steel and wrought iron start to decrease and at about 500 °C the yield point approaches the normal working stress for members subject to full design load. Cast iron retains its properties up to about 300 °C above which its strength rapidly declines, reaching about 50 per cent of its original value at about 600 °C. For all three metals the strength properties continue to decline with increasing temperature. Young's modulus also decreases with rising temperature.

There is thought to be a risk of cast iron columns shattering, if hit by the cold jet from a fireman's hose, after having been heated by in a fire. Instances of this actually happening are not well documented. There are, however, documented cases of serious cracking in cast iron beams and columns resulting from fires and firefighting. The cracks have usually started at flaws or points of previous mechanical damage.

6.7.2 Performance of Structures in Fires

As mentioned above, the strengths of the metals reduce significantly above 200–300 °C. The maximum temperature that the metal of a structural member attains in a fire depends on a number of factors:

1. The intensity of the fire: obviously, the hotter the air surrounding the member becomes, the higher the temperature of the metal in the member.
2. The duration of the fire: the longer the member is exposed to the heat, the hotter the metal becomes.
3. The mass of metal in the member: for a greater mass of metal per unit perimeter area, exposed to the fire, the temperature rise for a given heat input will be less (this is often referred to as the 'heat sink' effect). This means that members with solid cross-sections or with thick flanges will, for a given fire, take longer to reach the critical temperature at which their carrying capacity has been reduced to the point when they can no longer carry their loads.
4. The exposure to the heat of the fire: when structural members are encased in concrete or clad with insulating materials, it will take substantially longer for

the heat to penetrate to the metal, and hence for the metal to reach the critical temperature.

Of the factors above, 1 and 2 depend principally on what is known as the fire load, together with the geometry of the room and the ventilation. The fire load can be defined as the nature and amount of combustible material, stored in the building and/or forming part of the building finishes. This is independent of the metal structure and of any protection to it and is governed solely by the use and nature of the building. The periods of structural fire resistance, stipulated in the Building Regulations, broadly reflect these variables by defining building designations, compartment volumes, etc.

Factors 3 and 4 depend on the structure and on its protection, if any, and are not, in principle, influenced by the use of the building. However, users may have outfitting/decoration work carried out which may damage any applied fire protection. It is considered by some that a contributory factor to the collapse of the World Trade Center towers was that the, not very robust, fire protection to the steelwork had been damaged, prior to the attack, by such users' work.

Requirements for fire resistance are usually quoted as the number of hours for which the elements of the structure would resist failure under their design load in a British Standard fire exposure test. Such a test involves exposing the loaded element to a fire which produces a 'standard' rise of temperature with time. This does *not* reproduce conditions in a real fire, but it enables assessment of relative levels of fire resistance.

The period, required by the regulations, varies with the size of the building and with the category of its use (from a practical point of view, it is as often as not the safety of the firefighters that is the governing factor). An existing building with continuing and unchanged use will normally be accepted in its current state, subject to adequate means of escape. The requirements for fire resistance will however be imposed on any major refurbishment, particularly if it envisages a change of use.

As inferred under 1–4 above, the time that a structure will be deemed to adequately resist a fire depends not only on the structure, but also on the finishes and the use of the building. This means that a given structure may be sufficiently fire resistant for one type of new use, but not for another. It should also be remembered that the authorities may require the structure to be *incombustible* in areas such as escape exits, regardless of the structural fire rating.

With the exception of warehouses, it is rare for structures to be loaded to their design capacity when a fire reaches its maximum intensity. If the metal does not significantly exceed 500–600 °C during the fire, the structure is unlikely to collapse under its load. If expansion has not caused damage to other structural elements, such as loadbearing walls, and members have not buckled due to restraint of the expansion, the question that remains is whether the margin of safety has been permanently reduced.

6.7.3 Appraisal of a Structure Following a Fire

The carrying capacity of a structure, that has been exposed to fire, may be affected by:

- 'Macro'-physical damage, such as buckling or distortion of wrought iron or steel beams or cracking of cast iron columns.
- Metallurgical changes, which impair the strength properties of the metal. The heat may have 'annealed' the steel, removing the 'warm-working' effect of the rolling, and thus reduced the yield point.
- Damage to bolts or rivets, overloaded by forces created by restraint of expansion.
- Heat relaxation of connection devices such as rivets and, in newer structures, high-strength friction-grip bolts, the holding power of which depends on their clamping action.

If a beam is buckled in its entirety, it is usually unfit for its purpose and straightening will rarely be a practical proposition. If the compression flange is buckled, the carrying capacity of the beam, unless subsequently encased in concrete, will have been drastically reduced (tension flanges rarely buckle on their own). Columns of wrought iron or steel, which have buckled, will have lost most of their carrying capacity, apart from the fact that the buckling will have been accompanied by distortion of the supported beam structure.

Longitudinal cracks in a cast iron column have the effect of detaching a part of the cross-section from the remainder for the length of the crack. This reduces the stiffness of the column over that length and hence its load-bearing capacity. Even if this turns out to be insignificant, per se, such cracks are liable to propagate, if left to themselves, and should therefore be 'stitched', as described in Sec. 6.5.6. Whilst buckling is visible with the naked eye, cracks in cast iron should be searched for by tapping and/or by ultrasonic techniques.

Even if no 'macro defects' can be seen, the question of metallurgical changes remains. Tests, carried out by British Steel Corporation, show that as long as mild steel has not been heated to more than about 600 °C, its mechanical properties are practically unaffected. For cast iron the same applies and for wrought iron some tests even seem to indicate a slight enhancement of tensile strength after exposure to high temperatures (due to the great variability of wrought iron, this does *not* mean that higher loadings can be allowed on wrought iron beams after a fire!). Some high-tensile steels show a reduction of their mechanical properties after prolonged exposure to high temperatures.

Where the post-fire mechanical properties are in doubt, there are instruments available for non-destructive in-situ measurements of hardness. There is usually a good correlation between the hardness and the tensile strength and, for extra confidence, the hardness results can be calibrated against measurements on unaffected parts of the structure, or against laboratory tensile tests on samples taken from the affected parts.

In light structures such as roof trusses, the connections usually form better heat sinks than the members themselves. Bolts and rivets are therefore subjected to less severe temperatures than the metal in the members. Hence, if the members are left intact after a fire, the connections should not have been weakened. The exceptions to this are the main supporting connections to the stanchions, as these may have been overloaded and the bolts/rivets partially sheared by 'expansion forces'.

The forces, generated by restraint of thermal expansion, are generally more likely to damage connections between heavy beam and stanchion sections and, similarly the joints between portal frame rafters and columns. Rivets and ordinary older, so-called 'black', bolts were normally designed to carry their loads in shear, without any enhancement from their clamping action. Being made of material similar to the mild steel of the members, their mechanical properties after a fire would be affected to the same or to a lesser degree as those of the members. High-strength friction-grip bolts may however loose most of their clamping effect at only moderately high temperatures, e.g. 250–300 °C; they should therefore always be replaced after a fire if there is any evidence, e.g. more than minor blistering of paint, that the steelwork has been subjected to substantially raised temperatures.

The cost of replacing bolts will in any case be a very minor item of a fire reinstatement and should always be carried out if there is the slightest doubt about their residual strength, regardless of the type of bolt (which may in any case not be known). Replacement of rivets poses a problem for preservation purists, as riveting is no longer a practised trade and site riveting would contravene today's safety regulations. Special bolts with rivet-like heads, that can create visual similarity on one side, are, however, available (trade name: Huck-bolts).

6.7.4 Fire Protective Treatments

A proposed change of use of a building may require its structure to have a longer period of fire resistance than it may already have. There are only a few treatments, that will effectively improve the fire resistance of an iron or steel structure. The first action to be taken should therefore be to establish if, taking account of the fire load resulting from the new use and the nature of the structure, it may satisfy the requirements in its existing condition (see Sec. 6.7.2).

Concrete encasement or cladding with fireproof, insulating, boards is a very effective way of protecting a metal structure from the high temperatures generated by fires. Concrete encasement means bringing a wet trade into the building and there are difficulties in arranging formwork, that will allow pouring the concrete to a full storey height. Unless the concrete is intended to serve the additional purpose of increasing the carrying capacity of the stanchions, dry fire casing will therefore be the preferred choice for an existing building.

Fire casing will however change the appearance of the structure and where this is of any consequence, other solutions are sought. It is worth noting that board-type materials can now be cast into specific shapes; for example, half-round 'shells' can be made for protective cladding of round (and tapering) columns.

Certain paints and pastes are available, which, when heated by a fire, swell up to form a porous non-flammable, insulating, coating. These so-called 'intumescent' substances provide a measure of insulation to the structure against the heat from the fire, but have limitations: Those of paint consistency provide only a thin layer of insulation, which however does enhance the period of fire resistance of larger members quite effectively. Most cast iron columns can be treated in this way to achieve one hour's fire resistance. The pastes are capable of giving an hour's fire resistance to most structures, but, in order to do so, they have to be applied in thicknesses of some 5–7 mm. They are then vulnerable to mechanical damage (see comment on World Trade Center under Sec. 6.7.3). The thickness also means that fine detail, such as decorative features on cast iron columns, tend to be obliterated.

It has from time to time been suggested that the fire resistance of existing hollow columns could be increased by filling them with concrete or grout. There are unfortunately a number of problems combining to make this a less than ideal solution: Grout has, as opposed to water, a fairly low specific heat and hence does not form a very good heat sink; the best that could be achieved under ideal circumstances would be 3/4 hours enhancement of the fire-resistance period of the untreated column. There are also the practical difficulties in getting the filling into the column without making a great hole in it; grout injection from the bottom might provide the answer to that, although this would require a vent hole in the top to ensure complete filling.

Even if this were successfully achieved, the inside surface of the column is likely to be rusty and may, in the case of cast columns, be rough and have remnants of the core former adhering. This, combined with the gap, created by the shrinkage of the grout, will impair the heat transfer from the metal to the filling, thus rendering the heat sink effect of the filling ineffective. (More detailed information is given in the report by British Steel Corporation's Swinden Laboratories entitled *The Reinstatement of Fire Damaged Steel and Iron Structures*. Parts of this appear however to be biased towards demonstrating the claimed superiority of steel structures over concrete structures in terms of fire resistance.)

SEVEN

CONCRETE AND REINFORCED CONCRETE

"Highpoint", Highgate, London: This block of flats was constructed in the late 1930s, using slip-forming for the walls—a technique not used again for dwellings until 20 years later. The external walls have had to be repaired because of reinforcement corrosion, which was due to the thin cover, considered adequate at the time of construction (*Architect: Berthold Lubetkin, Engineer: Ove Arup*).

Concrete as a construction material did over the last decades of the twentieth century acquire a bad reputation in the popular mind. This was not entirely unfounded; the 1960s and 1970s saw a great amount of ill-considered and/or badly executed use of the material as structure and, particularly, as cladding to buildings. Various, initially promising, concrete-like materials were also used in construction before all their characteristics were properly understood. All of this resulted in premature needs of drastic repairs or, in some cases, demolition after a useful life of only 15–20 years. Most of the demolitions were prompted by consideration of factors, which were not inherently structural, and reinforced concrete, when properly executed, is an excellent form of construction, which has the added advantage that it can be moulded to almost any desired shape and size. This was a property that was (and is) much appreciated by architects and there are a great number of buildings with reinforced concrete structure, which have considerable architectural merit. Some of them date from the first decade of the twentieth century, but many of them of are much more recent. These are all worthy of conservation and in Britain a few of them have in fact been listed, although some very good early examples were allowed to be demolished for reasons of financial gain.

Successful conservation of such buildings requires assessment of the load-carrying capacity of the structure and of the significance of any apparent defects for their continued serviceability.

7.1 PROPERTIES AND BEHAVIOUR OF MATERIALS

7.1.1 Composition of Ordinary Concrete

Concrete is a man-made 'conglomerate', generally consisting of stones and sand with a binder. This can be of lime, sometimes with a pozzolan as in Roman concrete, or it can be of Portland cement as in modern concrete.

For the purposes of the following, 'ordinary concrete' is taken to mean the Portland cement-based concrete, as used nowadays for the majority of construction work.

The rubble-and-mortar filling, found in cores of mediaeval walls in northern Europe, could be described as a primitive concrete. The lime, used as binder, was usually only capable of hardening by exposure to the carbon dioxide of the atmosphere. In walls several metres thick, the air could not penetrate to the core if the mortar was dense, so the 'concrete' inside these walls is sometimes not hardened. In others, the poor quality is due to very sparing use of lime in the mortar.

The limes used in Roman concrete and in engineering structures up to the mid-1800s, are commonly described as 'hydraulic', due to their ability to harden under water. Their essential property is that, due to their content of substances other than calcium hydroxide, they harden in the absence of air, as opposed to the pure limes. These substances, which are sometimes natural impurities and sometimes deliberate additives such as crushed brick, are known collectively as pozzolans.

Portland cement was originally developed as a synthetic substitute for the 'natural' cement, obtained by the burning of Portland stone. It is nowadays manufactured from a slurry mixture of lime and clay, which is fired in a long cylindrical, inclined, rotary kiln, from which it emerges as spheroidal clinkers that are subsequently ground. The chemical composition of Portland cement is complicated; it can be varied to modify some of its properties such as the rate of strength gain and the resistance to certain chemicals such as sulphates.

The stones and sand that, together with the binder, make up the concrete, are known collectively as 'aggregates'. These are usually from natural sources, either sedimentary or marine deposits or obtained by the crushing of rock. Certain late nineteenth and early twentieth-century concrete floor structures were made with crushed brick or coke clinker in lieu of natural stones; in these latter, the unburnt residue of coal in the clinker, which contains sulphur, can cause corrosion of any embedded steel (see Sec. 6.6.4).

Since the 1950s inert lightweight aggregates, manufactured by bloating clay or sintering fuel ash in a kiln, are sometimes used in lieu of natural stones; this reduces the bulk density of the concrete, at the cost of a moderate reduction of strength and modulus of deformation. Similarly in some parts of the world, where stone is not available, crushed brick is still used as coarse aggregate. Aggregates should be free from any substances, which may cause deleterious chemical reactions with the cement in the hardened concrete or with any embedded metal.

The amount of water that is required to make the concrete workable is usually far more than what is necessary to complete the reaction with the cement, the so-called hydration, which transforms it from a powder to an adhesive, surrounding and holding together the individual aggregate grains. The excess water evaporates, leaving air-filled pores, which reduce the strength of the concrete and allow the ingress of air and moisture.

Older compaction methods, such as hand tamping, required more water than modern vibration compaction to ensure that forms were properly filled and reinforcement properly surrounded; older concretes are therefore sometimes more porous than more recent ones, although a greater cement content often compensated for the poorer compaction.

7.1.2 Mechanical Properties of Ordinary Concrete

Ordinary concrete is strong in compression. For concrete produced in the first half of this century, the crushing strength of a 150 mm cube would have been about 15–20 N/mm^2; from 1950 the strength of structural concrete in Britain was gradually increased and today it usually ranges from 35 to 70 N/mm^2. These high strengths are partly due to advances in cement technology and partly due to the availability of good natural aggregates. With modern man-made lightweight aggregates, the strength ranges from 25 to 40 N/mm^2. In regions where poor aggregates have to be used, strengths as low as 10–15 N/mm^2 are still common.

The tensile strength of concrete is low—roughly one-tenth of the compressive strength. Being a brittle material, the resistance against shear is governed by the tensile strength and by any compressive stresses acting on the section being considered.

7.1.3 The Role of the Reinforcement

The purpose of the reinforcement is primarily to resist tensile forces. Its second most important function is to limit cracking. It is also used to increase the load-carrying capacity of columns.

In order to utilize the tensile strength of the steel in bending members such as slabs and beams, the reinforcement *and its surrounding concrete* has to stretch beyond the tensile limit of the concrete and this will therefore crack. This is of no structural significance, as the tensile forces are resisted by the reinforcement.

Cracking is inherent and unavoidable in reinforced concrete structures; it does not necessarily indicate structural deficiency.

Cracks at right angles to the reinforcement, which are wider than about 1 mm, may however be a symptom of overstress in the reinforcement. Wide cracks, in a 'crazy-paving' pattern, may be a symptom of alkali–silica reaction (see Sec. 7.3.7).

7.1.4 Shrinkage, Creep and Moisture Movements

Cement, being a lime-based material, shrinks during the hardening process, and during any further drying-out, and the concrete shrinks with it. Shrinkage, *per se*, would not be harmful, but would merely lead to shortening of the structural members; these are however usually not free to shorten, but are restrained by being connected to other components of the building. This restraint of the shrinkage sets up tensile stresses in the concrete and these may cause cracking.

This is again not a cause for alarm; the cracking is the concrete's way of relieving itself of stress and if reinforcement is present and capable of resisting any tensile forces, the structure is perfectly safe. As often as not, once drying out is complete and the environment of the building has become stabilized, such cracks will not move any more. The same applies to the surface crazing that can occur if the surface of the setting concrete dries out too quickly, or cools down too quickly. (Shrinkage and other structurally harmless cracks are dealt with in detail in the Concrete Society Technical Report 22: '*Non-structural Cracks in Concrete*' and some are illustrated in Fig. 7.5.)

During the drying-out process another phenomenon takes place: where the concrete is subjected to compressive stresses, it shortens with time, even though the loads are not increased. This is known as 'creep' and could be described as stress-assisted shrinkage. In slabs and beams, shrinkage and creep will shorten the compression zone; the effect of this shortening strain is similar to that of an increased bending moment (see Sec. 1.2.3) and results in increased deflections. As the loads are not increased, this type of deflection is however *not* a sign of overstressing; it may however cause cracking of partitions, such as those of blockwork, which have not been designed to accommodate such movements.

Creep and shrinkage cause shortening of columns and load-bearing walls, and this can sometimes cause problems with masonry infill and cladding.

Like most building materials, concrete will, after its initial drying-out shrinkage, respond to variations of the relative humidity and temperature of its environment, increasing temperature and/or humidity causing expansion, and decreasing temperature and/or humidity resulting in contraction, or shrinking.

7.1.5 Special Cement-based Products

In Sec. 7.1.2, mention has been made of concrete made with light-weight aggregates. The properties of this are so close to those of concrete made with natural aggregates that it has here been included under 'ordinary concretes'.

There are however concrete-like materials, which have densities in the range of 400–1200 kg/m^3, as against the 2300–2500 kg/m^3 for natural aggregate concrete. These owe their development to the emergence of the cavity wall, which led to a demand for a cheap material, which needed only moderate strength, but preferably provided some thermal insulation, and which could be constructed like masonry to form the inner leaf. In Britain, the first of these materials was the *breeze block*, made of crushed clinker from coal-fired furnaces bound together with cement. In Scandinavia, clinker was not abundant and the need for thermal insulation was greater. This led to the development in the late 1930s–early 1940s of the *aerated concrete* block. Aerated concrete is made from a sand–cement 'slurry', to which aluminium powder is added. The alkali in the cement react with the aluminium to form hydrogen bubbles which give the mixture a spongy structure. Alternatively, soap powder is added and the mixture is 'whipped' to produce a similar result. The 'frothy' mix is usually factory cast in large blocks, which are then wire-cut to produce building blocks or slabs of the desired dimensions.

In ordinary concrete, the coarse aggregate, which accounts for 70–80 per cent of the total volume, resists shrinkage. Aerated concrete has no coarse aggregate and would therefore undergo far greater shrinkage than ordinary concrete. To counteract some of this, most of these products are therefore autoclaved, i.e. cured at high temperature and pressure.

Aerated concrete is, as mentioned, widely used for blockwork masonry, but during the 1960s–1970s it was also used (with varying success) in the form of lightly reinforced storey-height wall panels and even as reinforced floor planks (see Sec. 7.2.5).

Another material, popular in the 1960s was '*Wood-Wool*'. This consisted of wood slivers, about 50–100 mm long and 2–5 mm wide, bound together with a cement slurry and made into 'planks' about 0.6 m wide and in varying lengths. These had a very open texture. Some had the edges reinforced with pressed steel channels and were used to support felt roofing. Unreinforced, they were used as permanent formwork to concrete floor slabs over parking areas, etc. under buildings, in order to provide some thermal insulation to the storey above. The material was also made into blocks, to be used as void formers in lieu of hollow clay blocks (see Sec. 7.2.5).

Finally, whilst *Glass-Fibre Reinforced Cement*, or *GRC*, is not known to have been used for primary structures, the material enjoyed a considerable vogue as cladding during the 1970s. It was usually produced by spraying a mixture of a (usually white) cement–sand slurry and glass fibres into a mould. In the earlier applications the resulting panels were used as single skins, backed with traditional blockwork. Later, sandwich panels, incorporating various insulating materials between two skins of GRC, were developed. These suffered severe problems (see Sec. 7.3.10) and whilst, at the time of writing, some have been repaired, most of the GRC sandwich panels have had to be replaced with different cladding.

7.2 ASPECTS OF STRUCTURAL APPRAISAL

7.2.1 General Problems of Appraisal of Concrete Structures

Common to all structural appraisals is the fact that the properties of the materials are initially uncertain. What is special about concrete, and particularly reinforced concrete, is that the initial uncertainty is so much greater than for other materials.

It is usually possible to visually distinguish between oak and softwood and between cast-iron and steel, but all concrete is grey and shows imprints of the formwork on the surface, regardless of its strength. Furthermore, except in cases of severe spalling, caused by corrosion, it is impossible to assess what the reinforcement, if any, is.

Even when construction documents are available, one comes up against the fact of life that, what was built was not always what was shown on the drawings: concrete strength is not always as specified and reinforcement may not be in accordance with the drawings. The information that can be obtained from the original construction drawings and specification will however result in such a reduction of the investigation work required for the assessment, that no means should be spared in tracking them down.

7.2.2 Stability and Robustness

In buildings, constructed with load-bearing masonry walls and reinforced concrete floor structures, the stability depends to a large degree on the wall configuration. The concrete floors, however, act as horizontal diaphragms, transferring lateral loads, e.g. wind-loads on façades, to walls that can resist them, such as end gables. When, as part of a refurbishment, it is desired to create a new two- or three-storey space, this may limit the size of floor opening, that can be permitted.

Some buildings may have a reinforced concrete frame, which is designed to carry vertical loads only, but rely for their stability on the masonry (usually infill) walls to stair wells and lift shafts. If a refurbishment scheme envisages re-location of these vertical means of movement, their function as 'structural cores', or 'strong-points' must be considered, both from the point of view of their permanent structural replacement, but also in the temporary condition; they should not be removed, before their replacements have been constructed.

Many (usually tall) reinforced concrete building frames rely for their stability on concrete core walls, enclosing stairwells, lift-shafts and service ducts. For these, the same considerations apply as above. Occasionally, the core walls are made to work compositely together with façade columns, by means of connecting concrete walls, located on plant-room floors. Such walls cannot be removed without some structural replacement, no matter how inconvenient they are for the new plant.

Similarly, some tall blocks of flats have pairs of cross-walls, coupled by beams that span a central corridor. These coupling beams are essential for the resistance against wind loads.

A properly detailed in-situ reinforced concrete structure will be inherently robust, having enough reinforcement to 'tie it together'.

With any part of the structure being of pre-cast construction, this cannot be taken for granted and the provision and adequacy of transverse and longitudinal ties must be checked.

7.2.3 Procedure for Assessment of Load-carrying Capacity

1. If continued or new use does not impose greater loads than *are known* to have been carried in the past *without signs of overload*, the structure can be deemed to be adequate, subject to its condition being sound.
2. One should then check for corrosion damage and take steps to protect against future deterioration.
3. If past loading is *not* known, the reinforcement amounts and arrangements *must be* established in sufficient detail to allow rough calculations of load-carrying capacity to be carried out.
4. All possible places for records of design and construction that may give some information, should be searched. (If it is a nineteenth- or early twentieth-century construction, an indication of a possible patent system may be helpful.)
5. Reinforcement details, shown on drawings, must *always* be checked on site by local opening-up in selected, structurally significant, places.
6. If no drawings can be found, more extensive opening-up will be necessary. (Early design regulations may give an indication of the loads for which the structure was designed and the permissible stresses generally used at the time of construction, but total reliance should not be placed on such non-specific information.)
7. For initial, 'back-of-envelope', checks, permissible working stresses under *unfactored service loads* may be assumed to be:
   (a) 120 N/mm^2 for mild steel reinforcement (180 N/mm^2 for high–tensile reinforcement)
   (b) 3–6 N/mm^2 for concrete, depending on the age of the structure and the quality of workmanship.

8. Later, if accurate measurements of structural dimensions, reinforcement amounts and -positions, thicknesses of screeds, partition weights, etc. have been carried out and/or the strengths of the materials have been *reliably* ascertained by testing, final calculations may be made with higher permissible stresses, or reduced partial safety factors. (See Secs 2.3, 2.5 and the Institution of Structural Engineers' report: 'Appraisal of existing Structures'.)

7.2.4 Early Concretes and 'Patent' Systems

The properties of early cements, and hence early concretes, may sometimes differ from what is familiar nowadays. Early reinforced concrete structures were also sometimes designed and constructed with proprietary systems, which used reinforcement components and/or bars, which are no longer in production and which may appear highly confusing, when first uncovered.

If the year of building can be established, even if only approximately, it may give a clue to the concrete and the system, or systems, that may have been used in the construction and hence facilitate assessment of the load-carrying capacity of the structure. The following chronology may therefore be helpful:

1824: Joseph Aspdin's patent for Portland cement.
1845 (approx.): First reliable Portland cement.
1851: Marc Isambard Brunel's reinforced brickwork cantilever.
1854: William B Wilkinson uses flat iron reinforcement.
1855: Francois Coignet's first (British) patent system.
1861: Coignet publishes 'Betons agglomerés'.
1867: Joseph Monier's patent for tanks, pipes and box culverts.
1887: Wayss & Freytag (Köenen) publish 'Das System Monier': first design textbook.
1892: Francois Hennebique's patent.
1890–1910: Emperger, Bauschinger, Mörsch and Considère's research.
1901: CWS warehouse in Newcastle-on-Tyne.
1911: Royal Liver Building: Liverpool.
Mid-1930s: German planetarium domes and early shell roofs (Flügge et al.).
1931: Boots factory, Nottingham (Owen Williams).
1938: 'Isteg' twin-twisted high-tensile bars (Germany) welded reinforcement mesh, Hoyer 'system' of pretensioning. Highpoint in Highgate, London (Ove Arup).
(Isteg bars and welded mesh may have been used in Germany for some years prior to 1938.)
Late 1940s: High-tensile, deformed bars; square-twisted in the UK, hot-rolled in Scandinavia and Germany, high-tensile, smooth round bars in the UK, Shell roofs widely used.

Some of the special types of reinforcement, used in the early 'patent' systems, are illustrated in Fig. 7.1.

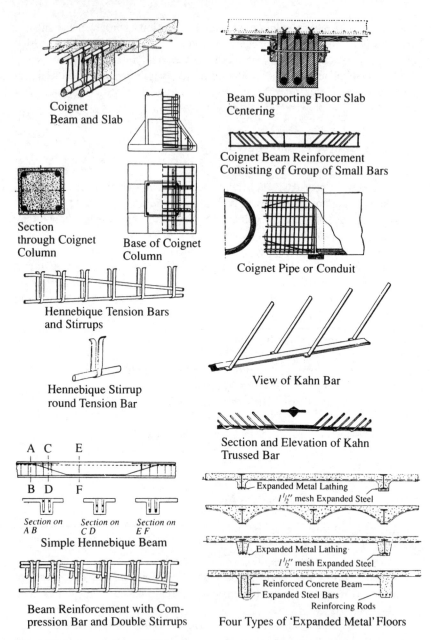

Coignet
Beam and Slab

Section
through Coignet
Column

Base of Coignet
Column

Hennebique Tension Bars
and Stirrups

Hennebique Stirrup
round Tension Bar

Simple Hennebique Beam

Beam Reinforcement with Compression Bar and Double Stirrups

Beam Supporting Floor Slab
Centering

Coignet Beam Reinforcement
Consisting of Group of Small Bars

Coignet Pipe or Conduit

View of Kahn Bar

Section and Elevation of Kahn
Trussed Bar

Four Types of 'Expanded Metal' Floors

Figure 7.1 Some early 'patent' reinforcement systems (*from Cassell: Reinforced Concrete, 1920*).

7.2.5 Special Forms of Construction

From about 1930–1970 some, seemingly 'orthodox', in-situ reinforced concrete structures incorporated floor slabs of what was commonly known as 'hollow pot' construction. Whilst having flat soffits, structurally they were in effect ribbed

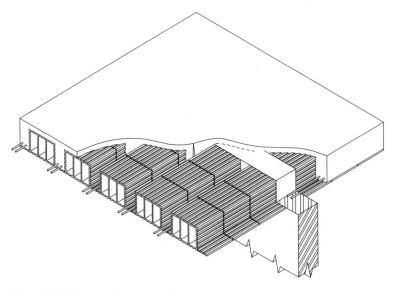

Figure 7.2 Hollow-pot floor construction.

slabs, the spaces between the ribs being filled with void-formers of hollow blocks of fired clay laid end-to-end (Fig. 7.2). Sometimes blocks of wood-wool were used in stead of hollow clay blocks (Fig. 7.3).

This form of construction resulted in a structural depth giving adequate bending strength and stiffness for fairly long spans with significantly reduced weight. As the design codes of the time did not generally require shear reinforcement

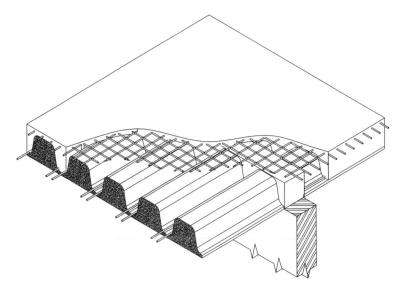

Figure 7.3 Wood-wool former floor construction.

in such ribs, they may however by today's standards be sensitive to concentrated loads.

The earlier examples of hollow pot floors were cast on traditional formwork, the width of the ribs being defined by clay slip tiles laid between the blocks. To economize on formwork, a variant was developed, in which the bottom part of the ribs were prefabricated in the full length, so as to only require few intermediate temporary supports within the span. They were sometimes of inverted Tee section, sometimes of 'X'-section, usually pre-tensioned, and the 'pots' were shaped to be self-spanning between the ribs (see Fig. 7.4). Most of the 'X'-sections were made of high-alumina cement concrete, in order to obtain what was at the time very high concrete strengths and to allow rapid turnover in the precasting works. The long-term strength loss of some of this concrete did in some cases cause problems (see Sec. 7.3.4).

The next logical step was to combine the rib and the void-former into hollow-core pre-tensioned 'planks', capable in the temporary condition of bridging the whole span on their own and only requiring the profiled narrow 'valleys' between them to be filled with a small aggregate concrete. These planks were proprietory products, marketed under trade names by the makers who provided design tables. Sometimes the design required a 'structural topping' of similar concrete to provide the necessary compression zone (see Fig. 7.5). Where no structural topping was required, contractors would sometimes fill the valleys at the same time as they carried out the finishing screed, using the same sand-cement mix for both.

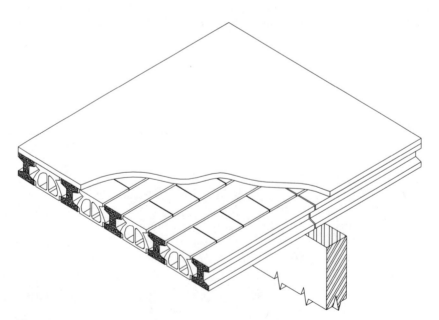

Figure 7.4 'X'- section pre-tensioned 'joist' and hollow block floor construction with in-situ 'topping'.

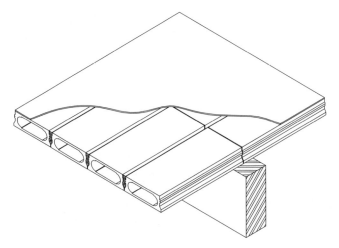

Figure 7.5 Floor construction of hollow-core pre-tensioned planks.

Sometimes, when a high standard of surface finish was desired or local conditions advocated a minimum of in-situ formwork, the beams and columns of the frame were pre-cast. They were then joined together on site, either with obvious 'mechanical' devices, resulting in structural 'hinges', or with hidden in-situ fillings of pockets, in which the reinforcement of the joined members interlocked, so as to provide continuity.

The demand for dwellings in the 1950s–60s, led to the adoption of various systems of 'large panel pre-cast construction'. Walls were factory made in full storey heights and often in full room-size lengths, and floors were made of wide, multiple hollow core, panels. They were then placed by tower cranes, with horizontal joints made of in-situ 'stitches' of fine aggregate concrete and with vertical open, 'self-draining' joints between facade panels.

7.2.6 Assessment of Concrete Strength

A full structural appraisal requires knowledge of the properties of the concrete, primarily the strength and the durability, and also of the nature, disposition and strength of the reinforcement.

A first rough indication of the concrete strength may be gained from specified mix proportions and test requirements. These may be found as notes on the drawings or in the contract documents (notes on drawings may be a more reliable guide, as these are more likely to have been seen by the site operatives). A more tenuous clue may, in the absence of specific project information, be found in building regulations and codes in force at the time of construction. These will often quote mix proportions to be used *and* the corresponding strengths that should be obtained from tests on samples. When using strength data, derived directly or indirectly from documents predating the 1960s, it is worth bearing in mind that the specified strengths normally referred to tests on samples of the concrete 28 days after casting and that the cements, then in use, would usually in the course of 1 or 2 years produce a strength gain in the structure of up to 25 per cent over the 28 day strength.

It is also worth remembering that the definitions of the specified strengths and the test samples vary: Until the 1960s British practice was to require a minimum strength of three 6 in. cubes from each sample, whereas later codes authorized a statistical manipulation of the cumulative test results. German practice was to use 20 cm cubes and in America 6 in. diameter cylinders, 12 in. long, were the norms. As the strength, obtained from a test, depends on the shape and size of the test specimen, as well as on the material being tested, quoted strengths may need multiplication with a conversion factor in order to be compatible with the basis of any assessment calculations.

For structural members, other than columns, concrete strength is often not critical and an initial assessment on the basis of a conservative value of strength, obtained from documents alone, may suffice if a crude in-situ 'test' with a hammer indicates that the concrete is quite hard. In most cases, however, concrete strength has to be confirmed by some form of physical testing.

The most direct measure of strength is obtained by taking samples of the concrete from the structure and testing them. The samples are almost invariably in the form of cylindrical cores, cut by means of a rotating steel barrel with a diamond-impregnated cutting edge, flushed and cooled by a small flow of water which can fairly easily be recycled and contained. (Air flush, using tungsten-impregnated cutting edges, may be possible where water has to be avoided at all cost, but it produces clouds of dust and is very expensive, due to the rapid wear of the cutting edge.)

The cores should ideally be 150 mm in diameter and about 160 mm long, when cut; this size will give the most consistent results. For most existing buildings, the weakening of the structural members that removal of such cores creates, is however unacceptable and a diameter of 100 mm with a length of about 110 mm tends to be the preferred size. Where even this is too great, 75 mm diameter, 85 mm long cores can be used, but the results of the tests will show greater scatter, the smaller the cores. (The ideal length-to-diameter ratio for testing is 1:1; the 10 mm excess of length, quoted above, is to allow for the ends of the core to be ground smooth and perpendicular to the axis; an alternative end preparation by capping is acceptable, if not quite as good, and does not require the extra length.)

Core test results may be used as a basis for calculation in two ways: (a) to arrive at the general strength level of the concrete used in the structure, or (b) to establish the strength of the concrete in the most critical structural member(s). Calculation (a) requires a large number of cores, taken in random locations, whereas for (b) one or two cores suffice, if they are taken where non-destructive testing has indicated that the strength is lowest.

BS 6089 and the Concrete Society's Technical Report No.11 give detailed guidance on sampling and testing procedures, as well as the use of the test results in assessment of load-carrying capacity, including the conversion factors necessary for translating 'raw' core strengths into 'design strengths' or permissible stresses.

Non-destructive testing on its own can only give comparative strength values, but if calibrated against tests on cores, taken from the structure being appraised, they can be used to arrive at the actual strengths of the structural members to be assessed. The non-destructive strength assessments are usually based on either (a) the ultrasonic pulse velocity or (b) the elastic surface rebound.

The ultrasonic pulse velocity is measured by placing a transducer, or probe, about the size of a stethoscope head, either side of the concrete member. The probes are connected to a 'black box', which via the one probe transmits an ultrasonic pulse to the concrete surface and then measures how many microseconds it takes for the pulse to reach the other surface, and hence the other probe (see Fig. 7.6). The thickness of the concrete, through which the pulse travels, is measured, and when divided by the time measured by the 'black box', gives the velocity of the pulse through the concrete; this is usually in the range of 3.5–4.5 km/sec. The pulse velocity can be shown to be proportional to the square root of Young's modulus for the concrete and this, in turn, has been found to be proportional to the square root of the compressive strength.

The pulse velocity is a very sensitive indicator of variations in strength (40 per cent variation of strength shows as 10 per cent variation of pulse velocity); the instrument ('the black box') is easily portable and if both sides of the concrete can be reached with the probes *and the wires connecting them with the 'black box'*, it is a very convenient method for estimating relative strengths and, if calibrated against a few cores, absolute strengths. The most popular instrument in the UK is known as the 'PUNDIT' (portable ultrasonic non-destructive integrity tester). The method is however *not* operator-proof and the presence of reinforcement, connecting the bars at the two faces of the concrete, can act as a short-cut path for the pulse and thus give false results. To give reliable results, it requires to be used by skilled and experienced operators.

The elastic surface rebound is measured by the 'Schmidt-Hammer', a torpedo-shaped instrument that has a plunger at one end (Fig. 7.7). An internal

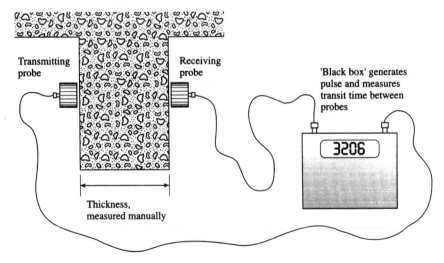

Figure 7.6 Principle of ultrasonic pulse velocity measurement.

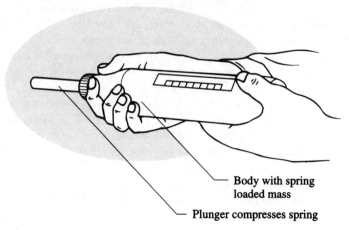

Figure 7.7 Elastic surface rebound hammer.

spring is compressed by pressing the plunger against the concrete surface; further pressure releases the spring, which then drives a mass against the plunger. The plunger makes a (very shallow) dent in the concrete; most of the denting is elastic and recovers, 'kicking back' at the plunger and the mass. The amount of 'kick-back', which is recorded on the instrument, is a measure of the surface 'hardness' or elasticity of the concrete and, in turn, of the strength. By its nature, this instru-ment measures the properties of the concrete close to the surface, where they may have been affected by the nature of the formwork, the curing treatment immedi-ately after removal of the forms and the subsequent environment; these factors will not have had an influence on the body of the concrete. For this reason the Schmidt-Hammer is frowned upon by many experts. It does, however have the virtue of being simple and robust and only requiring access to one face of the concrete and hence only one person to operate it, even in the cases of floor slabs and walls. As long as some simple precautions, described in the instructions, are followed, it can give useful readings of relative strengths. (The conversion factors on the instru-ment scale, giving absolute strengths, should normally be treated with caution.)

To give reliable absolute strengths, both the Schmidt-Hammer *and* the PUNDIT *have to be calibrated* by taking readings at points from which cores are subse-quently taken and tested.

The intrinsic form of hollow-pot and woodwool former slab construction makes it very difficult to test the concrete strength. If small areas of floor finishes can be removed, it may however be possible to locate the ribs by means of the Schmidt-hammer and take small diameter cores.

7.2.7 Assessment of Reinforcement Strength, etc.

A first, and often adequate, estimate of reinforcement strength can be obtained from the standard specifications and codes of practice, in force at the time of construction. For appraisal purposes, the strength properties from steel specifications are preferable to permissible stresses, as the margins of safety, implied in some early design codes may be unduly conservative by today's standards.

It must however be remembered that early steel specifications tended to only require a minimum ultimate tensile strength, whereas nowadays design stresses are based on the yield stress. Some knowledge of the ratio of yield stress to ultimate strength is therefore essential. For example, the British standard BS 785 (1938) specified the values given in Table 7.1 in tons per square inch (approximate metric equivalents in brackets).

The permissible working stresses (to be used with *unfactored loads*) for mild steel were about 18 000 lbs/in² (124 N/mm²). (Similarly, 1940s German mild steel reinforcement was specified to have a tensile strength of 3700 kgf/cm² (365 N/mm²), its yield point was about 240 N/mm² and the permissible stresses allowed in design codes ranged from 1200 to 1400 kgf/cm² (118–137 N/mm²), the higher values prompted by war time and post-war restrictions on supply.)

In the rare event that accurate values of reinforcement strength are required, samples will have to be removed for testing. Apart from the physical work involved in removing the concrete cover and cutting the bars, this is not a great problem in slabs, where removal of a few lengths of bar near the supports will not cause serious weakening. In beams, however, great care is necessary to avoid structural weakening and bar removal may be hampered by the links.

In the same way as non-destructive testing is used to locate the weakest areas of concrete, in-situ hardness testing, requiring removal of only small patches of cover, can be used to locate the weakest bars, thus reducing the number of bar samples to be removed (see Sec. 3.5.3). Determination of just the strength of the reinforcement thus requires very limited opening-up. In hollow pot and wood-wool former slabs this procedure will however be hampered by the slip-tiles/block lips. More extensive removal of cover is necessary to confirm the arrangement of bars, shown on the drawings, if these are available; but if no drawings can be found, finding the arrangement and sizes of reinforcement bars becomes a major task.

There are electromagnetic instruments, known as 'covermeters', which will indicate the presence and orientation of bars and their depth below the surface (up to about 75 mm), and one make at least, is claimed to indicate the diameter of the bar as well. Full use should obviously be made of such instruments, but a substantial

Table 7.1 Steel tensile strength in Tons/in² (N/mm²) (*From BS 785: 1938*)

	Mild steel	Medium tensile steel	High-tensile steel
Ultimate tensile strength	28–33 (433–510)	33–38 (510–587)	37–43 (572–664)
Yield point (minimum) For maximum diameter 1 in.	None specified	19.5 (301)	23.0 (355)

(For larger diameters the specified yield points were slightly less.)

amount of opening-up will still be necessary to confirm the disposition and size of just the essential 'structural' reinforcement.

Where serious corrosion of reinforcement has taken place, it is necessary to assess the remaining cross-sectional area of the bars. This is best done if sufficient concrete can be removed all around a length of bar to allow calliper measurements across several diameters, after removal of the rust. As 'wholesale' application of this procedure is usually impractical, the fact that rust takes up about eight times the volume of the steel, from which it has been formed, can be used to estimate the reduction of the effective bar diameter; for example, if 3 mm rust is seen to have formed on one surface of the bar, the radius of the bar will have been reduced by about $3/8 = 0.375$ mm and, by inference, the loss of diameter is 0.75 mm, which for a 16 mm bar is equivalent to a 9 per cent reduction of cross-sectional area.

7.2.8 Assessment of Durability

The durability of a reinforced concrete structure can be endangered by two different processes: (a) disintegration of the concrete material as such and (b) corrosion of reinforcement with consequent spalling and cracking of the concrete. The causes of both processes can be either attack by aggressive substances in the external environment, *or* chemical reaction of deleterious ingredients in the concrete. Both are dealt with in detail in Sec. 7.3.

A benefit of core sampling for strength testing is that the density of the concrete can be obtained by weighing the core(s) prior to strength testing; this will give an indication of the porosity, which has an influence on the durability. It is also possible, after strength testing, to carry out a chemical analysis on the crushed remains of the core to determine the (approximate) cement content and the presence of deleterious substances, e.g. sulphates and chlorides, etc., which can seriously affect the durability.

A quantitative assessment of the permeability of the concrete, which affects its resistance to external attack, can be obtained by the Initial Surface Absorbtion Test (ISAT), or better by the Figg Permeability Test; these are described in detail in the Construction Industry Research and Information Association (CIRIA) Technical Note 143 (1992).

The risk of spalling, due to reinforcement corrosion, depends on the depth to which the concrete cover to the reinforcement has carbonated (see Sec. 7.3.1). This can be detected by application of phenolphthalein, a chemical indicator, to a *freshly cut* section of a removed lump, or to drilling dust, *immediately after DRY drilling*. Uncarbonated concrete will then show up by a distinct purple-red colouring, whilst the carbonated concrete retains its grey colour. With care, the drilling method allows testing for each, say 5 mm, penetration of the drill, whereas the 'lump' will show the actual depth of carbonation.

Ongoing corrosion, that has not yet caused spalling, can be detected by electrochemical potential or resistivity testing. Potential measurements are carried out by measuring the (milli-) voltage, created between a reinforcement bar, exposed in one place, and a copper-coppersulphate electrode, placed in contact with the

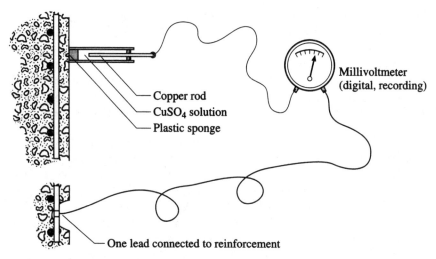

Figure 7.8 Principle of potential measurement.

concrete surface on a large number of points, located on a grid (see Fig. 7.8). Contours of the voltage are then drawn, and by opening up at the locations of high and low readings, some indication of the seriousness of the problem is obtained.

7.3 COMMON DEFECTS, THEIR DIAGNOSIS AND SIGNIFICANCE

7.3.1 Reinforcement Corrosion, due to Excessive Depth of Carbonation

When Portland cement has just hydrated, it contains among its many constituents calcium hydroxide, $Ca(OH)_2$. This is alkaline and where it surrounds reinforcement steel, it protects it against corrosion, that would otherwise result from contact with oxygen and water. However, when the concrete is exposed to air, the calcium hydroxide reacts with carbon dioxide in the atmosphere to form calcium carbonate:

$$Ca(OH)_2 + CO_2 \rightarrow CaCO_3 + H_2O$$

This process, known as carbonation, proceeds from the surface into the concrete at a rate that reduces with time and which results in a depth of carbonation of about 5–10 mm after 10 years. The carbonation neutralizes the alkalinity, so when carbonation reaches the interface between the concrete and the steel, reinforcement corrosion will take place *if* oxygen *and* moisture can penetrate to the surface of the steel. The rate of progress of the carbonation depends on the permeability of the concrete and this, in turn, depends on the cement content, the amount of water, relative to the cement, and the degree of compaction that the concrete has received.

In dry situations, the consequences of thin and/or permeable concrete cover to reinforcement are usually not too serious, provided no other substances that promote corrosion (such as chlorides) are present. For concrete exposed to the weather or to internal humid conditions (such as may be found in kitchens and bathrooms in temperate climates and in non-air-conditioned rooms in humid tropical conditions), the result is often that the rusting of the reinforcement with the accompanying expansion causes extensive cracking and spalling of the concrete cover.

Prior to the 1950s, reinforced concrete building structures were often clad with masonry and hence protected from the weather. The hand compaction, then in use, required a fairly wet mix in order to be effective. The strength gain of the Portland cement, available at the time, was relatively slow, so fairly high cement contents were used in order to achieve the specified 28 day strength. Where compaction was conscientiously carried out, this resulted in a moderately dense concrete which could give fairly good protection to the reinforcement.

In the late 1940s, compaction by mechanical vibration was introduced and the water content could therefore be reduced. At that time, strength was considered the main criterion for concrete quality and as the then specified, usually fairly low, strengths could be obtained with less cement when the concrete was vibrated, cement contents were often reduced, leading to more permeable concrete covers. The thicknesses of cover, stipulated by the then current design codes were moreover inadequate by today's standards and were often not properly maintained.

In the late 1950s, externally exposed concrete structures and cladding became fashionable and many structures with inherently inadequate protection of the reinforcement were built and required extensive remedial work after only 15–20 years. The cracking and spalling resulting from this 'atmospheric' corrosion of the reinforcement is unsightly and may in serious cases pose a risk of falling fragments of concrete. If caught in time, it is however rarely significant in terms of structural safety and it can usually be reasonably successfully repaired.

7.3.2 Reinforcement Corrosion, aggravated by Chloride Attack

Chlorides do not attack concrete as such; their presence can however destroy the protection against corrosion of reinforcement, that is normally provided by the alkalinity of the concrete. They act as catalysts for the corrosion and are thus *not* consumed in the corrosion reactions, which therefore continue unabated. Chlorides are hygroscopic, that is, they absorb water vapour from the air, even in environments usually considered dry, and in doing so, may create conditions that encourage corrosion. This is however considered rarely to be significant.

Chlorides may be introduced into the concrete from its environment, or they may have been part of its original ingredients.

Sea salt consists mainly of sodium chloride. In locations near beaches, salt spray from breaking waves may be carried by the wind (sometimes several kilometres inland) and deposited on the surface of the concrete, from where the chlorides find their way into the body of the concrete. In places with limited sources of freshwater, such as Gibraltar and Hong Kong, sea water is used for street cleaning which may cause salt spray and/or it is used for flushing lavatories and may then find its way to the concrete through overflowing cisterns or leaking pipes. The consequences of the latter can be particularly disruptive to the occupants of the building as remedial works have to be carried out inside (as opposed to spray damage which, being external, can be repaired from the outside).

Chlorides in the concrete mix may originate from two sources. In some countries beach sand has in the past occasionally been used as fine aggregate for concrete, without prior washing to remove the salt. The chloride content of such sand is particularly high if it has been dug up between the high water- and the low water-line on a beach in a hot sunny climate, where evaporation during each receding tide increases the concentration of salt. (Sea-dredged and beach aggregates *can* however safely be used in concrete, *provided* they are adequately washed before use.)

The other principal source of chlorides in the concrete mix has been the use of calcium chloride as an admixture to accelerate the setting and hardening of the concrete in temperatures at, or below, the freezing point. In the 1950s it was thought that a dosage of 2 per cent $CaCl_2$, by weight of cement, would not result in any harmful effects, but experience led to a ban on the use of calcium chloride in reinforced concrete being imposed in Britain in 1977.

Whilst the rust from 'ordinary' atmospheric corrosion tends to crack and spall off the cover at a fairly early stage, the substances produced by chloride-assisted corrosion are often able to 'infiltrate' and spread into the surrounding concrete for quite a time before cracking shows up the problem. It can then often be identified by a characteristic reddish-brown staining. By then, serious loss of reinforcement cross-section may however have occurred. (In one case, all that was left of 6 mm link bars to a column was the red stain.)

As the chlorides are not neutralized in the reaction, the consequences of the corrosion of the reinforcement are usually far more serious than those, where corrosion is due merely to carbonation of inadequate cover. This is because of the difficulties of achieving a lasting remedy. Whilst structures with surface attack from spray can usually be repaired to last another decade or two, corrosion due to chlorides in the mix can sometimes be practicably beyond redemption.

7.3.3 Acid and Sulphate Attack, etc.

Portland cement is susceptible to attack from acids that may occur in the environment or may be accidental spillage originating from the use of the building. Sulphates also react with the calcium compounds in Ordinary Portland cement. The reaction produces compounds, which occupy a greater volume than the original constituents and the resulting expansion can cause disintegration of the concrete.

Where the attack is due to acids from atmospheric pollution, the result is a gradual disintegration of the surface of the concrete, so that sand grains can be rubbed off with a bare finger; in south-east England it is sometimes accompanied by a yellow discolouration, due to the sand losing its 'skin' of cement. Apart from the gradual loss of surface and hence cover, the effects of this are rarely of great structural significance.

Sulphates are present in certain soils, notably London Clay, and in the groundwater in these soils. Where such groundwater impinges on foundations and basement walls, the concrete, if made with Ordinary Portland cement, will be attacked and may disintegrate to a significant depth below the surface. The sulphates are consumed in the reaction and the attack is therefore likely to be most severe where there is a groundwater flow, which replenishes the sulphates as they are consumed. Sulphate attack, per se, rarely causes structurally serious damage, although the 'crumbling' of the cover may affect the bond of the reinforcement bars and open the way for corrosion. However, some early basement walls, constructed of poor quality, porous, concrete did suffer severe damage, as the groundwater would seep through the concrete and evaporate from the interior faces of the walls. The evaporation would draw fresh groundwater to the outside of the walls, leaving an increasing concentration of sulphates and other salts on the inside. These salts crystallized and disrupted the concrete. Although the end result may appear the same, chemical analysis should be carried out to ascertain whether acid or sulphate is the main attacking agent, as the remedial action may have to be different.

Between the two world wars, a cement of different composition: high-alumina cement, was sometimes used to combat the problem. This cement does however produce other, sometimes more serious, problems (see Sec. 7.3.4). Sulphate-resisting Portland cements have however since been produced and have generally been used in underground works, where sulphates have been found to be present in the soil and/or groundwater.

'Ordinary' sulphate attack is characterized by the formation of the mineral 'Ettringite'. In the late 1990s a variant of sulphate attack was identified, which affected concretes made with sulphate-resisting cement. In this, a different substance, 'Thaumasite', is generated. Whilst Ettringite attack makes the concrete crumble, Thaumasite tends to turn it into 'mush'. In addition to the conditions favouring ordinary sulphate attack, the formation of Thaumasite requires the presence of carbonates (e.g. limestone aggregate or sea shells in the sand) and very wet and cold conditions (below 15 °C).

7.3.4 Strength Loss of High-Alumina Cement Concrete

High-alumina cement is a cement manufactured mainly from Bauxite, an aluminium ore. Its composition is quite different from that of any Portland cement. One of its main characteristics is that concrete made with it, achieves its full strength (which is very high) after 24 hours as opposed to 28 days, or more, for Portland cement. This high early strength is unfortunately accompanied by an unstable crystal structure of the hardened cement, and this subsequently transforms itself to a more compact form, with the result that the cement matrix of the concrete becomes porous and the strength of the concrete is reduced.

The speed at which this re-crystallization, commonly known as 'conversion', takes place and the degree of strength loss depend on many factors, of which the principal ones are: the original water/cement ratio of the concrete, the temperature rise in the concrete during hardening and the temperature and moisture to which the hardened concrete is subsequently exposed. A concrete, from a wet mix, which had subsequently been exposed externally to solar heating (in Britain), was found to have its strength reduced from 40 N/mm^2 at 24 hours, to an average of about 10 N/mm^2 after less than 10 years. Conversely, concrete from pre-stressed, precast beams with a low water/cement ratio and hence a 24 hour strength of 65 N/mm^2, *from the same building*, but situated *in a dry internal environment*, were found to have retained a strength of about 35 N/mm^2.

The strength loss of High-alumina cement concrete (HACC) is aggravated if subsequent to conversion it is exposed to alkaline water, such as may originate from leakage through Portland cement screeds. The porosity of the converted HACC makes it particularly susceptible to this 'alkaline hydrolysis', which can lead to almost complete disintegration of the concrete.

The HACC can often be recognized by its dark, brownish-grey colour, but confirmation of this diagnosis and the appraisal of its load-carrying capacity is a subject for the specialist.

7.3.5 Clinker Aggregate Concrete

In the latter part of the nineteenth century and the first quarter of the twentieth century, crushed clinker or cinders from coal- or coke-fired boilers and furnaces were often used as aggregates for concrete, because they were cheap and plentiful and because their porosity gave the concrete a lower density than if it were made with natural aggregates. The rough texture of this aggregate required proportionally more sand and water to be used to make the concrete workable and this resulted in fairly low strengths; this was however not significant for filler-joist floor slabs (see Sec. 6.6.4), which were the main field of use of this type of concrete.

The potential problem with clinker concrete arises from the origin of the aggregate: the boilers and furnaces of the time did not have very efficient combustion and as a result the clinker often contained substantial amounts of unburnt coal or coke. These residues could make the aggregate and hence the concrete volumetrically unstable and in a couple of cases a carbon content of some 30 per cent by weight of the concrete raised questions from the Building Control Officer about the incombustibility of floors constructed of such concrete.

Steelwork embedded in clinker concrete is generally not well protected against corrosion, due to the porosity of the aggregate and hence the concrete. In the case of filler joist floors, if the top flanges of joists and beams are flush with the concrete, they have, of course, no protection at all. Serious problems can therefore arise if the building has been left open to the weather, because of neglected roof damage, or during protracted conversion work.

The corrosion of steel, embedded in clinker concrete, may be aggravated by the sulphur in the clinker producing sulphuric acid in wet conditions (see Sec. 6.6.4). In one case, a mezzanine floor over part of an 1890s swimming pool had been constructed as a filler joist floor of clinker aggregate concrete; during an investigation, it was found that the webs of some of the filler joists had rusted through completely.

7.3.6 'Destructive' Aggregates

Certain aggregates become unstable when exposed to the atmosphere. Boiler clinker have been mentioned above. Slag from some metallurgical processes as well as certain shales are also volumetrically unstable and expand when exposed to moisture and air, causing disruption of the concrete in which they have been used as aggregate. Other aggregates may cause local 'pop-outs': small cones with their apex at an aggregate particle get pushed off the surface.

Certain natural rocks contain inclusions of iron sulphides, known as pyrites. These are, for example, found in the slatey waste from tin mining in Cornwall, where they are known as 'Mundic'. The waste rock was used as aggregate in concrete blocks, which in the 1920s and 1930s were extensively used for construction of buildings, mainly for dwellings; these were always rendered.

The 'Mundic' blocks apparently perform satisfactorily, as long as they do not get wet. Water may however get access to the blocks through defects in the rendering, or through unprotected window reveals during window replacement. If that happens, a complex chemical reaction takes place, causing the concrete to expand and eventually disintegrate. There appears to be no cure, and the main protective action is to prevent water ingress.

7.3.7 Alkali–Silica Reaction

Portland cement contains varying amounts of alkali metals, i.e. sodium and potassium. Aggregate particles of certain amorphous forms of silica can react with the sodium and potassium ions and in so doing, they expand and/or form a gel and, if near the surface, the expansion disrupts the concrete.

Expansive reaction requires three conditions to be satisfied: There has to be particles of reactive mineral of a certain size in the aggregate, there has to be a certain minimum concentration of alkali in the mix and there has to be moisture present. In concrete exposed to the weather in a temperate climate, there will practically always be enough moisture for the reaction to proceed if the other two factors are present. The alkali concentration depends on the composition of the cement *and* on the cement content of the concrete. The changes in the cement chemistry, referred to in Sec. 7.3.1, were also sometimes accompanied by an increased content of alkali. The reactive minerals are found in aggregates from certain sources.

On surfaces of unreinforced concrete, Alkali–Silica Reaction (ASR) initially produces star-shaped cracks. As these propagate along the weak planes in the concrete, they form 'three-legged' cracks, which subsequently merge into a 'crazy-paving' pattern (see Fig. 7.9c). Where there is reinforcement near the surface, the cracks tend to become parallel with the heavier bars (due to the

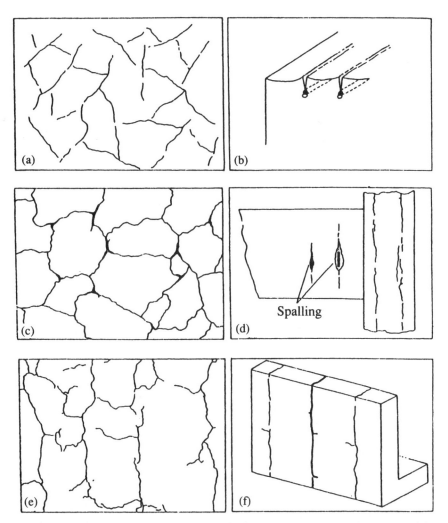

Figure 7.9 Visual defects on concrete surfaces and their probable causes: (*a*) Plastic shrinkage of the wet concrete; (*b*) Plastic settlement of the wet concrete; (*c*) Alkali–silica reaction; (*d*) Reinforcement corrosion; (*e*) Drying shrinkage, restrained by concrete behind; (*f*): Shrinkage, restrained by structure at the ends (*from Fookes et al., 1983*).

restraint against cracking that these provide). ASR cracking rarely causes any significant reduction of the overall safety of well detailed reinforced concrete structures.

The cracking is however often extremely unsightly and, by allowing easier access for air and water to the reinforcement, it can reduce the durability of structures exposed to the weather. In some cases there may even be a risk of fragments (up to the size of a hand) becoming detached and falling on passers-by.

7.3.8 'Hollow Pot' and/or Wood–Wool Former Construction

It can be difficult to adequately compact the concrete in the narrow ribs of these floors. Quite often, therefore, the workmen would use a very wet mix. The potentially lower strength, resulting from this, will however rarely be of consequence for the bending strength. The lack of shear links, mentioned in Sec. 7.2.5, could however make a floor slab sensitive to a heavy line-load, parallel to and near a support. The hollow clay blocks are usually capable of providing a certain amount of transverse load-shedding between ribs, so that moderate point loads can be accommodated.

If wood–wool formers have been used, another problem may have been caused: The wood–wool, being of a very coarse porous nature, may have absorbed the cement–sand fines of the concrete so quickly that only the coarse aggregate reaches the bottom of the rib to envelop the reinforcement bars. With little or no cement surrounding the bars, they are vulnerable to corrosion in damp conditions and the bond and/or anchorage of the reinforcement becomes very dubious.

A similar problem may be found when the steel fixers have omitted to put spacers under the reinforcement bars, so that they rest directly on the slip tiles or the 'lips' of the blocks. This, again leads to suspect bond and potential risk of corrosion.

Investigations of these problems are difficult because the slab soffits are made up of undersides of clay blocks and slip tiles or by wood–wool formers. A covermeter may indicate reduced cover, but assessment of the integrity of the rib concrete by non-destructive means may not be reliable. Examination of the ribs therefore requires the removal of the slip tiles or the protruding lips of the blocks or wood–wool formers.

7.3.9 Reinforced Aerated Concrete Wall Panels and Floor Planks

There have been cases of units, such as wall panels, displaying a 'crazy-paving' pattern of fine cracks after some years. This has been thought to be the result of incomplete autoclaving. The consequences usually affected appearance rather than structural safety. When, however, this effect was combined with optimistic design assumptions regarding load sharing between floor planks, the results were alarming: Whilst the dovetailed joints between the planks were filled with a sand–cement mortar, that may have been strong enough for the purpose of shear transfer between planks, the lips and/or corrugations on the plank edges were not. As a result, individual planks were overloaded and the incipient crazing opened up to gaping cracks.

7.3.10 Glass-Fibre Reinforced Cement (GRC) Panels

Despite being specially formulated, the glass-fibres in GRC are attacked by the alkali in the cement. This increases the bond between the fibres and the matrix, but the net effect is to significantly weaken the material and make it brittle. Whilst this is not significant in terms of wind and weather resistance of the panels, it should be borne in mind, where impact resistance of existing panels is required for a new use.

The cement–sand slurry in GRC is cement rich and the material, as sprayed, has a fairly high water content. In consequence, the inherent shrinkage, as well as the subsequent moisture movements, of the finished material are relatively high. In single-skin applications, of simple shapes and with suitable support conditions, this appears however not to have caused serious problems. Rigid bolted connections to the structural frame, which restrained shrinkage and moisture movements have however caused problems in the past.

Sandwich panels were made by first spraying the 'visible' skin against the mould, applying the insulating material to the inside of this and then spraying the backing skin, all these operations being carried out 'wet on wet'. This meant that gauging the thickness of the backing skin was difficult, and sometimes it has been found to be too thin for its purpose. Furthermore, the most popular form of sandwich panels incorporated an insulating 'filling' of a light-weight concrete, made with beads of expanded polystyrene as 'coarse aggregate', which had very high moisture movement. The GRC is not entirely waterproof and it is not vapour-proof and in some cases there was water ingress or condensation of water vapour, migrating from inside the building. This caused the filling to expand and crack the GRC.

7.3.11 Pre-cast Slab/Beam/Column Construction

In addition to the design and construction defects, occasionally found in in-situ reinforced concrete construction, there are several problems that can occur at the connections between pre-cast members. Whilst some of these are caused by plain bad workmanship, others have been caused by the designer ignoring the effects of temperature and moisture movements, which usually do not cause problems in run-of-the-mill in-situ construction, but which may impose significant forces between the pre-cast components. Neglect of the practicability of construction has sometimes also contributed to the trouble. The main 'trouble spots' are:

1. Corbels, Nibs and 'half-joints'.
2. Bearings where 'secondary' beams are supported on ledges of 'primary' beams, spanning at right angles.

The most common defect in the case of (1) is the lack of effective reinforcement to the corbels and/or half-jointed beam ends/slab edges. This is largely a problem of size; the vertical reinforcement 'loops', that make these connections viable for larger members, cannot be scaled down beyond a certain limit. Faced with this difficulty, some designers adopted reinforcement details that were not effective, and this did lead to some failures, particularly where pre-cast stair flights were supported on in-situ nibs on landing slabs.

Where corbels or nibs project from the bottom of the sides of beams, they have to be 'hung up' by sufficient reinforcement to transmit the reaction to the centre of the beam—another point sometimes forgotten by designers.

In the case of (2), designers were sometimes tempted to limit the horizontal depth of beam bearings to that necessary to keep within permissible stresses. This did in some cases result in the theoretical depth of the bearing being as little as 50 mm. With an accepted tolerance on column verticality of ±15 mm, this could result in a bearing of only 20 mm, if the tolerance was taken up at one end only. As a result, the beam end would only be supported on the unreinforced cover of the supporting ledge. This is then likely to shear off at the slightest provocation. Even where a full 50 mm bearing has been achieved, creep shortening of pre-tensioned beams, or rotation of beam ends, consequential upon flexure, due to varying temperature gradients, are known to have caused the bearing ledges to be pulled off their supporting columns or edge beams.

7.3.12 Pre-cast Large Panel Construction

The most serious problems are found in the horizontal joint between the storey-height load-bearing wall panels. These joints were intended to transmit the vertical loads through a filling of 'dry-packed', damp-earth-consistency, sand–cement mortar. This mortar filling has sometimes been found to be imperfect, or even missing, leaving the upper wall panel (and the panels above that) supported on just the two 20 mm diameter levelling bolts, used to align the panel during erection. Related to this, there was often no horizontal tying between wall panels and floor planks, resulting in serious lack of robustness.

External wall panels were usually of sandwich construction. They had an inner concrete load-bearing leaf, a layer of cork- or foamed plastic insulation and an outer weather-resisting leaf. The two concrete leaves were connected by inclined non-ferrous tie bars, which penetrated the insulation layer. Phosphor bronze was usually specified for these ties, but occasionally 'Manganese Bronze' was used. This is in reality a high-tensile brass (and therefore cheaper) which can be susceptible to stress-corrosion. However, the more important problem with these inter-leaf ties has been that not enough were put in.

7.3.13 Pre-stressed Construction

Post-tensioned members may have had their tendons imperfectly grouted and consequently corroded. The most common troublespots are the high parts of the ducts at the vent pipes and just behind the anchorages, if during grouting, sufficient time for 'bleeding' was not allowed. The presence, and seriousness, of the grout voids is very difficult to diagnose. Where tendons are all in one vertical plane, radiography may be of help, but where two or more tendons run parallel, shadow-effects make this method almost impractical (the exclusion zones, attendant on radiography, also makes it very cumbersome and expensive). Pulsed radar can be effective, but not if the tendons were placed in metal ducts. Even when gamma-rays or radar can be used, ascertaining the degree of corrosion, if any, needs very careful opening up of the duct(s), or drilling a small hole to insert a 'borescope'. Corrosion may also occur if anchorages on the outside of a building have not been given adequate weather protection.

Shortening of the concrete, due to creep (and/or shrinkage, if the post-tensioning was carried out too soon), may cause relaxation, i.e. loss of pre-stress. This will normally not affect the ultimate strength, but may lead to cracking and/or excessive deflection. Where the vertical structure is excessively rigid, the force of the post-tensioning may have been partly 'absorbed' in bending the vertical structure, leaving the beam with reduced pre-stress; subsequent shrinkage may then cause cracking.

Most pre-tensioned members are cut to length with a diamond saw. If it is possible to inspect the end face, any subsequent 'pull-in' of the tendon ends can be observed. Any pull-in over 3 mm may indicate loss of bond.

7.4 REPAIR AND STRENGTHENING TECHNIQUES

7.4.1 Repair of Damage caused by Reinforcement Corrosion

Unless the corrosion has already caused structurally significant loss of cross-sectional area of reinforcement, the main aim of repairs is to prevent further deterioration. In some cases there is also a strong desire to restore the original appearance.

If significant reinforcement loss has taken place, strengthening may become necessary (see Sec. 7.4.3), but in the case of chloride attack this may not be a practical, nor an economic proposition.

To prevent further deterioration, the reinforcement has to be given adequate corrosion protection. This is usually provided by a combination of a protective coating on the bars and a reinstated, more effective, cementitious cover to the reinforcement. Sometimes an impermeable coating is applied to the face of the reinstated cover, in order to provide additional protection.

When the sole cause of corrosion has been inadequate cover *and no chloride attack has been involved*, the following procedure will normally suffice to give a life of the repair of 20 years or more:

(a) All loose concrete is removed by manual hacking, pneumatic hammering and/or grit blasting. (Where isolated bars would otherwise end up with inadequate thickness of cover, they can usually be pushed in after removal of some of the concrete behind them.)

(b) All rust is cleaned off the reinforcement, exposed by (a), down to bare metal, preferably by grit blasting.

(c) Within about an hour of (b), the reinforcement is given a rust-protective coating. (This may be followed by a second coating, dusted with sand whilst wet, to improve adhesion of the new cover.)

(d) A bonding agent is brushed on to the concrete, followed by the application of a Portland cement mortar, usually modified by an admixture of a waterproofing latex emulsion. This new cover may be applied by trowelling or as sprayed mortar ('gunite' or 'shotcrete'), finished off by trowelling. It is often given greater thickness than the original cover.

(e) The repair mortar is carefully cured, i.e. kept moist by various means, for several days.

Blast cleaning creates large amounts of dust, although this can be minimized by wet blasting and/or the use of a vacuum attachment surrounding the blasting nozzle. In the gunite (or shotcrete) process only 30–50 per cent of the material emerging from the nozzle stays on the treated surface (where it adheres very well); the rest rebounds and ends up on the floor or working platform.

The use of waterproofing agents in the repair mortar means that, even if the sand and the cement in the repair come from the same source as those in the original structure (and this is often impossible to ascertain and achieve), it is extremely difficult to achieve a colour match.

As the weathering characteristics of modified repair mortars are usually different from those of the parent concrete, one may also have to chose between colour matching at the completion of the repair, or matching after, say, 5 years of weathering.

Board-marked, exposed concrete poses special problems of matching the appearance of repairs with that of the original surface.

In the case of the rear wall of the 'listed' Beethovenhalle in Stuttgart, where atmospheric corrosion had spalled the cover on isolated areas of the vertically board-marked concrete, the cover was removed in rectangular panels, corresponding to the width of one shutterboard or a whole number of boards and with carefully cut vertical and horizontal edges. Where the reinforcement had been too near the surface, the concrete was cut back and the bars were stapled back or, where they were not structurally essential, they were cut out. The sand for the repair mortar was taken from the source of the original sand, identified from the contract documents, and the repair cement was supplied from a single day's production to ensure uniformity of colour. Bonding agents and polymer admixtures were deliberately omitted, in order to give the repair patches, as far as possible, the same vapour transmission from inside the building, as well as the same surface absorbtion from the outside, as the original concrete. The repair mortar was applied by trowel and finished by pressing sawn shutterboards against the surface.

The repairs, described from (a) to (e) above, do not stop any corrosion that may be going on in adjacent areas and which has not yet caused spalling of cover. Where access is not too difficult, the most economical course of action may be to wait until sufficient further spalling has appeared to justify another repair campaign.

As mentioned in Sec. 7.2.8, there are techniques for detecting corrosion that has not yet caused spalling and for determining the depth to which carbonation has progressed. These methods are not exact, but they can assist in determining the extent of the potential problems and help to indicate when repairs *will have* to be undertaken.

In the case of chloride-aggravated corrosion, *all* chloride-contaminated concrete in contact with the reinforcement has to be removed and a barrier against chloride migration be provided. This ideally means cutting away not only the cover but also some of the concrete *behind the bars*, before treating them as in (b) and (c) above. The amount to be removed should be sufficient to allow enough new concrete to be placed to resist the migration of chlorides back on to the steel. For beams and columns this may result in significant structural weakening during the repair works, necessitating temporary supports. For slabs with closely spaced bars in both directions, the operation becomes almost impossible to carry out in a satisfactory manner.

As it is rarely possible to guarantee that there will be no pinholes or thin spots in the coating applied to the bars (see c above), and as chlorides do tend to migrate, repairs to structures, where the chlorides were incorporated in the original concrete, often do not last more than 5 years. Where the chlorides originate from salt spray, the prospects can be better.

7.4.2 Alleviation of the Effects of Acid/Sulphate Attack and Alkali–Silica Reaction

Short of cutting out and replacing the affected concrete, there is not anything that can be done to repair the acid or sulphate damage. Preventing continued attack can, in principle, be done by applying an impermeable coating or membrane to the concrete.

Where the concrete is exposed to acids from polluted air, it is also exposed to view (usually intentionally, as part of the architectural design) and the application of a coating will significantly change the appearance of the surface, as no transparent, non-shiny, effective coatings have so far been developed. This change may however be a price worth paying for a longer useful life.

Where concrete in a basement wall is being attacked by sulphates in the ground, it would in theory be possible to apply a coating or membrane to the face of the wall. The necessary excavation would however be a vast and costly operation and there is no way that the treatment could be applied to the underside of the basement floor. Treatment of footings and stub columns is equally impractical.

A practical way of slowing down such an attack may be to divert the groundwater (which usually is responsible for most of the sulphate attack) away from the basement by means of deep drains. Care must however be taken not to dry out any clay under the floor or under any adjacent foundations, as this could cause settlement (see Sec. 8.2.4).

In one case of a basement in silty ground with very acidic, but stagnant, groundwater, the remedy proposed was to re-excavate the soil immediately adjacent to the basement walls and then backfill with substantial amounts of slaked lime to neutralize the acid.

In the case of ASR-cracking, the only way that the reaction can be stopped is to dry out the concrete and prevent subsequent moisture ingress by means of an impermeable coating. This may be impractical for external structures, such as bridges where not only direct rain, but also seepage from the roadway (through the often inadequate waterproofing) has to be excluded.

In one case of a building with an exposed concrete structure, suffering from ASR, a preliminary proposal was made to provide a ventilated overcladding to the external faces of the frame members, so as to exclude driving rain and reduce the relative humidity of the air next to the concrete face. This was, in the event, not carried out and there have been no other documented cases of this treatment being applied.

7.4.3 Traditional Remedies for Reinforcement Deficiency

Where, due to the original design, or due to severe corrosion, the reinforcement is found to be inadequate for the loading that the structure is desired to carry, there are several ways of enhancing the carrying capacity of the structure. These do however either entail drastic surgery or they leave external strengthening devices that have to be protected against the elements and, in some cases have to be disguised.

Every effort should therefore be made to ascertain, by the principles described in Secs 2.2, 2.3 and 2.4, whether the structure, as existing, can be shown to be adequate for the proposed use, *before* any decision is made to carry out strengthening works.

The available strengthening methods are in principle:

1. adding new reinforcement;
2. applying post-tensioning tendons;
3. attaching strengthening plates or sections to the concrete face.

To add new reinforcement, grooves are cut in the concrete, new bars are inserted and then grouted in. This is generally not too difficult in the tops of slabs, but in beams the, often close-spaced, existing reinforcement can make the insertion very difficult and for both beams and columns the existing links may have to be cut and bent aside, and then re-joined by welding or new links may have to be introduced.

Post-tensioning tendons (cables or rods) are usually applied externally to the member that is to be strengthened and special brackets or blocks for the anchorages have to be fixed to the member. In floors it is often possible to accommodate both cables and anchorages within the thickness of the screed which, when re-instated, will also serve as protection. For beams, the tendons can be positioned either side of the downstand and held in a 'curved' alignment by deflectors in the shape of heavy bars placed in holes drilled horizontally through the downstand. Anchorages may need steel brackets bolted to the sides of the downstand, or may be placed in pockets cut in the slabs associated with the beam or on cross beams. All these new exposed metal components will need corrosion protection *and also* fire protection and they will be visually obtrusive.

It may be possible to strengthen a beam by bolting steel plates or channel sections to the side faces of the downstand. The structurally most satisfactory way is to have through bolts, but these give rise to the practical difficulty of drilling the holes in the concrete so as to line up on both faces with the holes in the steel.

The plates or sections may be visually as obtrusive as post-tensioning tendons, but both types of strengthening may preserve more of the original structure than will the addition of extra reinforcement and they are partially reversible; something that may appeal to conservation purists. Plates and/or sections will (unless located in a dry environment) require corrosion protection and also fire protection.

7.4.4 Adhesive-bonded Strengthening Components

The provision of additional tensile capacity by adhesive-bonded external steel plates was developed during the 1980–1990s. The technique was mainly used on bridges. It required the concrete face to be flat and clean; the latter was usually achieved by grit blasting. To prevent 'peeling failure' of the glue, the ends of the plates were usually held to the concrete by expanding anchor bolts. Corrosion protection of the plates and bolts was of course necessary and if the extra margin of safety, provided by the glued-on strengthening plates, was to be maintained during a fire in a building, the fact that the adhesive would lose strength at a lower temperature than the steel, meant that a greater degree of fire protection was required for 'glued-on' strengthening devices, than for those relying on just steel and concrete.

Since the mid-1990s, a similar technique, using fibre-reinforced polymers instead of steel has been used to strengthen a number of building structures, as well as bridges. The most commonly used components have so far been carbon-fibre-reinforced resin strips (usually referred to as 'plates'), but Aramid ('Kevlar') and glass fibre sheets have been used in a few cases. These fibre composites have extremely high-tensile properties (ultimate strength 1500–3000 N/mm^2, E-value: 150–300 kN/mm^2). They are light in weight and quite flexible (most are between 1 and 2 mm thick and between 50 and 150 mm wide). These properties make for much easier installation than steel plates (no propping is usually necessary) and this largely compensates for their high material cost.

The disadvantages of the technique are that (a) the strips are vulnerable to 'vandalism' by other building trades; (b) there are, as yet, no experience of the long-term durability of the strengthenings and (c) whilst the fibres (with the exception of aramid) can withstand high temperatures, the resin matrix and the adhesive cannot—fire protection will therefore be required in most building applications. (a) and (b) mean that periodic inspections and perhaps some monitoring should be carried out—something that many building owners are reluctant to instigate, but which should be a price worth paying for the relatively painless process of strengthening; (c) is no worse than for strengthening with bolted-on steel.

In one respect, these fibre-composites behave quite differently from almost any other building material: their stress–strain diagram is straight right up to failure. This means that they cannot provide proper load sharing, if the strains in the strengthening do not match those in the structural member to be strengthened. This will however not be a problem when the aim is merely to allow increased imposed load on a floor, as long as strictly linear-elastic analyses are used for both the 'before' and the 'after' condition. The easy handling of the strips (some can be delivered as coils and cut on site) means that strengthening of slabs, to take increased imposed load, becomes a fairly simple operation. Installation is best carried out by specialist subcontractors.

Detailed information is given in The Concrete Society's Technical reports 55 and 57.

7.4.5 Remedies for Concrete Strength Deficiency

There are two ways of remedying concrete deficiency: to add extra concrete or to install external steel components. For slabs it is often possible to remove the non-structural screed and replace it with a structural concrete topping, finished with a floated surface of similar smoothness to a screed. The slab may need to be propped in order to minimize deflections and some preparation of the surface and/or installation of shear connector dowels may be necessary.

Where column size is not critical, the existing cover can be removed, a new reinforcement cage 'wrapped round' the column and a casing, 100–200 mm thick, cast around the column, either through a hole in the floor slab above, or through a 'letterbox' slot in the formwork. In both cases it is beneficial to stop the casing some 50–75 mm short of the top and fill the gap by ramming in a 1:3 cement:sand mortar of 'earth-dry' consistency, after the concrete in the casing has hardened and dried out.

An alternative way of providing the new concrete is to apply it by spraying ('shotcrete'); this also allows smaller thicknesses (down to 75 mm) to be applied. The problem of 'rebound' of sprayed material should however be borne in mind (see Sec. 7.4.1).

Bolted-on structural steel sections (most often channels) can be used to strengthen columns as well as beams (see Sec. 7.4.3). For stocky columns, the sections are provided with end plates top and bottom, which are dry-packed to provide load transfer. For slender columns, where the shortfall is only significant over the middle part of the height, it may be possible to effect load transfer simply by bolting the sections to the concrete, mainly near the top and bottom.

The glued-on metal plates, described in Sec. 7.4.3, can be equally used to make up a modest shortfall in compressive strength. As the glueline may be less strong in direct tension than in shear, buckling of the plates may be a concern; some additional bolting along the length may therefore be desirable. A moderate shortfall of compressive strength may be remedied by the 'glued-on' fibre composites, described in Sec. 7.4.4. The necessity for corrosion- and fire protection of steel- and fibre-composite components remains the same as for beams (see Sec. 7.4.3).

For some columns, wrapping with fibre-composite sheet will increase the lateral confinement of the concrete and thereby the axial strength; this can be particularly effective for round columns.

7.4.6 Remedies for Wood–Wool Former Problems

The only effective cure is to remove the open-textured rib concrete and replace it. 'Guniting' may be the most practical way of doing that. Where the problem is severe, reinforcement anchorage and shear capacity may be jeopardized in the temporary condition. This means that the repair may have to be done one or two ribs at a time, with adjacent ribs being propped.

7.4.7 'Structural' Crack Filling

Cracks in concrete structures are sometimes due to a structural cause in the past, that has since been eliminated. These cracks may cause a structural weakening, for instance if they occur near the ends of a beam, where they reduce the shear capacity.

Low viscosity epoxy resins that can seal and penetrate even very fine cracks to a considerable depth are available. Some of these have strengths that equal or exceed the strength of concrete. The low viscosity is usually accompanied by a short setting time, once the resin and the hardener are mixed (short 'pot life'). To overcome this, special injection heads are used, to which resin and hardener are pumped separately and mixed just before being fed through the nozzle. (Low viscosity can also be achieved by diluting the resin with solvent, but this leads to porosity and reduced strength of the hardened resin when the solvent evaporates.)

7.4.8 Temporary and palliative Measures

It may sometimes be thought expedient to postpone radical treatment, such as described under Sec. 7.4.1. In that case measures should be taken to prevent fragments of concrete falling and injuring people. This may involve safety nets and/or screens ('scaffold fans'). Where the cover has already spalled off in places, it may be worth knocking off any loose bits and applying paint to the exposed steel so as to slow down the rusting, but only if it can be done from a cradle, as opposed to a scaffold.

Occasionally in the past there have been treatments marketed that, without discriminating the cause of the problem, purported to cure reinforcement corrosion by a simple process of cleaning only the exposed areas of the bars, applying an epoxy-based paint and reinstating the cover in some 'magical' (and expensive!) resin mortar.

Where the corrosion is due to chlorides in the original concrete mix, such treatment will only offer a brief respite, before the chlorides in the concrete, remaining behind the bars, make their effect noticed again. In the case of atmospheric corrosion, if the treatment is only applied in patches, it is likely to retain moisture, which may aggravate corrosion in adjacent areas.

7.5 FIRE PROBLEMS AND REMEDIES

7.5.1 Behaviour of Concrete Materials and Structural Members in Fire

With the exception of some limestones and some sintered lightweight materials, concrete aggregates contain minerals with chemically bound crystalline water. When the concrete is exposed to the heat from a fire, this water boils and the resulting vapour pressure bursts the minerals and causes rapid, inwardly progressing, spalling of the concrete surface. This exposes the reinforcement to the heat which causes it to soften when its temperature approaches 500 °C (see Sec. 6.7.1).

The concrete itself begins to deteriorate at about 200 °C and loses most of its strength above 300 °C even if it has not spalled. The strength loss is accompanied in most concretes by characteristic colour changes: at about 300 °C they turn pink, above 450 °C a yellow-grey colour appears. These colour changes can afterwards be useful indicators of the temperatures that the concrete and the reinforcement have attained during a fire.

The structural performance of a continuous reinforced concrete floor during a fire is often modified by the effect of the thermal expansion. As the high temperatures are usually localized at any one time, the expansion tends to be restrained by the surrounding structure and this puts the 'hot' part of the structure into compression. This compression may enable it to develop arch action which, at least partly, compensates for the softening of the reinforcement.

No such enhancing effects are however available for columns.

7.5.2 Assessment of Structural Members after a Fire

Sometimes the effects of a fire on a reinforced concrete structure may be visible as gross deflections and spalling; on other occasions all that can be seen is blackening by soot. It is however essential for an assessment of the carrying capacity after a fire to ascertain any changes of the mechanical properties of the materials that may have been caused by the heat. The nature and degree of these changes depend on the temperatures that the materials will have reached during the fire.

The colour changes mentioned above, can be of considerable assistance in such an assessment: Concrete that has turned pink will have reached a temperature of about 300 °C and will have lost practically all of its strength; any reinforcement partly or wholly surrounded by such pink concrete will likewise have reached 300 °C, but no more, and will therefore not have suffered any significant permanent loss of strength.

Concrete that shows a dirty yellowish grey colour, and has pink concrete behind, will have reached 450–500 °C and will of course be structurally useless, but any steel inside such concrete will have reached a similar temperature and its strength after the fire will then depend on its nature.

Reinforcement which relies mainly on its alloying composition for its mechanical properties, e.g. mild steel and hot-rolled high-tensile bars, will have suffered very little reduction in ultimate tensile strength and yield point (the stress at which the bar elongates without any increase of the applied force). Cold-worked high-tensile reinforcement, such as twisted square bars, twisted twin bars and twisted round, ribbed, bars (recognizable by the spiral pattern of the ribbing) may have been partly or wholly annealed and hence had their yield point and strength reduced to that of mild steel. When there is doubt about the remaining strength of reinforcement, it may be possible to carry out in-situ hardness tests on exposed portions of bars (see Sec. 6.3.4); these tests can, if calibrated on unaffected bars, give a fairly reliable indication of the yield point.

Pre- and post-tensioning wires, cables and strand may owe their strength partly to alloying, partly to cold drawing and partly to heat treatment during manufacture. Assessment of their remaining strength after a fire requires specialist advice.

A point to bear in mind when appraising a, seemingly lightly, fire-damaged structure, is the following: Were there any materials in the building (whether forming part of the finishes or being stored at the time of the fire) which by the heat of the fire may have been turned into corrosive substances, forming part of the soot deposit on the concrete? (For instance, polyvinyl chloride (PVC) will produce chlorine when burning and this, when combining with water from fire-fighting turns into hydrochloric acid). Where this may have been the case, steps must be taken to remove such substances, before they migrate too far into, and damage, the possibly porous concrete and attack the reinforcement.

7.5.3 Reinstatement after Fire Damage

The first and essential step is to remove damaged concrete. This is similar to the procedure for other concrete repairs, but in the case of fire damage (as opposed to, for instance, reinforcement corrosion due to carbonation) it is essential that *all* the affected concrete is removed, because it will have lost its strength and any repair-mortar or -concrete will therefore not bond structurally to the sound concrete underneath.

The weakening of the structure, that such removal may cause, should be borne in mind and scaffolding material should be held in readiness for propping to be put up speedily, if it becomes necessary.

Once the unsound concrete has been removed, the strength of the reinforcement has to be assessed, as described in Sec. 7.5.2, and the load-carrying capacity of the member can then be calculated on the assumption that only the damaged concrete is replaced. If this is adequate, replacement can proceed by one of the methods, described in Sec. 7.4. If the reinforcement has been weakened to the extent that the member would not be adequate for the intended *new* use (as opposed to the original design), strengthening by one of the methods, described in Sec. 7.4.3, will have to be carried out, prior to re-concreting. (Detailed information on assessment and repairs of concrete structures can be found in: Concrete Society Technical Report No. 33; Assessment and Repair of Fire-damaged Concrete Structures.)

EIGHT

FOUNDATIONS

York Minster exploratory excavation 1967: The imposition in the fifteenth century of the 16 000 tons Central Tower on the eleventh-century foundations, intended for a much less heavy crossing, caused differential settlement which distorted and cracked the foundation masonry.

'The root of the trouble' with old buildings is sometimes to be found in their foundations, but this is by no means always so. Remedial works to foundations tend furthermore to be expensive and very disruptive to the users of the building and they do, in some cases, have unforeseen side effects. They should therefore not be undertaken lightly. Only after a careful study of the structure as a whole, taking full account of the principles of foundation behaviour, can it be established that foundation settlement has occurred *and is continuing to occur.*

8.1 PRINCIPLES OF FOUNDATION BEHAVIOUR

8.1.1 Settlement, its Nature, Significance and Effects

No foundation carries load without settling!

The soil under a strip footing or a column base is subjected to pressure from the load from the superstructure acting on the foundation and will, like any other material, compress and in doing so, allow the foundation to settle. Piles are subjected to compressive forces from the superstructure load and will like any other structural compression members shorten and, in doing so, allow the pile-cap to settle. Even a pillar founded on rock will exert pressure on the rock, which will, in consequence, like any other material, compress and make the pillar settle, albeit by an almost immeasurably small amount.

Settlement is an inherent part of foundation behaviour and is not, per se, a sign of deficiency or danger.

'Uniform settlement' does not cause damage to the superstructure: the building sinks, usually a small amount, 'on an even keel' and all that may happen is that service pipes, going into the building are bent a small amount. Larger uniform settlements will still not harm the building, but where there originally were seven steps up to the entrance door, there may now only be five or three. These very large settlements occur in areas with very soft soils, e.g. in Mexico City.

Settlements that are unexpectedly large, i.e. much more than is normally predicted and observed on adjacent buildings in the locality should, however, be viewed with suspicion, as they may indicate that the soil beneath the foundations is overstressed and near to failure.

If an originally horizontal plane at some level in the building becomes slightly inclined, but remains flat, this can be described as 'planar settlement'. This again will be harmless, except in the case of free-standing towers that are tall in comparison with the width of their foundation, e.g. The Leaning Tower of Pisa.

What does cause superstructure damage is 'differential settlement', i.e., some footings or bases go down more than others, so that what was a flat horizontal plane through the building becomes a dished, domed, or twisted, surface. This causes masonry to crack and floors to become uneven.

In the case of ancient buildings, differential settlement was usually not a serious problem; buildings were either flexible enough to accommodate it (timber frames—bricks in lime mortar), or it was accepted that the masonry would crack (but remain stable) and would be repaired when the movement stopped, after a generation or so.

Modern construction and finishes are stiffer and brittle and much less capable of accommodating settlement. 'Permissible bearing pressures', quoted in modern regulations and codes, are therefore often set at a level to limit settlements rather than to provide a rational factor of safety against failure.

8.1.2 Factors governing Foundation Behaviour

The behaviour and performance of foundations are influenced by the following factors:

1. the load from, and the nature of, the superstructure;
2. the nature, and dimensions, of the foundations;
3. the nature and history of the ground strata (soil or rock) under the foundations;
4. the groundwater, its levels and fluctuation;
5. any interference with 1, 2, 3 or 4 after the initial construction.

Factor 1 is fixed, once the superstructure is built. It may however be changed by subsequent alterations. Apart from its weight, the stiffness and strength of the superstructure may affect the response of the foundations to local variations in the underlying strata.

A continuous, rigid and strong superstructure, such as a solid masonry wall with few and small openings, may be able to bridge across a local area of weak soil under the footing (Fig. 8.1). Conversely, an articulated superstructure, such as a colonnade, with individual lintels spanning between columns, will offer no resistance to the settlement of a column founded on a soft spot (Fig. 8.2). The damage in the latter case will however be less than that caused to a rigid wall that is not strong enough to bridge or cantilever over differential settlement, or which has been weakened by subsequent alterations.

The remaining factors 2, 3, 4 and 5 will be discussed in detail in the following.

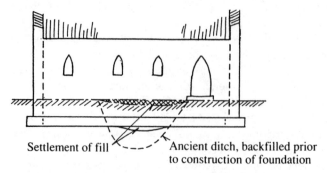

Settlement of fill   Ancient ditch, backfilled prior
to construction of foundation

Figure 8.1 A rigid and strong superstructure bridges over 'soft spots' and resists differential settlement.

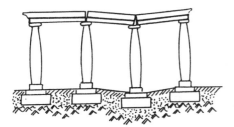

Figure 8.2 A flexible superstructure has to follow differential settlement.

8.1.3 Foundation Types

If a building stands on a reasonably level area of rock or sufficiently strong soil, no foundation in the usual sense of the word is necessary. The walls can rise directly from the bottom of a trench that need only be deep enough to ensure that the bottom course of masonry is not too easily dislodged. Except where the rock comes right to the surface, it may be necessary to dig down a little way to get to good bearing strata, but that is all.

There are also many buildings dating from as late as the nineteenth century that have no more foundations than that, even on soils where today we would construct quite elaborate foundations. Buildings on such 'non-existing' foundations often display signs of past, substantial differential settlements, of a magnitude that would be considered unacceptable today. *Such settlements in buildings of that age do however not justify intervention, if they are no longer increasing.*

On more compressible soils, the load from a wall has to be distributed over a greater width than the thickness of the wall. This requires a strip footing. Similarly, the load from a pier or column is transmitted to the soil by a pad footing with a horizontal area, larger than that of the pier or column. Strip footings and pad footings, collectively known as spread footings, transmit their load in direct bearing on the soil immediately below.

Strip and pad footings for old buildings were sometimes constructed with corbelled brick footings or with large stone slabs as a first course. They might alternatively consist of large stone blocks laid directly on the soil (see Fig. 8.3*a*). Where the ground was suspect, the foundation masonry was sometimes reinforced with timber baulks (for example: the Norman masonry foundations of York Minster had embedded oak timbers 200 × 250 mm at 1.5 m spacing in both directions in a double grid, four layers in all). Lowering of the groundwater (by, for instance, introduction of street drainage) may cause trouble in such foundations due to rotting of the timbers.

Brick foundations sometimes had a bottom 'course' of lime mortar into which had been rammed broken and crushed brick. The latter would act as a pozzolan, enabling the mortar to harden in the absence of air and gain a fair amount of strength.

If the upper layers of soil are too weak or too compressible, deeper excavation may enable spread footings to be placed on good bearing strata, but this

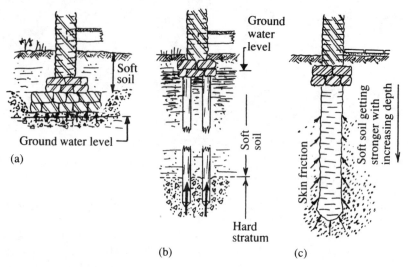

Figure 8.3 Foundation types: (*a*) strip footing, (*b*) end-bearing timber piles, (*c*) cast in-situ friction pile or caisson.

solution is limited by operational constraints on depths of excavation (trenches have to be planked to prevent cave-in; spoil has to be lifted to a great height) or by the presence of a groundwater table, which makes deeper excavation impractical unless the site can be temporarily de-watered; an option generally not available before the late nineteenth century, although baling had been used earlier in the construction of caissons for bridge piers.

In the past, land was often reclaimed from river foreshores, etc., by placing fill on top of the, already present, silty deposits. The fill was often firmer than the silt, or it became so in time, with shallow founded, low, buildings being initially constructed on it. When, in the first half of the eighteenth century, more ambitious edifices were to be constructed, the builders would hit water when attempting to excavate for deeper foundations. They would then place timber grillages, just below the water, to help even out the expected settlement and act as permanent formwork for rubble fill below the bottom course of masonry. As long as the timbers remained under water, all was well. When subsequently the water level drops, for whatever reason, the timbers rot, their contribution to the bearing area is lost, the ground pressure increases and new settlement occurs.

As mentioned in Sec. 6.2.5, the footings for the stanchions of some steel-framed buildings, from the first half of the twentieth century, consist of steel beam grillages, encased in mass concrete (see Fig. 8.4). The grillages would often consist of several layers, each at right angles to the one above, 'stepping out' from the stanchion until the required foundation area was achieved.

Generally, however, when spread footings cannot reach sound bearing strata, piles or caissons are used to bypass the weak layers. Where the soft overburden rests on a hard layer, the load is transmitted mainly in end bearing by the pile or

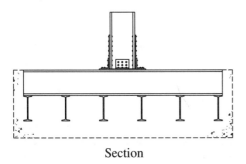

Section

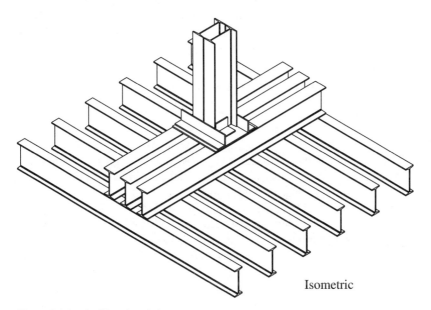

Isometric

Figure 8.4 Steel grillage foundation.

caisson. If the soil gradually improves in strength with depth, the load is largely transmitted by friction between the shaft and the soil.

Traditional piling consisted of timber piles driven through the soft strata and carrying the load by end bearing (Fig. 8.3b). In earlier construction they were capped with a timber grillage platform from which the masonry would rise. Pile tops and grillage would generally be kept below groundwater level to ensure their durability. When mass concrete replaced stone as a foundation material, the piles would be cut as far as possible just below the water level and the concrete would be poured over and around the pile tops.

Caissons were originally shafts, hand-excavated, often inside a (timber) sheet-pile enclosure, and subsequently filled with masonry or concrete. (The foundations of the Taj Mahal are described as being hand-dug 'wells', lined with brickwork and filled with rubble and mortar.) Later, pre-cast concrete rings were used to line the excavation. Bored, cast in-situ concrete piles (Fig. 8.3*c*) are formed in a similar way, but using mechanical means of soil extraction. A driven steel tube is used to line the hole where ground conditions make it necessary. Cast in-situ piles transfer their load, largely in friction. Since the 1950s they have largely superseded driven concrete piles as foundations for buildings, except where very soft deposits overlie hard strata. During the 1960s *under-reamed* cast in-situ piles were introduced for carrying very heavy loads, mainly in clay soils. The boring 'tool' for forming these has a pair of 'wings' at the bottom end. These are pushed out when the prescribed depth has been reached, so as to form a bell-shaped enlargement at the bottom of the pile. This enlargement will carry a substantial proportion of the load, particularly in the ultimate condition, whilst the remainder of the load is resisted by the 'skin-friction' of the shaft. These under-reamed piles made it practical to construct tower blocks in places where the ground consists of slow-draining clay (see Sec. 8.1.4).

8.1.4 Ground Strata and their Behaviour

Foundation strata can be roughly classified as rock, gravel and sand, clay or silt.

Rock is inherently very strong, but it may be fissured or weathered and it may have been deposited in relatively thin layers, interspersed with layers or veins of softer material. Foundations on rock usually have ample reserves of strength and their settlements are very small and whilst weathered granite can turn into a 'sensitive clay', the main hazard is instability of the rock mass as a whole.

Certain rocks, such as limestone, are slightly soluble in rainwater, which is acid due to its natural content of atmospheric carbon dioxide; others, such as rock salt, are highly soluble in pure water. Fissures may allow rainwater to percolate down and dissolve limestone to the extent that underground caves and tunnels are formed. These can gradually enlarge to the point that the 'roof' of the cavity becomes too thin and caves in. One form of these solution cavities is known as 'swallow holes' from the way that smaller buildings may disappear into them; they occur amongst other places in south-eastern England, where the chalk is near the surface.

Alternatively, fissures in rock may allow rainwater to penetrate to lower veins of silt, clay or shale. These are inherently planes of weakness, and if they slope down towards a valley, due to folding of the strata after deposition, the softening and lubricating effect of the water on these veins may eventually result in a landslide of the whole rock mass above. The *nature and inclination* of the rock strata under a historic building, particularly if situated on a slope or escarpment, should therefore always be investigated. Figure 8.5 shows some possible hazards for buildings founded on rock, but note that the juxtaposition in the same locality of landslide due to clay softening *and* the presence of solution cavities in limestone is not very probable.

Gravels and Sands are composed of relatively coarse grains. These have a small specific surface (surface per unit volume). The grains rest on each other at points of contact and the voids between them are relatively wide so that air and water can move freely through them. (Deposits with a minimum grain size of 2–3 mm are usually classified as gravel, whilst the grain size for sand ranges from 2 to 0.06 mm.)

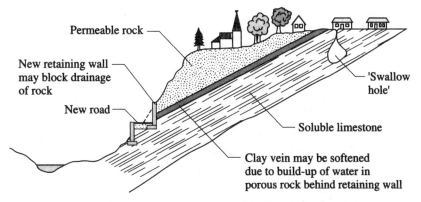

Permeable rock

New retaining wall
may block drainage
of rock

New road

'Swallow
hole'

Soluble limestone

Clay vein may be softened
due to build-up of water in
porous rock behind retaining wall

Figure 8.5 Some possible hazards for buildings founded on rock (the juxtaposition in the same locality of landslide due to clay softening and solution cavities in limestone is not probable).

The capacity of gravels and sands to carry foundation loads depends on their internal friction. It is therefore increased by any vertical load ('surcharge') imposed on the area immediately outside the footing. The 'effective' strength of such soils therefore increases with the depth of the footing below the surface.

Water reduces the strength, partly because it acts as a lubricant on the soil below the footing, decreasing the internal friction, and partly because it reduces the effective density of the surcharge material. A strong flow of water, across the site of a building, can wash away the fine particles from the foundation stratum, if it has mixed grain sizes, and leave it as a weakened 'honeycombed' material. Excavation below the water table inside a sheet-pile or diaphragm wall can create an upward flow of water turning fine sand into 'quicksand' and washing out material from under adjacent buildings.

Settlement of foundations on sand or gravel depends on the elastic deformation of the grains as they are stressed. It therefore tends to be almost instantaneous as the load is applied. Sometimes settlement is also caused by consolidation of loose-bedded material subsequent to construction (Fig. 8.6). This may occur if the

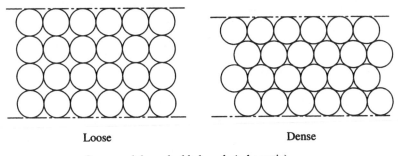

Loose

Dense

Loose and dense-bedded sands (schematic)

Figure 8.6 Consolidation of loose-bedded sand causes settlement.

ground is subjected to vibration, e.g. from work on an adjacent site, such as pile driving, or from heavy lorry traffic. (Fine wind-blown sands of uniform grain size, such as Loess, can be meta-stable: they may sustain load when dry, but collapse if subsequently inundated.)

Fluctuations in the moisture content of sands and gravels do not cause any significant shrinkage or swelling.

Clay consists of very fine (less than 0.002 mm) particles that are flakey in shape. This gives them a very high specific surface which means that water in the very fine pores is far harder bound to the particles than it is in sand. The behaviour of clay is largely governed by this pore water. Unlike sand, clay can shrink significantly when drying, and swell when wetting up. (In some ways its behaviour is similar to that of a cellulose car-washing sponge: pressure, e.g. from a foundation, squeezes out the pore water and conversely, if the clay is unloaded, e.g. by excavation of overburden, it takes up water, and in doing so, swells or 'heaves'.)

The load-bearing capacity of clay depends on its cohesion and on the pore water pressure. Settlements can go on several years after the application of load, before they reach their final value. This is because the elastic deformation of the particle 'skeleton', which governs the volume change of the clay, cannot proceed any faster than the displacement of the pore water, which is a slow process; everything else being equal, the finer the clay particles, the slower the water movement.

So-called 'sensitive clays' are particularly prone to heave, when they are unloaded by excavation for a deep foundation. When load is then applied by the new foundation, the pressure slowly drives out the pore water until the settlement, relative to the original level, equals the consolidation due to the net pressure increase minus the heave.

'Laminated clay' has a plywood-like structure with alternating thin layers of clay and fine sand/silt. The sand/silt bands connect with 'open' water or water-bearing sandy deposits. This means that pore water can move fairly rapidly, in which case settlements reach their final values fairly quickly (in months rather than years as would be the case for ordinary clay).

Settlement of foundations on clay can take place without any change of the load on the foundations, if the moisture content of the clay is changed. This can be caused, for instance, by the growth of trees and may be aggravated by a succession of dry summers (these occur roughly on an 11-year cycle, coinciding with the increases and decreases of the Sun spots). Removal of trees can, conversely, cause clay heave. Settlements and heave, caused by moisture changes of clay, are usually differential and often cause damage to the superstructure.

Clays are *not* susceptible to consolidation by vibration.

Some clays have high contents of sulphates. These attack concrete and limestone, particularly if the sulphates are present in groundwater that continuously flows around the foundations.

Silt has grain sizes in between those of sand and those of clay (0.06–0.002 mm). It has some of the undesirable properties of both. The grains are too fine to develop solid point contact through the water, so the internal friction is low, but at the same time the grains are too large to allow the 'electro-chemical' bond of clay, so the cohesion is also low. Silts often contain organic matter such as peat.

Peat, being extremely compressible, aggravates the poor performance of silt as a foundation stratum by giving rise to large (occasionally gross) settlements. Under certain conditions it undergoes anaerobic decomposition, which can produce, in turn, various organic acids and then methane gas. The acids attack lime and cement, whilst methane is highly explosive in a wide range of concentrations.

8.1.5 The Influence of Groundwater

As mentioned in Sec. 8.1.4., the presence or absence of groundwater significantly affects the properties and performance of foundation soils. It can also affect the actual foundation structures.

If water depletion of clay is induced by outside influences, such as for instance, pumping from an aquifer under the clay, settlement will be caused, but at the same time such 'under-drainage' will in fact increase the strength of the lower layers of the clay. Conversely, excessive recharging of the aquifer can weaken the lower layers of clay; this can reduce the bearing capacity of deep piles and caissons.

Silts, particularly those containing peat, are very susceptible to volume changes caused by variations of moisture content. Where river foreshores, etc. have been reclaimed by filling, and buildings 'founded' on timber grillages, a lowering of the groundwater table will initially encourage rotting of the timbers, as described in Sec. 8.1.3. If the water table is lowered further, the silt will shrink and further settlement will occur.

Timber piles that are totally submerged will usually last for centuries, because the organisms (mainly fungi) that cause rotting, are inhibited by the very low oxygen content of the water. If the water table is lowered so as to expose the tops of the piles to air, and particularly if the water level fluctuates so as to alternately wet and dry the timber, it creates conditions that encourage the fungi, thus accelerating the rotting of the pile tops, leading to eventual failure of the foundation. Timber grillages are similarly susceptible to rotting if the water table is lowered. In this case the foundation will not fail, but will sink as the gap, left by the timber, closes up.

Conversely, if the groundwater is acidic and its level rises, it will attack mortar, limestone and concrete in the foundations. Failure due to acid attack is however far less common than failure due to pile rotting.

8.2 CAUSES OF FOUNDATION PROBLEMS

As with any other form of structural distress, a correct diagnosis of the cause of apparent foundation malfunctioning is a pre-requisite for effective treatment. However, before investigating foundations in great detail (and at great cost!) one should ascertain that any distress symptoms above ground do not have their cause in the superstructure itself.

Malfunctioning of foundations may be caused by one or several of the following.

8.2.1 Original Deficiency

It is frequently assumed that anything built a 100 years ago or earlier was 'designed' by sound rules-of-thumb and built with solid conscientious workmanship. This is not so: 'Jerry-building' has a long inglorious history, not least in the area of foundations where any shortcuts could be quickly covered up. Some shortcomings may be due to honest ignorance of soft strata underlying the apparently sound ground. Some may represent the best that could be done in the circumstances, e.g. where the depth of excavation was limited by groundwater before really good bearing strata was reached, or where the building had to be partly founded on new footings and partly resting on the footings of an earlier building, which were too massive to remove with the means available at the time.

There are however equally many instances where, on investigation, it is found that the foundation 'design' and execution were based more on the 'market economy' of the day, not to say sharp practice, than on sound building/engineering judgement. For example, the buttresses to the East Wall of York Minster were found to have practically no projecting footings; the footings did in fact taper inwards with depth, having a boat-shaped cross-section, and they rested on made ground.

There may also be instances in which the foundations, per se, are adequate and the actual bearing stratum perfectly sound, but where the site and the surrounding area are part of an unsuitable formation, e.g. a hillside subject to progressive slipping of clay strata at some depth.

Original shortcomings of foundations can therefore not be ruled out as a cause of structural distress, *but*, conversely, a foundation, that today may appear to be unacceptably undersized, may have completed its settlement at some time in the past and will carry its load without any further settlement, unless something or somebody interferes with the soil, the groundwater or the superstructure.

8.2.2 Overloading, subsequent to Original Construction

This is really another instance of poor design and construction. The difference from Sec. 8.2.1 is that the culprits are not the original builders, but the people who built the extension or the new superstructure, perhaps several 100 years later. Any record that indicates rebuilding or enlargement on the site of a previous edifice may help to ascertain this cause, but should obviously be verified by physical investigation.

A prime example of 'post-completion' overloading is the central tower of York Minster, rebuilt in its present form in the first half of the fifteenth century on the Norman foundations from 1080 AD and raising the ground pressure to about 530 kN/m^2 as opposed to the 290 kN/m^2 elsewhere under the church, with drastic differential settlements as a result.

A modest increase of the load on a foundation, that has completed its settlement, need however not have serious consequences: District Surveyors in London used to allow a 10 per cent increase, without the need for a full site investigation, if the foundations were reasonably sound. (District Surveyors were autonomous building inspectors for London with wide-ranging powers, originally appointed by Parliament after the great fire in 1660, and functioning until the 1980s.)

An increase in foundation loading can occur, without any apparent building extension or reconstruction, due to a change of use leading to heavier floor loadings. In many cases these heavier floor loadings are of short duration (e.g. in lecture rooms) and their effect is therefore likely to be far more significant for the superstructure. A conversion, in which large openings are cut in load-bearing walls, can however cause significant overloading of the foundations supporting the walls at either side of the opening.

8.2.3 Deterioration of the Foundations

As mentioned earlier, tops of timber piles, timber grillages supporting masonry off piles, and timber 'reinforcement' of masonry footings are all liable to rot if exposed to air and subjected to wetting and drying. These conditions are most likely to arise if, for some reason, the groundwater level has been lowered.

Sulphate-bearing and/or acidic groundwater will attack the mortar in masonry foundations and if the groundwater flows across the site, there will be a constant supply of the attacking substance. If the product of the chemical reaction is soluble, the mortar may, in theory, lose all its binder with time and become weakened to the extent that the masonry crumbles. There are however no generally published cases of this having caused damage to a superstructure, whereas there are many instances of problems due to rotting timbers.

8.2.4 Change of Soil Conditions and/or Water Regime

There are several ways in which load-bearing characteristics of the strata under foundations can change with time:

1. Sands which in the past have been above the level of the groundwater may, even if firmly bedded, lose some of their internal friction, and hence some of their bearing capacity, if the groundwater rises and 'lubricates' the grains.
2. Loose-bedded sands, that have remained stable for centuries during which the roads surrounding buildings have supported nothing heavier than an oxcart, may consolidate under the influence of vibrations from heavy lorries. This consolidation may extend under the buildings and cause settlement of their foundations; this will be more severe nearest to the roads.
3. Clays may lose some of their bearing capacity, if they are 'wetted up', due to a rise of the groundwater level.
4. If the moisture content of clay is reduced, due to a succession of years with low rainfall, or due to a lowering of the groundwater table, the clay will shrink and the volume reduction will lead to settlements. This effect is most pronounced with 'fat', or 'sensitive' clays.
5. If water depletion of clay is induced by outside means such as, for instance, pumping from a deep aquifer under the clay, settlement will be caused, but

such 'under-drainage' will in fact increase the strength of the lower layers of clay. Conversely, where decline of local industry leads to cessation of pumping, excessive recharging of the aquifer may occur. The consequent wetting-up will weaken the lower layers of clay and thus reduce the bearing capacity of caissons and (particularly under-reamed) piles. This has, for example, been found in parts of London.

6. If, in silty soils, an increased flow of the groundwater is caused by some interference such as water extraction or drainage in the neighbourhood, the finest particles may be washed out of the soil, with settlement as a result.

7. Where river foreshores, etc. have been reclaimed by filling, and buildings 'founded' on timber grillages, a lowering of the groundwater table will initially encourage rotting of the timbers, as described in Sec. 8.1.3. If the water table is lowered further, the silt will shrink and further settlement will occur.

8. As mentioned in Sec. 8.1.4, limestone may be dissolved by carbon dioxide in water. This may result in 'swallow holes', sometimes several metres deep. These and the less dramatic manifestations of solution cavities tend to occur suddenly, without warning. Similarly, the wetting-up of sloping clay or silt layers in sedimentary deposits does not give any visible warning of a possible impending landslide.

Some of the above-mentioned changes take place slowly, sometimes over many years. The foundations and superstructure of a building, constructed before the onset of the changes, will remain intact for several years, bridging or cantilevering over the more softened, or shrinking, patches of soil without showing any differential settlement, so that the owner is not aware that anything is amiss. At some stage, however, stresses in the structure exceed its strength and the structure adjusts itself to the new ground condition by settling and cracking and thereby relieving the stresses. When this happens, one must *not* think: 'It has settled 20 mm in a week, so in six months it will have gone down half a metre, we must do something quickly!' Whilst this is a typical scenario for damage from rotting timbers in the ground, it can be due to other causes.

During the extreme drought in the summer of 1976, the house of one of the authors developed a crack at one corner overnight. This was due to desiccation of the clay under the foundations. The crack width was monitored during the following 12–18 months and was found to vary by no more than ±0.5 mm. At the next redecoration, linen-backed lining paper was applied over the affected area, and this has remained intact since.

8.2.5 Mining Subsidence and Brine Pumping

Underground mining leaves cavities, where the minerals have been extracted. Whilst modern practice is tending towards returning most of the waste, this filling cannot be sufficiently compacted to completely prevent some collapse of the workings; this leads to a 'wave' of subsidence on the surface which follows the progress of the mining underground. The deformation of the surface is not just vertical settlement; across the convex part of the wave there is also a stretching (tensile strain) of the surface.

Where financial considerations are given complete priority, as they universally were in the past, back-filling is not carried out and the workings are allowed to collapse to their full height with resulting greater subsidence on the surface. (As the collapsing 'roof' material does not completely compact in the workings, the surface subsidence, however serious, does not correspond to the total volume of the workings.)

There have been instances of old wells and mine shafts which have been simply closed near the surface with a timber platform, with a few metres of fill on top. The timber eventually rots and whilst the fill may arch over the shaft, any disturbance may precipitate a collapse.

Brine pumping, whereby salts are extracted in solution from underground deposits, will lead to the formation of underground cavities. The resulting subsidence can be worse than that caused by even non-waste-return mining, partly because the direction and extent of the progress of the cavities are not known, nor controllable, as they are in mining. The subsidence can also be sudden, without warning.

8.2.6 Tunnelling and/or Adjacent Excavations

Railway tunnels and sewers were, in the past, lined with brickwork. Nowadays these and road tunnels are lined with concrete walls in the case of cut-and-cover construction, or with cast iron or pre-cast concrete segmental linings in the case of bored tunnels, so as to keep the final cross-section clear. There is however always a certain amount of ground loss involved in their construction, because the 'bore' has to stand unsupported until the lining segments have been installed and the gap between them and the soil has been grouted. This will cause a wave of movement of soil towards the tunnel and this, in turn, will result in settlements of any foundations in the vicinity. The novel so-called 'New Austrian Tunnelling Technique' provides an initial sprayed concrete lining to limit the 'stand-alone' time of the soil, but this does not completely eliminate the problem.

Similarly, construction of a deep basement, other than in rock, is nearly always accompanied by some, usually small, inward movements of the enclosure and the supports to the excavation (sheet piles or slit-trench wall). The soil will tend to follow the movements of the excavation supports and this can cause settlement of the foundations to adjacent buildings.

If a basement excavation is carried out in fine sands below the water table, the excavation usually has to be de-watered by pumping from inside the enclosure. If the cut-off of the enclosure is not deep enough below the bottom of the excavation, the pressure difference will cause the water to flow under the enclosure and upwards inside the excavation. In doing so, it will take with it some sand from outside the enclosure and hence from under any adjacent buildings, causing them to settle. In moisture-sensitive soils, settlement may also be caused by the pumping;

this will lower the groundwater outside the excavation, unless the water extracted from within the enclosure of the excavation is returned to the ground outside.

8.2.7 Precautions in Case of Construction Works adjacent to Historic Buildings, etc.

When any mining, tunnelling or deep excavation in the neighbourhood of a historic building is proposed, the person responsible for that building should obtain a technical description of the new works from its promoters. This should be sufficiently detailed to allow an assessment to be made of the likely effects on the historic building.

In Britain, The Party Wall Act makes the promoter responsible for notifying the neighbours (including all tenants of multi-occupancies), and for paying for independent professional advice, but pre-empting the promoter's notice may serve as a useful reminder to him, and gain a bit of time for the custodian of the historic building to find and brief his expert. A basic problem in this situation is that the promoter/developer's design team will have people who have expertise in new structures and geotechnical predictions of movements, but who know little or nothing about historic buildings and how they respond to ground movements.

As a rough preliminary guide, any tunnel, basement, foundation or pile within a plane sloping down at 45° from the bottom of the foundations of the existing building is a potential hazard. The opposite does unfortunately not always apply: Works *outside* such a plane *may* constitute a threat to the building.

There are computer techniques available that make it possible to predict the approximate amounts of vertical and horizontal ground movement, likely to result from a given extent and method of tunnelling or excavation. Techniques have also been developed which allow estimation of the response of certain forms of masonry walls to the predicted ground movements. The use of these techniques may help to ascertain whether proposed works pose a real threat to a historic building.

An *independent*, experienced, civil/structural engineer should therefore be employed to examine drawings, specifications and method statements for the new works, in order to assess the probability and likely severity of damage to the historic building and, if appropriate, suggest alternative, less damaging layouts, forms of construction and/or methods. Such an expert will be more aware of the effect of the new works on the upper strata, in which the historic building is likely to be founded, whereas they may be ignored by the developer's team because they will be dug out to form the basement, and be bypassed by the new foundations.

Immediately prior to the start of the actual construction operations on site (but in the case of tunnelling, before the tunnel reaches a point at a distance away, equal to its depth below the surface), a condition survey of the historic building should be carried out *and agreed with the promoters of the new works*. This will help to prove or disprove whether any defect, subsequently discovered, has been caused by the new works.

Any pre-existing cracks should be monitored well in advance of the commencement of the works, so as to establish their response to normal environmental fluctuations. All cracks or other defects, whether pre-existing or occurring during the works, should be monitored during the execution of the works and for some time after completion (see Secs 3.3.4 and 3.3.5). The intervals of the readings should be adjusted so as to provide frequent readings when the effects of the new works are likely to be most serious, whereas longer intervals will be appropriate for monitoring residual movements after completion. The period after completion, during which readings have to be taken, will depend on the nature of the works and on the nature of the soil; for tunnels in clay soils the period may have to be over a year.

If vibrations, likely to be caused by the construction operations, give cause for concern, arrangements should be made for their intensity to be measured at various points in the building and the results should be compared with published 'permissible' limiting values (see *BRE Digest* 353 and *ASCE Journal of the Geotechnical Division*, Feb. 1981).

The custodian of the historic building should, if at all practicable, obtain from the promoters of the new works a legal undertaking to keep certain measurable building movements (usually settlements) and vibrations below the levels which are considered not to cause any significant damage to the building.

8.3 METHODS FOR FOUNDATION AND SOIL INVESTIGATIONS

If some defects in a building have been clearly shown *not* to be caused by some superstructure problem, but by settlement *that is continuing*, the foundations and the soil below have to be investigated. The methods, by which this can be done, are described in the following.

As with any investigation, costs are usually low for desk studies and modest for simple procedures that can be carried out on site by the architect or engineer in person, assisted by a local builder as necessary. Methods involving specialist contractors and laboratories are likely to be more expensive, but compared with the cost of a potentially overelaborate foundation strengthening, it will be money well spent, unless the simple procedures have given conclusive answers.

8.3.1 Geological Information and other documentary Evidence

Geological maps show the nature (and sometimes the thickness) of the strata underlying an area. The main information is usually about the deeper 'solid' layers, but for some areas the upper, so-called 'drift' strata, which may govern the foundation performance, are shown in some detail. More detailed information may be found in descriptions compiled by national or local geological societies. In Britain, the most comprehensive collection of maps and monographs is held by the Institute of Geological Sciences at Keyworth in Nottinghamshire. Historic topographic maps may identify old ponds, streams and gravel pits, etc., and even street names such as 'Loampit Lane', 'Spring Gardens' and 'Lower Marsh' can give useful clues.

The information from these sources refers to the area in general, but may quote specific results from borings for water extraction, etc., marked on the map, which will be indicative of the local conditions. The local building control officer may also have useful information on conditions likely to be encountered in the area.

Additional, and often very valuable, information can sometimes be found in archives: design drawings, or failing those, contemporary descriptions of the original construction or even the builder's accounts may indicate whether difficulties, due to poor foundation strata, were encountered during the works and how they were overcome. Similarly, descriptions of subsequent restorations or alterations may give useful indications, not only of the ground conditions, but also of the foundation construction.

If construction works, carried out adjacent to the building, were preceded by a site investigation, the records of such trial pits or boreholes (see Secs 8.3.2 and 8.3.4, below) will usually give the most reliable information on the underlying strata. Documentary evidence should however be corroborated by physical site investigation; this can be by trial pits, hand probing, boreholes or a combination of these techniques.

8.3.2 Trial Pits

This is the name given to excavations, usually hand-dug, with a plan area of between 1.5 and 5 m^2, taken down to, or slightly below, the bottom level of the foundations. They have two advantages over other site investigation methods: they allow observation of the soil in bulk, and they allow inspection of the foundations.

The strata in the sides of the pit and immediately below the bottom can be probed and sampled with hand tools. It is also possible, if so desired, to take samples of the soil with a minimum of disturbance. For soft ground, a simple probe can be improvised from a 10 or 12 mm reinforcing bar and a piece of plumbing copper tube makes a simple sampler.

The disadvantage of trial pits is that the depth of excavation, and hence of visual examination, is limited by the groundwater level and by the problems of getting the excavated material to the surface.

Single trial pits can be misleading if they inadvertently are located where modifications to the foundations have been carried out; a prime example of this is where local underpinning (see Sec. 8.4.2) has been carried out, but past excavations for drains, etc. can also result in misleading information being deduced, unless checked by inspection of one or more additional pits (Fig. 8.7).

Where, otherwise inexplicable, settlements have occurred, a clue to their origin may be found by excavating a trial pit, where settlement has *not* occurred. This is often the only way of ascertaining the past presence of timber grillages, such as described in Sec. 8.1.3.

Excavation of trial pits usually requires the services of a builder or small general contractor. The excavated material may not compact adequately, in which case thought must be given to the backfill material. Depths greater than 2–2.5 m increase the problems of supporting the sides and removing the soil to the extent that trial pits may become uneconomical.

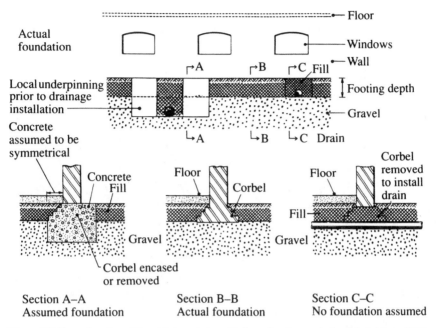

Figure 8.7 Examples of possible misinterpretation of information from a single trial pit (*from CIRIA 'Structural Renovation of Traditional Buildings', Report 111, 1986*).

In any but the most stable strata it is *essential* that *all*, except the most shallow, pits are properly planked and strutted. Examination of the soil *has* to be done in the bottom of the excavation or through modest gaps (about 0.5 m wide) in the planking to the sides. Convenience gained by omitting these safeguards has in the past been be paid for by lives.

If the site was previously used for potentially noxious industrial activities, such as a tannery, metal foundry, gasworks, or similar, one should check beforehand whether the soil is likely to be contaminated, so as to pose a risk to health and safety.

Findings from trial pits should be carefully recorded on sketches and colour photographs. It is good practice to retain samples of the materials found for later reference. If the findings are to be used later to convince a Building Control Officer, it is advisable to invite him to see the pit for himself.

8.3.3 Probing with Hand Tools

The hand-auger, or 'post-hole auger' has a shaft, consisting of short (about 1 m) lengths of steel rod or tube, connected with threaded joints to enable any desired length to be produced. At the top end of the shaft there is a 'T'- handle to enable it to be rotated and at the bottom end it has a short helical corkscrew-like bit, or alternatively, a hoop-like 'spoon', both of which can loosen and lift soil as the auger is 'screwed' into the ground (Fig. 8.8).

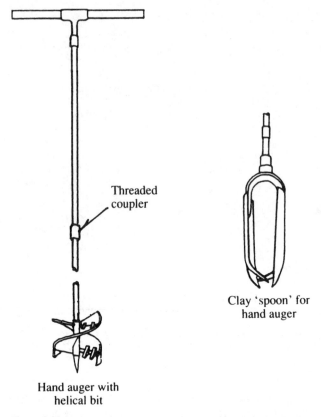

Threaded coupler

Clay 'spoon' for hand auger

Hand auger with helical bit

Figure 8.8 Hand-auger (*after Stroud and Harrington, 1985*).

With this tool, holes of 10–15 cm diameter and up to 4 or 5 m deep can be produced and the nature of the soil, brought up by the spoon, can be examined. A qualitative indication of the relative strengths of the penetrated strata is also obtained from the resistance encountered when turning the handle and pushing the auger into the bottom of the hole.

The 'Scandinavian' weighted probe is similar to the hand-auger but in place of the spoon has a conical or pyramidal point at the bottom end (see Fig. 8.9). Just below the handle a flange or cross-piece supports a standard weight (usually 50 kg). The hole is started by pushing down on the handle, the weight is then attached and the handle is merely turned and the penetration resulting from a standard number of half-turns (usually 20) is recorded at successive depths. In this way a quantitative indication of *relative* (inverse) strengths at the various depths is obtained.

Cone penetrometers are similar to the weighted probe, but have an obtuse angle cone at the tip. Some types have a smaller weight that slides between two stops at the upper end of the rod. This allows the weight to be dropped a standard height and the number of drops, or 'blows', required to produce a standard depth

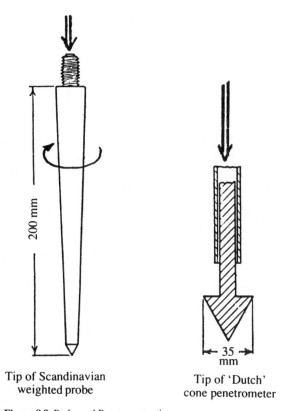

Tip of Scandinavian
weighted probe

Tip of 'Dutch'
cone penetrometer

Figure 8.9 Probe and Penetrometer tips.

of penetration at the bottom of the hole, without rotation, is recorded. Other types, mainly used for soft strata, rely on static loading. Both types give quantitative values for *relative* soil strengths and in some circumstances the tool can be calibrated to give indications of absolute strength.

The hand-auger is easily transportable and often requires only moderate physical effort to operate. It enables a fair number of holes to be bored to a depth that is significantly deeper than practicable for trial pits, over a reasonably large area, within a day. It will enable rough identification of the nature of the soil strata and, if supplemented with a weighted probe, or better still with a cone penetrometer, can give approximate soil strengths.

The disadvantages are that it only samples the soil in the boreholes, it cannot go to great depths, particularly in granular soils below the water table (the soil surrounding the hole tends to flow into it as the spoon is withdrawn) and stiff and stony soils can be too difficult to penetrate by muscle-power only. In glacial deposits, if isolated boulders are encountered, they may be mistaken for bed-rock. Hand-augers do not enable standard 100 mm diameter undisturbed samples to be taken, but smaller sizes may be possible.

8.3.4 Mechanically Drilled Boreholes

Where deep strata have to be investigated and/or undisturbed samples have to be obtained for laboratory tests, mechanically drilled holes are used. The most common machinery for this is the tripod rig, which usually folds up to become a trailer, easily towed behind a van.

The boreholes produced with this equipment are usually 15 cm in diameter and depths of 30 m are not uncommon. In granular soils the holes are lined with steel tubes; this enables samples to be taken without contamination from higher strata, using a baler just below the bottom of the water-filled liner tube. Undisturbed samples of clay, usually 10 cm in diameter, can be taken by means of special sampling tubes that can be sealed to prevent moisture loss from the sample during transport to the laboratory.

Impact, 'Standard Penetration Tests' can also be carried out; these count the number of 300 mm drops of a 50 kg weight that are required to drive a standard sampling spoon, at the end of the drill rod, 300 mm into the bottom of the hole. The number of drops, or 'blows' can be translated into bearing capacity, but some corroboration from laboratory tests is usually required, particularly for cohesive soils.

In medium to soft clays it is also possible to measure the shear strength (which is the major parameter governing the bearing capacity) in-situ, by means of the vane apparatus. This has a rod, at the bottom of which is a vane consisting of four small rectangular 'paddles', parallel with the rod; this is pushed into the undisturbed soil at the bottom of the borehole. At the top is a torsion head by which a measured torque is applied to the rod. The torque is gradually increased until the 'paddles' twist off a cylinder of soil. The torque at which this occurs is a measure of the shear strength of the soil.

The depth to the groundwater table can easily be measured in the borehole and if it is kept open, with the liner tube retained, seasonal fluctuations of the water level in granular soils can be monitored. Where there is artesian pressure in water-bearing strata below clay, it is also possible to measure this by sealing the hole at a higher level, with a so-called piezometer tube going through the seal to the bottom of the hole.

Where rock has to be investigated, heavier rotary rigs, similar in principle to the core drills for sampling concrete, are used (see Sec. 7.2.4); these extract cores from the holes, which can subsequently be examined for potentially soft, lubricating, veins, signs of slip surfaces, etc. Both types of boring are carried out by specialist contractors and the tests on the samples by specialist laboratories.

The information obtained by these techniques, is the most accurate that can be achieved, but it is still limited to the area of the borehole and that immediately surrounding it. Several boreholes, disposed in a pattern across the site are therefore required, so that the variation in thickness and properties of the various strata within the area of the building can be ascertained. The results require careful interpretation by an experienced geotechnical engineer.

8.3.5 Laboratory Tests

Results of laboratory tests can show the strength properties of samples of cohesive soils, from which safe bearing capacities can be calculated. The sensitivity of clay soils to changes of moisture content can also be ascertained, and this will indicate the behaviour in case of possible changing site conditions, such as, for instance, new drainage.

Samples of soil and groundwater can also be analysed to ascertain the presence or absence of aggressive substances such as sulphates, or organic acids, that may attack mortar or concrete in the foundations. It is even possible to ascertain the origin of the water: traces of chlorine indicate a leak from a water main, certain organic matters suggest a leaking sewer, etc.

It is also possible to test the compressibility of soil samples. The results of such 'consolidation tests' on samples, taken from strata at successive depths below the surface, are normally used to calculate the probable settlements that a new building may suffer. In simple terms, this is done by calculating the vertical stresses, due to the weight of the building, at the depths of the strata from which the samples were taken and then multiplying the corresponding tested compression strain of the respective samples by the thickness of the strata.

In special circumstances the results can however be used to 're-construct' the settlement history of an existing building, if its weight at various phases of building can be estimated and its foundation geometry is known. This was done in the case of the central tower of York Minster as an aid to explaining the causes of the measured differential settlements and the observed superstructure distortions: The plan shape and depth of the foundations had been established by exploratory excavations. The superstructure geometries and hence the weights of the various parts: tower, nave walls, etc. of the earlier phases of construction had to be estimated, partly by reference to the features of similar churches from the eleventh and twelfth century, as neither remains nor records were available.

This enabled a graph to be produced, showing settlements as a function of time, and, despite the sketchy nature of some of the information, a remarkable correspondence with the observed differences in levels was obtained (see Fig. 8.10).

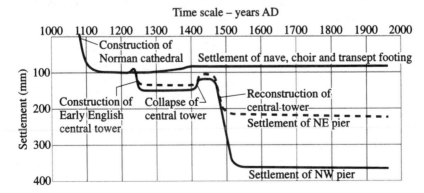

Figure 8.10 York Minster: Reconstruction of settlement history (*from Dowrick and Beckmann, 1971*).

8.4 PRINCIPAL METHODS OF FOUNDATION STRENGTHENING

When it has been established beyond doubt that the only (or the best) way of remedying the defects of a building is to strengthen its foundations, there are a number of techniques available. Each of these has its particular field of application, its advantages and disadvantages, and all of them involve a certain amount of risk. These factors have to be carefully weighed up, before the final decision is made. Particular care should be taken *not* to accept the guarantees offered by contractors, operating proprietary techniques, without reading 'the small print' which may invalidate the guarantee, if certain unforeseen conditions are encountered during the work. The advice of an engineer with specialist geotechnical knowledge *and* with experience in foundation strengthening, as well as familiarity with the structural characteristics of old structures, should always be sought, regardless of the method envisaged.

8.4.1 Soil Improvement by Grouting

When foundation failure or inadequacy is due to loose-bedded gravel or cavities caused by washing out of fine sands (such as may happen due to bursting of pressurized water mains), it may be possible to strengthen the soil and/or fill the cavities by grouting. This is done by ramming or drilling (usually vertical) pipes into the ground, so that their lower ends are located on a (usually square) grid, covering the area to be treated. Sometimes the pipe ends are arranged at two or more levels to enable sufficient thickness of stratum to be treated. Grout is then injected under pressure, through the pipes into the ground, where it sets and, by filling the voids and 'glueing together' the soil particles, increases the bearing capacity of the stratum.

Grouting can be carried out either with cement-based grouts or with chemical grouts.

Cement-based grouts may contain cement replacements, such as pulverized fuel ash (PFA) or ground granulated blast furnace slag (GGBFS); these have finer particles than the cement and they slow down the rate of hardening, both of which improve the penetration of the grout. Admixtures of various kinds may also be used for the same purposes.

Chemical grouts usually consist of two solutions which, when separate, remain liquid, but which, when combined, react to form a solid substance. Being solutions, chemical grouts have greater penetrating power than cement-based grouts (which are suspensions). They require strict observance of mix proportions and mixing procedures and some have to be injected as separate liquids in a prescribed sequence. They are by their nature more expensive than cement-based grouts.

The feasibility, procedure and materials to be used in soil grouting depend not only on the nature of the soil, but also on the groundwater regime. The results of a thorough site investigation should therefore always accompany enquiries to contractors, and only those who have experience in this type of treatment should be approached.

In order to obtain penetration, grout has to be injected under, sometimes considerable, pressure. This means that the technique cannot be used near the surface, as the pressure would here merely 'blow off' the top of the soil before the grout had penetrated. Grouting with excessive pressure can also lift parts or the whole of buildings with undesirable results.

Conversely, special, dedicated, high-pressure 'compensation grouting' procedures, with carefully designed injection patterns and grouting pressures, are sometimes used to counteract potential damaging settlement that might be caused by tunnelling. This requires the ground movements to be carefully monitored, so as to keep the grouting 'in step', as the tunnelling progresses.

Soil grouting generally requires access for plant and personnel both inside and outside the building. As it involves mixing and pumping on site, spillages are almost inevitable.

There is also no simple way of knowing where the grout may flow, once it emerges from the injection pipes. Precautions *must* therefore be taken to check that nearby sewers and cable ducts (with possible leaky or cracked joints) are not being filled with grout and that adjacent buildings are not being affected.

8.4.2 Underpinning

Underpinning is often used as a common description of any process that carries loads down to a greater depth than that of the original foundation. For the purposes of this section, it will however be used in its narrower sense of putting a deeper foundation under the original strip footings, so as to transfer the bearing pressure to a deeper stratum.

The procedure for continuous underpinning of a strip footing is generally as follows (see Fig. 8.11):

1. A narrow (about 1.2×1.2 m) pit is excavated alongside the original foundation, down to the desired depth.
2. From this pit, a heading (short tunnel) of the same width is excavated under the original footing across its entire width, down to the desired depth.
3. Vertical formwork is erected across the heading.
4. Concrete is poured into the heading, behind the formwork, so as to fill the space and create a new deep foundation block between the original underside of the footing and the new, excavated, bearing surface.
5. When the concrete has hardened, steps 1–4 are repeated at another location along the footing, leaving an undisturbed space equal to one, or preferably two, pit widths between the first and the second block.
6. When the whole length of the original footing has been completed in this hit-and-miss way, the gaps between the new, deeper, foundation blocks are then treated as above, keeping the width of the headings to about 1.2 m (Fig. 8.12).

In step 4 there is the problem of ensuring that the new concrete is in full contact with the underside of the original footing. This can be overcome by stopping

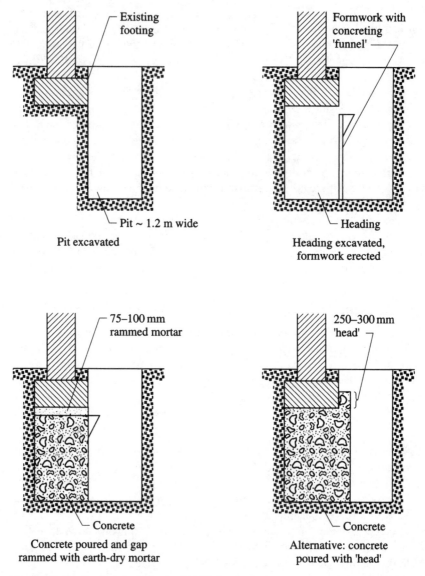

Figure 8.11 Construction stages for one 'block' of underpinning.

the concrete some 75–100 mm below the final level and filling the gap by ramming in an earth-dry mortar after the concrete has set. (Alternatively, the formwork is erected some 150 mm away from the face of the footing and taken up above the interface, so as to enable the concrete to be poured with a hydraulic head to force it up against the old bearing surface.)

The pits and the first group of headings will, for health and safety reasons, usually need planking and strutting (the sides of later headings are formed by the

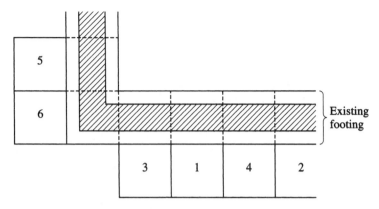

Figure 8.12 Underpinnig 'block sequence' plan.

concrete blocks of the first groups). The underside of the old footings may need temporary support.

For pad footings a similar hit-and-miss procedure of excavation and concreting may sometimes be followed. This can be risky, due to the small linear extent of such foundations. For lightly loaded piers or columns it may be possible to provide temporary supports, but these may suffer settlement themselves and 'minipiling', as described in Sec. 8.4.4, is usually a better option for piers and columns on pad footings.

A variation on the direct underpinning is to form isolated blocks going down to firm bearing either side of the wall (or pier) and connect the blocks by beams, going under, and supporting, the original footing. This 'indirect underpinning' is very similar to the use of piles and pile-cap beams, described below (see Fig. 8.14), and shares with that some of the disadvantages, such as the need for access to both sides of the wall (there is no equivalent to the tension piles).

The advantages of direct underpinning are:

1. It needs access from one side only and can therefore, for external walls, be carried out without disturbance of the interior of the building.
2. It preserves the original footing more or less intact. (If the original foundation masonry is of poor construction, stones or bricks in the bottom course may however drop off, if temporary support is not provided.)
3. It requires no special machinery.

One disadvantage of underpinning is that it generally does not increase the bearing area and hence *does not reduce the bearing pressure*. On the contrary, the bearing pressure is increased by the difference in weight between the new concrete and the soil, which it displaces.

This has significant consequences because the vertical stress (ground pressure) from any footing is dispersed with depth (see Fig. 8.13), so that *before underpinning*, the intensity of the vertical stress at the deeper level is significantly smaller than that at the underside of the original foundation. Underpinning transfers the load from the

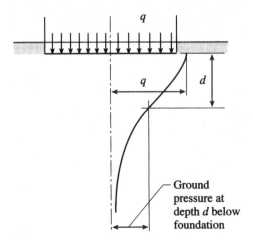

q

q   *d*

Ground
pressure at
depth *d* below
foundation

Figure 8.13 Variation of vertical stress with depth below foundation.

underside of the original footing, where the soil has been consolidated by the pressure, to a deeper level, which has hitherto been only lightly stressed, and this increase in stress then causes the deeper layer to consolidate with consequential settlement.

Underpinning will therefore, in principle, always cause some, usually slight, further settlement. It should, therefore only be used if the strength and stiffness of the deeper strata are better than those at the level of the underside of the original footings.

A further disadvantage of underpinning is that it involves fairly large volumes of excavated soil. A constructional limitation of underpinning is that the depth to which it can be taken, and hence the depth of good bearing strata that can be reached, is limited by the practicalities of safely supporting the sides of the excavated pits, so as to ensure the health and safety of the workmen, and also of raising the excavated material to the surface. Underpinning has nevertheless a long history of largely successful applications and, given favourable conditions, i.e. soil that significantly improves with depth in carrying capacity (and/or has better volumetric stability under fluctuating moisture regimes at greater depth) and can stand without too much planking and shoring of the excavation pits, it can be a very effective method of foundation strengthening.

When a basement has to be constructed immediately adjacent to, and deeper than, the foundations of an existing building, underpinning can also be used to support the existing building at, or below, the level of the new basement floor. In that situation, the projection of the existing footing into the adjacent basement space is cut off. If the depth of the new basement is too deep for a single under-pinning, the process may be repeated, e.g. a first underpinning down to, say, 2.5 m, followed by basement excavation to 2.5 m, then a second underpinning of 2.5 m and finally basement excavation to 5 m (this presupposes that the soil at the depths of the intermediate and of the final underpinning can support the load from the existing building when transmitted through the reduced width).

Underpinning should normally be carried out to all of the foundations of the building. Partial underpinning can lead to the situation that the original and the underpinned foundations respond in different ways to subsequent ground movements, such as may be caused by climatically induced variations of the moisture content of the soil. This has, in several cases, resulted in damage to the superstructure.

8.4.3 Foundation Strengthening by conventional Bored Piles

When ground conditions do not allow underpinning; either because the depth of good bearing strata is too great, or because the ground does not allow safe excavation of the underpinning pits, foundations can be strengthened by the use of bored piles (Fig. 8.14).

For a wall, the method in its basic form employs pairs of piles, with one pile either side of the existing footing. The piles of each pair are connected by a beam, constructed in a miniature heading or 'tunnel' under the footing. This beam transfers the load from a length of the wall, equal to the spacing of the pairs, to the tops of the piles. This layout is the most economic in terms of piles, as each pile only has to carry about half the load of the length of wall between pairs of piles. It does however involve work on both sides of the wall. The beam can be of reinforced concrete or a steel H-section, subsequently encased. The latter option reduces the health and safety risks in working under the old foundation, and it

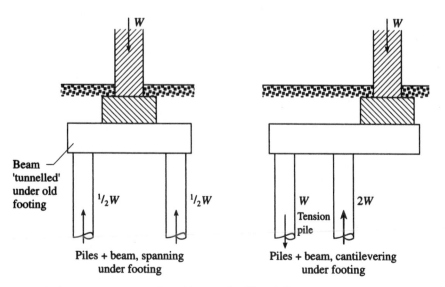

Figure 8.14 Foundation strengthening with conventional bored piles.

allows wedging, or jacking, to pre-load the new footings and counteract the effects of the beam deflecting.

Where access to both sides of the wall cannot be permitted, it is possible to have both piles of a pair on one side and to cantilever the beam over the pile, nearest to the wall, extending it under the existing footing. In this case, equilibrium requires that the pile furthest from the wall will have to act as a tie-down, and in consequence the load on the pile nearest to the wall will be greatly increased.

For instance, if the distance from the central plane of the wall to the nearest pile is equal to the distance between the piles in a pair, then the outermost pile will have to take a tension equal to the weight of the length of the wall supported by the pair, and the pile nearest the wall will have to carry a load equal to twice the weight of the same length of the wall. This means that both piles of each pair will have to be substantially longer and/or of greater diameter.

Whether carried out from one side or both, *foundation strengthening by piles has to be on an 'all or nothing' basis* (unless the untreated part of the building is directly founded on rock). If this is not done, the untreated part of the building will move up and down with subsequent ground movements, caused by variations in the groundwater regime, whilst the piled part will be on a 'hard spot' and hardly move at all; the result being superstructure damage. (An exception may be permissible, if there is no risk of heave. The piles may then be designed to allow slight settlement of the piled portion of the building, at the same rate as that of the untreated part. This is, however, a very delicate balance to strike.)

Both of the pile layouts, described above, require the wall to span between the pile-cap beams and these have, for practical and economic reasons, to be some distance apart, when conventional bored piles, of 350–500 mm diameter, are used. The pile-cap beams also have to be 'tunnelled' under the wall in question and this can, as mentioned above, be a hazardous operation. In the case of delicate superstructures, it should be borne in mind that some types of piles are formed, by dropping a heavy 'clay cutter' into the bore. This may cause ground vibrations.

8.4.4 'Minipiling'

To overcome the problems in the use of conventional piles and beams, a special type of piling was developed, initially by the Italian firm of Fondedile S.A. The process is now available from a number of specialist contractors.

This method uses special machines, resembling oversize core drills, to drill inclined holes, of a diameter between 170 and 250 mm, through the existing masonry, starting above ground if necessary, and through the footings and the soft soil layers, into sound strata. The holes may have liner tubes of steel or plastic where the soil strata or the groundwater conditions make it necessary. Reinforcement is introduced into the holes, which are then grouted or, for the larger diameter piles, filled with concrete, whilst any liner tubes are

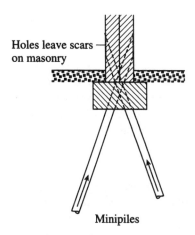

Holes leave scars on masonry

Minipiles

Figure 8.15 Foundation strengthening with 'minipiles'.

simultaneously withdrawn. In this way piles are formed that bond into the masonry (Fig. 8.15).

There are several advantages of this method:

1. The piles, being smaller than the conventional bored piles, can be positioned closer together and so do only require the wall to span very short distances.
2. No pile caps nor pile cap beams are required.
3. For pad footings to isolated piers or columns, no temporary undermining, such as is necessary for underpinning, is needed.
4. The piles with their reinforcement are grouted into the fabric of the existing construction and so help to bind it together. The grout may also penetrate soft patches and thus help to consolidate the ground.
5. The machines, unlike ordinary bored piling rigs, are generally quite compact and can be moved into, and operate in, cramped spaces like basements with low headroom. They can core through most materials, except timber.

There are however also real and/or potential disadvantages:

1. In order to neutralize the horizontal components of the inclined pile reactions, the piles have to be arranged symmetrically about the central plane of the wall and drilling therefore has to be carried out from both sides of the wall or pier. For small buildings, imposing only moderate loads on the piles, one-sided drilling is sometimes used; this requires the soil to resist the resulting horizontal forces and the piles may be subjected to bending in addition to the axial load.
2. Where the holes are started above ground level, they will disfigure the original masonry and durable restoration of the face, by patching with 'sawn-off' bricks, slip tiles or cut stones, can be difficult.

3. If the original masonry is poorly bonded, it may be doubtful if it can withstand the stresses from the load transfer, without bursting. In this case, some preceding 'cross-stitching' and grouting may be required.
4. Some operators tend to use water jetting to progress their liner tubes through the soil. This can, in some conditions, cause washing out of fines from under the foundations to be strengthened, thus causing further settlement of the building being treated. Adjacent buildings may also be affected. If the piles are being formed by grouting, there is the additional hazard of grout finding its way into adjacent sewers, etc.

To avoid any problems occurring, once the work has commenced, and particularly those arising from jetting, the advice of an *independent* civil engineer with experience of this type of foundation strengthening, should be obtained *before* any contract is entered into.

The 'all or nothing' rule, as stated for conventional piling, applies equally to minipiles.

8.4.5 Foundation Strengthening by Lateral Extensions

If problems of inadequate capacity or differential settlement are due to the ground pressure being too high under part of, or under the whole of the foundations, a viable option may be to increase the bearing area of the footings by extending them horizontally (Fig. 8.16).

This entails forming concrete blocks either side of the existing footing, with their bearing areas level with the existing one, and connecting the new concrete to the old structure in such a way that the load can be shared between the two.

This load-transferring connection is effected by 'bolting' the concrete blocks to the old footings, using steel rods having threaded ends with nuts and plates which bear on the outside of the concrete. The rods are placed in holes which are drilled through the existing masonry or brickwork, from side to side, and which extend as formed ducts through the concrete blocks. The rods are placed in the lower part of the composite concrete/masonry foundation and prevent the concrete

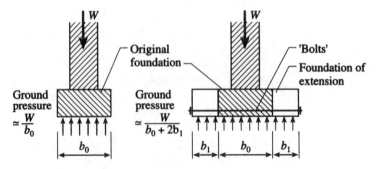

Figure 8.16 Reduction of ground pressure by lateral foundation extension.

blocks from moving away from the masonry. The upward forces of the ground pressure on the underside of the concrete blocks create, by lever action, a clamping force between the upper parts of the concrete blocks, so that part of the load from the superstructure can be transferred across the interface by shear and friction. To this end, the joints of the masonry/brickwork faces are often raked out, before the concrete is cast against them.

To provide the best possible load transfer capacity, the rods are usually tensioned, either by tightening up the nuts, or by 'pulling' the rods with a pre-stressing jack and running up the nuts (Fig. 8.17). This results in the concrete being pressed against the masonry over the whole interface, whilst in the case of unstressed rods, a slight gap would develop at the lower part, due to the elastic extension of the steel.

The rods have to be protected against corrosion; this can be done by wrapping them in waterproof, self-adhesive, tape. In that case it may be possible to de-tension the rods to inspect them and, if necessary, replace them, one by one, at a future date. The plates and nuts are usually recessed into the concrete, so that they can be protected by a substantial mortar covering.

Where the longest maintenance-free life is required, rods, nuts and plates are made of the most corrosion-resistant grades of stainless steel and the holes, in which the rods are placed, are pressure grouted as in post-tensioned concrete. This has the added advantage that bond develops between the rods and the surrounding masonry and concrete, particularly if continuously threaded rods are used. This aids the anchorage of the rods and reduces the risk of sudden failure in case of anchorage fracture. The disadvantage is that inspection and/or replacement would require the rod to be freed by concentric core drilling along it.

No footing does however carry load without settling: the new concrete blocks will simply act as so much dead weight until the whole composite foundation has undergone further settlement, to consolidate the soil and thus mobilize sufficient ground pressure under the concrete blocks to enable them to carry their share of the load. This may result in some further, although limited, damage to

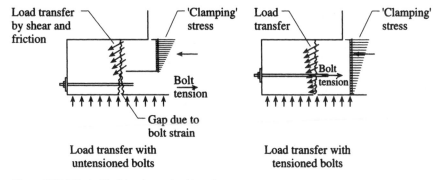

Figure 8.17 Effect of bolt tension on load transfer.

the superstructure. Even where this might be permitted, the process may on clay soils take a long time, during which final reinstatement of superstructure and finishes cannot proceed.

The remedy for this is to pre-load the soil under the new foundation extensions. This is done by casting the concrete blocks in two parts, separated by a small horizontal gap: the upper, thicker, part is bolted to the existing footing as described above; the lower part is separated from the existing footing and from the upper part of the block by suitable joint fillers. In the horizontal gap between the two parts, hydraulic flat-jacks are placed. (As explained in Sec. 4.3.4, flat-jacks are capsules (usually circular), that are originally dumb-bell shaped in cross-section. When pressurized with hydraulic fluid, they can exert very great forces—up to 15 N/mm² (1 ton/sq.in.) of their plan area. They are capable of expanding up to 25 mm at right angle to their plane.)

When the upper concrete blocks have been bolted or stressed to the original footing, the flat-jacks are pressurized; this exerts a downward force on the lower part of the blocks (commonly known as 'pressure pads') which forces them into the soil which is pre-loaded in this way (Fig. 8.18). The pressure in the flat-jacks, and hence the force they exert on the pressure pads, is adjusted to correspond to

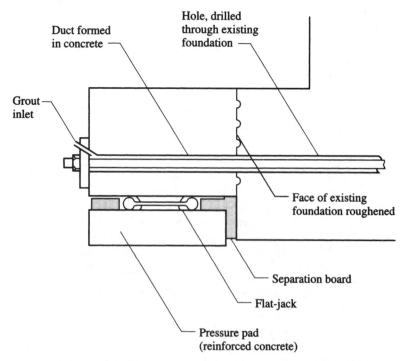

Figure 8.18 Details of pre-stressing of foundation extension and of pre-loading of soil beneath, by means of flat-jacks.

the proportion of the total load that is intended to be carried by the new concrete foundation extensions.

As the pressure pads are 'free floating' between the flat-jacks and the soil, the same proportion of the load will be carried by the soil under the pressure pads and hence the intensity of ground pressure under the combined footing is reduced by the desired amount without having to wait for the soil to consolidate.

On clay soils the pressurization of the flat-jacks may have to be carried out in stages. Each pressure increment then has to be maintained for some weeks, in order to allow the pore water to disperse and thus prevent failure due to excessive (although temporary) pore water pressure with its consequential reduction of the effective shear strength of the clay.

The first large-scale application, in Britain, of this method was carried out on the footings to the Central Tower of York Minster. The 16 000 tons fifteenth-century tower had been built on the footings constructed in the eleventh century for the original, much lower, central crossing. As a consequence, the ground pressures under the tower pier footings were averaging 530 kN/m^2, with a peak value of about 750 kN/m^2, whilst for the surrounding piers of nave, transepts and choir the pressure was 290 kN/m^2 (even 290 kN/m^2 would today be considered fairly high on the clay in question, but the slower mediaeval building process allowed a large proportion of settlement to occur during construction, and before the lime mortar had hardened).

Resulting differential settlements of up to 305 mm were measured between the central tower piers and the surrounding piers, and up to 115 mm differential between the tower piers themselves. Observations of cracks in the superstructure had furthermore shown that there was continuing movement: for instance, a 'new' sill to one of the lancet windows, installed in the 1930s had developed a 6 mm wide crack by 1966.

The scheme, finally adopted (see Fig. 8.19), added concrete blocks, triangular on plan, in three of the re-entrant corners of the strip footings of the original crossing. This enabled the centroid of the total foundation area (existing plus new) to coincide with the position of the resultant load from the pier and thus give uniform pressure under the combined foundation. The concrete blocks were stressed against the masonry by stainless steel rods, running diagonally, in two directions, under each pier. A special alloy, giving enhanced corrosion resistance and high strength, was used for the rods.

The lower pressure pads were 0.6 m thick and lightly reinforced; the main pre-stressed blocks were 2.1 m thick and reinforced at all exposed surfaces. In the gap between the two layers of each triangular block, seven or in some cases eight flat-jacks of about 350 mm diameter were positioned, protected by pre-cast concrete inverted troughs. (The number of flat-jacks depended on the plan-size of the triangle.)

To compensate for losses of pre-stress, due to compressibility of the masonry and relaxation of the rods, these were re-stressed two to four weeks after initial stressing. The losses were very small, due to the low compressive stress (0.86 N/mm^2), induced in the foundation masonry.

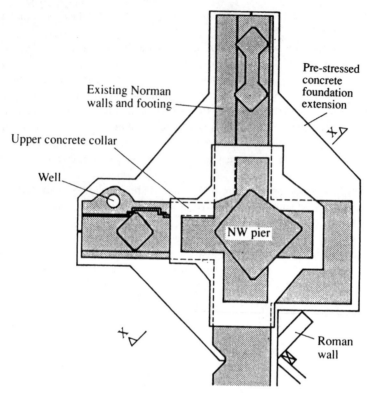

Pre-stressed concrete foundation extension

Existing Norman walls and footing

Upper concrete collar

Well

NW pier

Roman wall

Plan of NW pier foundations

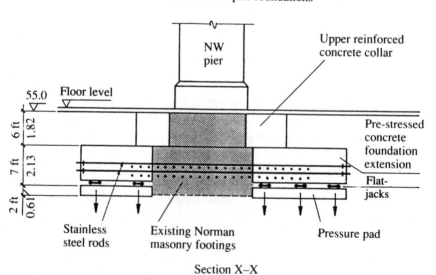

NW pier

Upper reinforced concrete collar

55.0   Floor level

Pre-stressed concrete foundation extension

Flat-jacks

Stainless steel rods

Existing Norman masonry footings

Pressure pad

6 ft   1.82

7 ft   2.13

2 ft   0.61

Section X–X

Figure 8.19 York Minster foundation strengthening (*from Dowrick and Beckmann, 1971*).

The flat-jacks were initially inflated to a pressure of 3.5 N/mm^2, corresponding to a ground pressure of 90 kN/m^2 and this pressure was gradually raised to 10.5 N/mm^2, initially in steps of 3.5 N/mm^2, but in the later stages by only 1.4 N/mm^2. At each stage, as the pressure pads moved down into the clay, the pressure dropped, but was 'topped up' as necessary until movement ceased. The final pressure, which was only attained after some 20 weeks, corresponded to a ground pressure of 250 kN/m^2.

CHAPTER
NINE

SEQUENCING OPERATIONS AND TEMPORARY WORKS

Qara Clise, Azerbaijan: The walls of this church were leaning out. Pending permanent remedial works, timber raking shores, on substantial concrete plinths, were installed to stop or slow down the movement. The two right-hand ones are properly braced horizontally, as well as vertically. The left-hand one may buckle sideways, if it has to resist a great force. Note the newer relatively flimsy steel tube access scaffold.

When a new building is to be constructed on a 'green field' site, the sequence of operations is usually fairly uncomplicated.

The client's requirements are stated, often in very general terms only. The architect then designs what the client never knew that he or she wanted. In doing this, the architect is nowadays usually aided and abetted by the structural engineer and one or more services engineers, so that the design not only gives delight, but also has a chance of providing firmness and commodity.

The design, when modified and agreed by the client, is worked up into contract documents. These form the basis of various contractors' tenders, from which the client usually chooses the lowest, or perhaps the one promising the quickest completion.

The chosen contractor then starts on site, provides what he considers the best temporary works for his purposes and completes the building more or less on time and more or less within budget.

In all of this, the only constraints on the design team are planning and building regulations; in addition, the structural engineer has to allow for ground conditions when designing his foundations. The contractor is usually free to do as he likes within the site, subject to health and safety regulations.

When work is to be done on an existing building, life is no longer so simple. The shape and size of the building may impose limitations on what the client can have. Its structure may limit the amount of alterations that can be made. In addition to the regulations mentioned above, the building may be listed, resulting in further limitations for client and design team. Floor-to-ceiling heights may circumscribe what services engineers can do and the structural engineer may have to worry about the loadings from the proposed new use. Worst of all, many of the physical limitations may not be known at the outset and require investigations, some of which may not be practicable before work starts.

The contractor is not much better off. The site *is* just the building, plus, if he is lucky, a backyard where he can put up his shed, so he has to modify his usual site set-up accordingly. The temporary works, that normally would seem obvious, may damage essential features, that have to be preserved, so they cannot be allowed, but have to be replaced by something else, and sooner or later something is almost bound to crop up, that he did not allow for in his tender.

All this means that the sequence of operations and the temporary works, in their widest sense, need special consideration when working on old buildings.

9.1 CONSTRAINTS ON OPERATIONS

Success in conserving historic buildings depends on:

- The client being aware of the special aspects of such projects, and how they differ from, say, the procurement of a new office building on a green-field site.

- The design team having experience to move the project forward in the pre-contract stages so that adequate investigations are undertaken and they can obtain all the necessary consents in time, so as to have the knowledge that the surprises on site will be limited and manageable.
- The contractor having experience, a skilled workforce and the resources to plan ahead and cope with the unexpected.
- There being good communication between the client, the design team and the contractor. Each needs to be aware of the roles and responsibilities of the others.

9.1.1 Constraints on Design and Investigations

On any building project, the design team works to a brief comprising the client's instructions as to what he wants from the building. When the project concerns an existing building, and if it is to be successful, the brief must include provisions that the work should be compatible with the existing fabric. Furthermore, in the case of a historic building, its architectural or historic features must be respected. This will generate some constraints—that is requirements that tend to preclude following what at first sight seems to be the easiest, cheapest, quickest route to achieving some aspect of the project. The constraints may be tangible and structural, resulting from the characteristics of the existing structure—or they may be historical, where parts of the construction, that might otherwise logically be removed during the course of the works, have to be preserved. There is an obligation on the design team to produce designs that can be implemented within the constraints, and the only effective way to demonstrate this is to illustrate both the constraints and the designer's assumptions about how the work would be undertaken on drawings or sketches.

The results of this exercise will then be clearly available for tenderers, and they can use the limited time available in the tender period to devise their method of carrying out the work and costing it realistically. Such drawings and specifications are also useful to re-assure owners, funders and those granting the various consents that what is proposed is feasible. They are also a most useful tool to anticipate what otherwise might crop up as a nasty surprise during the construction phase.

When this step in the design process is missed out, either because its importance is not recognized or in an attempt to save money by limiting design and investigation costs in the early stages, there are normally problems at the construction stage. Costs escalate, programmes slip and far more historic fabric gets destroyed and has to be reconstructed than should have been necessary.

There will always be some things that could not have been foreseen, but surprises that lead to major changes in the design while jobs are on site are best avoided.

Apart from the obvious restraints on design imposed by planning and listing, there will be limitations inherent in the nature of the building. For example, the existing floor construction may not allow archive loading, or removal of partitions to create 'open plan offices' may jeopardize stability. Sometimes the significance of the existing construction will be apparent from a fairly casual 'walkabout', but some significant points will only become apparent after a thorough inspection. Such inspection involving removal of floor boards, etc. may however not be allowed during the early design stage, if the building is still occupied. Previous experience on buildings of similar age and type may help to 'guess' what may be found, and thus inform later stages of design, but confirmation by inspection is essential. For instance, buildings from the middle of the twentieth century may or may not have asbestos-lagged central heating pipes; if this has not been allowed for and it is suddenly found during the works, the contractor will suffer a delay whilst the specialists have to be called in and the asbestos removed.

It should also be remembered during the design stage that any operation that may disturb or destroy bats' roosts or resting places, e.g. within a roof space, needs a licence; this will only be granted if the Department for Environment, Farming and Rural Affairs is satisfied that there is no alternative satisfactory course of action.

Other constraints may be less tangible at first, arising, as they sometimes do from the historical significance of some of the features of the building. It is vital that the structural designer understands the historical significance of the fabric as a whole at scheme stage in the design process. The architect will of course need this information as well, but not necessarily all that the engineer needs. For example, if the design brief relates only to the top two floors of a building, the architect may, quite understandably, not assign a very high priority to finding out details of the construction two floors below, nor whether the ceiling immediately below deserves special measures to protect it from damage. This information could however be fundamental to determining an appropriate sequence and temporary works. For example, there is no point in devising schemes for back-propping the floors of a building down to the basement if there are valuable painted plaster ceilings on intermediate levels that would be damaged by propping.

Inspections may sometimes need more resources than are available to a design team, even when assisted by a local builder, and therefore have to wait until a contractor is on site. For example, the full investigations of the foundations of York Minster required the moving of such quantities of soil that the contractor rigged up a miniature 'railway track' to take out the spoil by tipper wagons. In such circumstances, the pre-contract design must have enough leeway to accommodate the consequences of the later investigation.

The condition of the existing structure may also mean that some investigation has to be deferred. This can apply where it is suspected that there may be defects in hidden structural details but that cannot be verified without further investigations that might further reduce the stability. An example of this was a seventeenth-century house with timber floor joists spanning between timber beams. The walls and the beams were panelled. The building had distorted due to problems in the ground but there was also a suspicion that some of the stone corbels, on which the beams were supported, had fractured and that the panelling carried some of the beams. It was decided not to investigate further until a contractor was on site with resources to deal with the unexpected. This proved to be a wise decision, since some of the beams were indeed found to be supported by the panelling.

9.1.2 Constraints on Procurement

There is a tendency today for clients to want fixed-price tenders for construction. Given full and final information at tendering stage, this should in theory not cause any problems on new-build, when a competent contractor would know what risks to expect and price them accordingly. As clients and architects are, however, prone to changes of mind, these could involve the contractor in extra costs, which he will then try to recoup—a fruitful field for disputes!

As indicated above, when working on an existing building, the design team may, with the best will in the world, not know for certain what may be found, once work has started. This can have profound consequences for the procurement. Strategic decisions about budgets, overall programmes and the selection of the contractor inevitably often have to be made before all the details of the existing building are known. Informed judgements and assumptions based on experience are required, and it is vital to the whole process that these are clearly recorded and communicated between the various parties. Where records of the existing structure are limited and access for investigation is difficult, it is important that any partial or step-by-step investigations generate information relevant to determining the best sequence of construction and temporary works schemes.

The relationship between permanent works, sequence and temporary works is particularly critical to this process. The sequence of work and the temporary works, necessary for carrying out the work play a vital part in securing success. When things have gone wrong it is often because of the way the work was being carried out, not the work itself. This whole subject needs to be considered from the inception of the design, and carried through in the contract documents, although in all normal contracts the responsibility for ultimately devising and providing the means of implementing the design rests with the contractor. The reason for this is that there is usually more than one possible combination of plant and operations, which can be used to deal with any particular problem. The contractor, and nobody else, will know what plant he has available, or can procure, for this particular contract and how he can use it to carry out his operations, so as to provide a technically and economically sound solution to the problem.

The umbrella term 'temporary works' is often applied to this part of the design and construction process. This is misleadingly simple, as this is often a complex process of planning and sequencing work, with many strands to it.

9.1.3 Constraints on Construction

The contractor, working on an existing building, may face many restrictions on his activities. The site may impose its own limitations. If it is a stately home in the country, there will usually be a less well-kempt area that can be used for site huts and materials storage. If, however, it is a property in an urban street, fronting straight on to the pavement with other houses adjoining both flank walls, it is a different story. If in luck, the contractor may find a small back yard for some of his stores, but everything will have to be carried through the house. If it is a narrow street, loading and unloading will only be allowed at certain hours, etc.

In one instance, the site constraints dictated the entire project. During an internecine warfare between two, politically opposed, factions of emigrées from a 'far away' country, a bomb was exploded in the basement of one of two adjoining houses on a busy thoroughfare in London. The damage to both houses was considerable, the party wall in the basement had a 2.50 m wide section blown out and the only parts that would still be serviceable after (extensive) repairs were the brick walls, some partitions and parts of some floors. The buildings were of no historic or architectural merit, and the owner would have preferred to entirely demolish the interiors and start again, paying the extra over the insurance and get one modern property (on which he would probably be allowed an extra storey). However, this was at a particularly narrow part of the road, so no parking, loading or unloading was permitted between 7 am and 7 pm. There was a prestigious hotel situated just opposite, so no noisy activities were permitted between 10 pm and 6 am. There had been a backyard, but that had been roofed over and formed part of the shop premises, and the roof of this was fairly intact. The problems of demolition and disposal of rubble were such that repair, rather than re-build was decided upon.

The building itself may impose other constraints. It may have historic decorative finishes that have to be protected. Some of these may be susceptible to changes in air humidity, in which case window openings have to be sheeted with plastic. Floor finishes may have to be protected with hardboard, so as not to be damaged by workmen's boots, etc.

When scaffolding the outside of a building, the significance of the glass in the windows must be established. Important stained glass may require protection and be left in position, or be removed as a whole by specialists. Where the glass is plain and modern (e.g. where it was replaced after war damage), it may be perfectly acceptable to remove or even break individual panes, especially if this allows scaffold ties to pass through and avoids the need to drill restraining fixings into the external masonry.

9.2 TEMPORARY WORKS

'*Temporary works*' includes everything that a contractor constructs on site during the contract that is then either removed or becomes redundant on completion of the project. It includes all the site facilities, security protection, fire precautions, protection of the fabric from the weather and from impact damage, means of access for inspections and for construction work; also the more obvious shoring and strutting that may be required to compensate for stability that will be temporarily lost during the construction process, or to enhance the stability of a structure to withstand the additional loads which it experiences during works.

The activities and constraints that require temporary works are:

1. Safety legislation, etc.
2. Access for inspection
3. Provision of site facilities
4. Protection
5. Access for working
6. Demolition and structural alterations
7. Excavations
8. Interim stabilization.

9.2.1 Safety Legislation, etc.

The Construction (Health, Safety and Welfare) Regulations 1996 (Statutory Instrument 1996 No. 1592) applies to all works on site covered by this chapter. The regulations will no doubt be amended from time to time, and everybody involved should check up on the latest versions.

9.2.2 Access for Inspection

The principal risks, when inspecting a building, are of falling from a height or becoming trapped or incapacitated in a confined space. Most old buildings (as well as some fairly recent ones) were designed with little thought for such matters. Roofs with low parapets and crawl spaces with low headroom are commonplace. Thus even where the means of access exists, it may not be acceptable to use it under the current regulations.

Many buildings have had handrails installed to make access to roofs safe, but there can be difficulties in doing this in historic buildings, where the visual impact of a handrail above a parapet would be unacceptable. In these cases, the handrail may be set back from the edge of the roof, leaving an area to which access is not permitted, or there may be a wire system for use in conjunction with a harness. Harness systems need to be maintained and the wire fixings need periodic testing. It is important to make sure this has been done, that the equipment is complete and properly fitted, and that the method of operation—particularly how to get past the points where the wire is fixed to the structure—before using such systems. They are not designed to prevent falls, but merely to stop people from falling far enough to injure themselves.

One consequence of the coming into force of the 1996 regulations is that there is now available a vast variety of access equipment designed to comply with the requirements. The ladder and step-ladder still have a place, but for most purposes the simplest acceptable equipment is a light-weight aluminium tower, which may be fitted with outriggers and be on wheels. For a tower to be safe, there must be a suitable flat area immediately below the point where access is required.

Where there is no level surface below the point to which access is required, hydraulic mobile platforms can be the answer. These can either be mounted on a vehicle or have their own independent wheels. Some have a very narrow wheelbase when stowed for travelling, so they will fit through ordinary doorways. When in operation, they derive their stability partly from their self-weight and partly through extending out-riggers. They can be quite heavy, and the intensity of loading can be greater when they are rigged for travel than when they are in operation. A check should be made to see that floors will not be overloaded during delivery and operation. Before selecting a platform it is important to understand exactly what access it can provide. Some cannot achieve their maximum reach at all heights.

Where towers and mobile platforms cannot provide the necessary access, conventional scaffolding can be the answer. Another solution may be to employ specialist abseilers. These are trained rock-climbers, who use their specialist skills to set up ropes from which they can suspend themselves and operate 'hands free'. Using video cameras and two-way radios they can operate to the instructions of an engineer watching a monitor at ground level, and the whole proceedings can be recorded. They can also set up boatswain chairs so that anyone with a head for heights can be hauled up to look for themselves.

9.2.3 Provision of Site Facilities

Statutory Instrument 1592 lays down requirements for welfare facilities that are more comprehensive than were formerly commonly provided. Finding room for these can be a major challenge on constricted urban sites. Sometimes it is possible for facilities to be located within the building itself, in a part of the site that is not to be intensively worked on.

Where this is not possible and the site is very constricted, areas within the curtilage of the building such as flat roofs will need to be considered as sites for cabins. Sufficient information about the existing structure will be needed to judge whether cabins can be placed everywhere or to define strong points where additional temporary loads can be placed.

9.2.4 Protection

When all the moveable fixtures and fittings have been removed from a historic building there usually remain items such as panelling, stair handrails and balustrades which have to remain in position, but are not robust enough to survive the rigours of being on a building site. It is normal to box them in, and although there are rarely structural implications arising from that, it is as well for the engineer to be aware that it may happen, in case there is a need for investigations. These are then best done before things get boxed in.

There are usually structural engineering implications when it comes to protecting the building from the elements if that involves the provision of a temporary roof. When small domestic buildings with simple roofs have their coverings re-laid, it is quite common for roofers to operate without a temporary roof. They open up the roof and work on it when the weather is good and cover it with a tarpaulin when it is bad. This approach carries some risk that some water will get into the building, but where it is not listed and the only consequence of water ingress would be the cost of putting right the damage, the risk is often worth taking.

For most historic buildings undergoing work to their roof finishes, the risk of water ingress is not worth taking, and a temporary roof must be specified. This will also allow construction that is already wet to dry out, and ensure continuity of work regardless of weather conditions.

There are numerous proprietary systems of temporary roofs based on long span lightweight unit beams and flexible sheeting systems. These tend to be modular, and where the module does not suit the shape of the historic building, a more tailored temporary roof using conventional scaffolding and sheeting may be better.

Temporary roofs tend to generate greater horizontal forces due to wind than the roofs they cover. They need to be well above the old roof to provide the necessary working space, and the economics of scaffolding tend to lead to simpler bulkier profiles with a consequent increase in the area exposed to wind loads. These loads have to be transferred to the ground, and using the existing building may not be the best way to do this. The loads may be simply too great in overall terms, or they may need to be concentrated in places where there is no convenient place to transfer them. There will also be vertical reactions to consider. In some cases it may be possible to support the temporary roof off the existing structure. Often it turns out that an independent temporary roof is a better solution, particularly if the elevations are to be scaffolded in any case.

A temporary roof does not have to be complicated to warrant, at the very least, an engineering appraisal of the scheme. An interesting seventeenth-century barn suffered severe damage when it was overturned by the wind. The original roof had ridge vents and the sides were largely open, so the barn had survived many a gale. When the roof coverings finally deteriorated to the point that they were falling off, the whole thing was covered with plastic sheeting that was securely tied down, not to the ground, but to the posts supporting the roof just above ground level. This arrangement was made with the best of intentions in order to protect the building until a permanent repair could be affected. Sadly, and for want of an engineering appraisal, it had the opposite effect, ensuring that, in the absence of proper anchorage, the next high wind lifted the whole structure off the ground and turned it over!

9.2.5 Access for Working

Whilst it is possible to do simple light work from a tower or mobile platform, most high-level work now requires a fixed scaffold, even where, a generation ago, it would have been done off a ladder!

The key issue with historic buildings is to arrange for compatibility between the loads imposed by the scaffolding on the building and the building's capacity to take additional loads. These loads will normally comprise the dead and live loads acting vertically, and a nominal horizontal load to provide lateral stability. In some cases, there will be further horizontal loads due to wind.

In most cases, the vertical loads will be transferred directly to the ground, but where they must be carried on the structure this will need justification. It is important that the scaffolding designers are made aware at the outset that accurate predictions of the self-weight of the scaffolding will be required (not something that is done automatically in scaffold design offices) and of the locations where there is a need to limit the scaffolding loads. It is normal practice for the engineer in the client's design team to establish what loads the existing structure can carry, while the scaffold designer is responsible for determining the loads that are to be carried.

Where it is possible to use the existing building to carry the horizontal loads, the cost and complexity of the scaffolding will be reduced, but in some cases the scaffold will need to be a completely independent structure. This may be for directly structural reasons—for example, the structure cannot take the loads—or for conservation reasons—for example, drilling holes for restraint fixings in fair-faced masonry is not normally acceptable, and the elevation may not have window openings in suitable places or return walls with niches into which scaffolding can be braced.

In some cases an existing external elevation may have the capacity to accept horizontal loads acting inwards at floor level, but be unable to take loads acting outwards. In such cases, it may be possible to design scaffolding that is independent in one direction and not in the other.

Whatever the case, the issues need to be thought through and the restrictions must be defined well before tenders are sought. They must then be communicated clearly to the tenderers. If a completely independent scaffold is required that will almost certainly involve a raking scaffold held down with kentledge, and it is advisable to check, pre-tender, that there is adequate space within the site boundary for the footprint of the scaffold and normal site access. Where kentledge is provided this is often in the form of sandbags or concrete blocks, though steel plate is denser. Whatever the nature of the kentledge, its purpose must be made clear on site; otherwise people may mistake it for just another pile of building material and may remove it!

Scaffolds are either built with tubes and fittings tightened with a spanner or from proprietary kits of fabricated parts with 'patent' connections. Tubes and fittings are preferable where the form of the building is at all complex. System scaffolds are normally modular in terms of their setting out on plan and the heights achievable between boarded levels. They can be more economical than tubes and fittings if the requirements are straightforward and the scaffold module suits the building module. Otherwise the time and effort involved in making special pieces to cope with the non-standard setting out will cancel out the benefits of system scaffolds. All scaffolds must be certified as complete before they are used, and they have to be inspected regularly.

Whilst timber scaffolding was still used in Scandinavia and overseas in the 1950s, scaffolding will nowadays normally be galvanized mild steel tubes and fittings. Aluminium poles are lighter, but there is no difference in the weight of the fittings, and when there are high loads to be carried, aluminium poles may need to be closer together or be braced more frequently, which can cancel out any overall savings in weight. If the total weight of the scaffolding (including the boards and the design live load) becomes an issue, savings may be possible by considering the likely diversity of live load, since it is unlikely that all levels will be fully loaded at the same time.

The means by which lateral loads, and particularly pull-out loads, may be transferred from the scaffolding to the existing building will need to be defined prior to tender. In the absence of requirements to the contrary, scaffolding designers will assume that it is acceptable to drill in fixings for eyebolts on all elevations at a spacing of about 3 m horizontally and vertically. This may well be structurally unacceptable, especially in masonry that is not well bonded and/or is slender. But even where it is structurally acceptable, it may not be acceptable in conservation terms.

Firstly, it is notoriously difficult in some types of masonry to make good the holes left by such fixings without them being noticeable, and, secondly, if a building is scaffolded, say, every 50 years, and each time a new set of fixings is drilled in, the elevation will eventually become very scarred. In practice where the installation of fixings on a fixed grid is not acceptable, a satisfactory arrangement can usually be achieved by re-using existing fixings (after testing), by placing new fixings in positions where they are not visible, by taking ties through windows (having first removed precious glass!) and clamping to mobilize the thickness of the wall, and by bracing off niches and pilasters on return walls.

Where drilled-in fixings are accepted, they should be stainless steel, so as to avoid corrosion problems in the event that they are not removed on completion. They should be set slightly below the surface of the masonry, to provide space for a masonry plug. Fixings should be load tested, and the test rig should be designed not only to test if the fixing pulls out of the stone it is set into, but also whether that stone (or brick) is adequately held in by its neighbours.

When completed, a scaffold is often clad in some form of sheeting. Supporting the sheeting may be the prime function of the scaffold, as in the case of temporary roofs, or it may be a consequence of the operations to be performed from the scaffold. Some operations require controlled environments. Others use materials or generate debris that need to be contained. On prominent sites the cladding may be a requirement to make things look neat. It is important to establish whether the cladding is being procured separately from the scaffolding, in which case the type of cladding or enclosure must be made known to the scaffold designer. Systems vary in their behaviour under wind load. Some materials are perforated to reduce wind loads. Other materials, being impermeable, are attached by special toggles that are designed to fail in high winds, allowing the sheeting to flap rather than billow and pull on the scaffold.

9.2.6 Demolitions and Structural Alterations

Most building contracts involving structural alterations include elements of demolition and temporary situations where additional support or additional stability is required. In historic buildings, these will normally be found in parts that are, historically, of less significance. This reduced historical significance will often be in areas where the building has been altered quite a lot over the years, so that its historic integrity, and hence its historic interest has been lost. However this may mean that parts that are of historic significance now derive their stability from adjacent parts that are not historically significant. The structural implications of this must be recognized. Buildings have been known to collapse as a result of people thinking that just because something was not original or historically significant, it could be removed without further thought!

In other cases problems have arisen because no one realized that the order in which things were done was important, and that there needs to be planning and control over what activities occur simultaneously in different parts of the structure. However, once it has been recognized that there is an issue to be addressed here, the techniques for dealing with it do not require sophisticated plant or equipment and involve the minimum of engineering calculations.

The simplest form of shoring or propping uses 'Acrow' props. These can be thought of as scaffold poles that have been fitted with end plates and made adjustable by a threaded telescopic section. They are simple and quick to install and remove. Where something stronger is required, Acrows can be used in groups, laced together, or, to reduce congestion or where Acrows would be too slender, 'Tri-shores', stronger and heavier adjustable proprietary shores, having three tubes latticed together, are available, as are other adjustable props, made from perforated steel sections (see Fig. 9.1).

Structural steel H-sections with sizes of between 100 and 200 mm are used as spreaders below the bases of shores, or as 'needles' to temporarily support masonry over new openings while the permanent beams are installed.

The principle behind 'needling' is that masonry will arch between point-supports, and if the point-supports (the 'needles') are sufficiently close together, a flat arch can be generated between them. A beam can be inserted under this flat arch, masonry loads can be transferred into the beam by packing the gap, and the needles can be removed. The process works best in well-bonded brick masonry, where the bond enables the brickwork to arch in all sorts of directions, and the risk of even individual bricks falling out is low. It becomes progressively more difficult if the masonry is not coursed, or if the wall consists of two skins of coursed masonry with a rubble core. The presence of voids such as chimney flues will prevent or disrupt arching, as will pockets for floor beams, or door and window openings in a zone where arching is desirable. These can all be dealt with if they are recognized in time—rubble can be grouted or consolidated, openings can be braced up and redundant flues can be filled. This should be done in advance of the needling operation.

The temporary support system will need to be as stiff as possible, and that means the needles will be as short as possible, consistent with allowing room to manoeuvre the new beams into position. The temporary vertical supports may need to be taken down through the building to the ground—or to the nearest concrete floor. Inspections and work outside the site boundary may be needed here, and time must be allowed for obtaining consent from the road authority to do this.

When back-propping floors, Acrow props are normally the most appropriate form of shoring. The main risk associated with them is over-tightening. In extreme

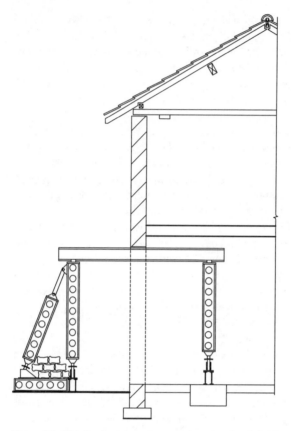

Figure 9.1 'Needling' using steel components; note kentledge to stabilize external 'dead-shore'.

cases, the props can lift a floor structure off its end-bearing and de-stabilize the supporting wall. When a floor is to be 'sandwiched' between two lifts of Acrows, its internal construction must be assessed to confirm whether it is strong or stiff enough.

Whilst the desirability of additional vertical propping can be quite easy to spot, it is not so easy to perceive when the lateral stability needs to be enhanced. For example, when a floor structure is being altered or repaired, its ability to provide lateral stability to an adjacent wall may be temporarily reduced. When a wall is being repaired and the repair involves cutting out masonry and replacing it, the wall will have reduced stability until the repairs are complete. Some measures to provide stability temporarily may be called for during the work.

9.2.7 Excavations

Statutory Instrument 1592 requires proper consideration to be given to the need to prop excavations. This is often interpreted as meaning that excavations over 1.2 m deep require propping. However, most historic buildings are founded less than 1.2 m below their lowest floor level, and any excavations may disturb their equilibrium. It is important therefore that the regulations do not get interpreted as meaning that excavations less than 1.2 m deep do not need support! Some of them do, and the need for this has to come out of an engineering appraisal. Excavations of all sorts—not just those for structure—have to be carefully considered, particularly those for drains and other buried services. The design and routing of excavations for services may fall outside the structural engineer's brief and may even be left to the contractor to decide on site. This may be inappropriate for the following reason. Historic buildings often contain retaining walls whose stability would be difficult to justify by calculation but are not visibly in distress. Such walls are nevertheless vulnerable to excavations at their toes, even if the excavation does not go below the bottom of the wall construction. The effect of excavating in these circumstances is to remove a block of ground that is providing passive resistance against lateral movement of the toe, and at the same time to increase the active height of the retained material. Taken together these can be enough to de-stabilize the wall.

In order to avoid this, it is necessary first to know enough about the existing structure and the proposals for new buried services to be able to recognize where there is the potential for a dangerous situation to arise. It is then a matter of deciding what action is required. There may be a need for temporary works or some re-routing of the services. Where temporary propping is deemed to be the solution during construction, it is important that the specification for back-filling the excavations is also reviewed, since it must be capable of providing passive resistance without undergoing much strain. Putting back the excavated material and compacting it as well as possible may not be enough.

Short telescopic shores, working on the same principles as Acrow props, are available for use with steel trench sheeting to support the sides of trenches.

Sometimes it is necessary to excavate directly below the foundations of a wall, either to form a route for buried services, or as part of a series of excavations for underpinning the wall. Before doing this, it is essential to check that the wall can arch over the excavation onto firmly founded construction on either side. The most reliable way to do this is to sketch the construction in elevation, and in proportion, draw in the proposed excavation, and note whether there are credible alternative load paths around the ground that is to be taken away. The necessary elevation drawings may not be already drawn, but to do so takes very little time and is far more reliable than looking at plans and trying to imagine what is happening in the third dimension.

9.2.8 Interim Stabilization

Sometimes it is found that a part of a building has moved and appears still to be moving and, whilst there is not enough money immediately available to implement a permanent remedy, something ought to be done to stop or retard the movement, until a proper repair can be carried out. The most common examples of this are leaning walls and the interim measures are 'raking shores'. These

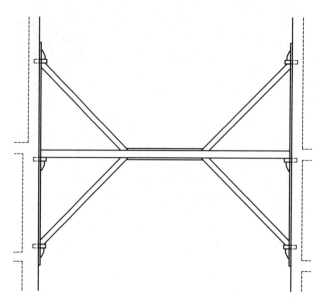

Figure 9.2 'Flying shore' using timber members.

used to be made from fairly heavy timbers, and this is still the practice when suitable steel components are more difficult to come by. As important as the design of the shores themselves, is the provision of proper bases, founded deep enough not to move when subjected to the forces from the shores (see chapter frontispiece).

Where, in a terrace, two flank walls are threatening to lean into the space left by a demolished, or bombed-out, house, so-called 'flying shores' are used (Fig. 9.2). These usually comprise a horizontal prop at the level of one of the upper floors, with raking props, bearing on the flank walls at floor levels above and below the horizontal prop. They have no support on the ground, so leave clear space for the re-building of the 'lost' house.

9.3 STRATEGIES AND TACTICS

9.3.1 General

Time and energy spent devising schemes which involve the minimum of work, often result in the best long-term solutions for old buildings. This is also the case for temporary works. It is easy, quick and often quite cheap to introduce so much scaffolding and shoring into a building that the weight of the scaffolding threatens its stability, and access for inspection and working space is compromised.

A better approach is first to undertake all the necessary structural repairs and reverse ill-considered previous alterations. This enhances the stability of the structure and can often be undertaken with the minimum of shoring. The improved structure then requires less shoring to safeguard it while more substantial alterations are carried out.

Similarly when partial demolition is being undertaken and there are inevitable repairs to be done at the junction between that which is to be demolished and that which remains, it is often preferable to do these as the demolition proceeds, rather than introduce lots of temporary shoring and return later to do the repairs.

Such an approach requires a complete integration of the different processes and trades involved. This is quite different from the procedures for major new buildings, which are currently designed so that contracts for site clearance, the sub-structure, the frame, the cladding, the services and the fitting out can successfully be let as separate packages, and where the biggest management challenge is to co-ordinate the delivery of materials from the different suppliers and subcontractors.

Despite the wide variety or circumstances leading to the need for temporary works, there is usually one underlying common purpose—they are only ever required in order to deal with some constraint or other within which the builder has to work. A good way for designers to proceed is first to consider the broad range of constraints that can exist, and then define which exist in the particular project in question. A review of possible ways of dealing with the constraints should come next, followed by a check to see whether sufficient is known about the existing building for the merits of different solutions to be compared. If sufficient information is not available, serious consideration should be given to obtaining it, particularly if cost and programme certainty is important at the construction stage—which is usually the case.

Projects that go wrong always teach more lessons than projects that go right. Thankfully only a small proportion goes spectacularly wrong, but when they do, it is always useful to study the reasons why. Time and again, in going over the evidence, it can be demonstrated that whilst there is normally an identifiable cause of any failure, more often than not it only occurred because the whole context of what was going on was likely to generate problems. In short, if the strategy for the building is wrong or poorly conceived, some sort of tactical mishap is likely to generate consequences disproportionate to the cause.

The following case histories should illustrate this general tenet.

9.3.2 Case Histories

Case History 1

An inner city site was to be re-developed. The existing buildings were a terrace of houses with ad-hoc extensions in what had once been back gardens. All were basically eighteenth-century houses, but as the area had been wealthy from the outset, most of the buildings had been refurbished and altered every 25 years for 250 years. As a result, only one of them was of sufficient architectural or historic interest to be listed. The planning consent therefore allowed the demolition of the unlisted buildings, but required that the one that was listed was to be retained. It had not been altered much, but neither had it been well maintained. There were known problems of rotten timbers and de-laminating brickwork. These would all have been repairable without destroying the historic interest.

The strategy adopted by the developer was to 'mothball' the listed building by installing temporary scaffolding and a temporary roof, so as to tie it together and keep the rain out, whilst the rest of the site was cleared and the new construction went ahead.

At 5.00 AM one day the whole of the listed building and the temporary roof collapsed into the street. Fortunately no one was injured, but unfortunately nobody witnessed the collapse and it was quite impossible to determine the precise cause. However, the way the work had been planned and carried out created a climate in which collapse, for whatever reason, was highly likely. In the temporary condition, the terraced house had become a detached house in the middle of an open re-development site. The scaffolding, that supported the temporary roof, was not independent of the building, so that *lateral loads from wind were carried by the building*. There was some rudimentary shoring and some scaffolding tying parts together, *but no permanent structural repairs had been done*. The protection made it difficult to inspect. Piling rigs had churned up the ground all round it, and this had turned to mud as it got rained on for the first time in 250 years. There had been some very high winds about 10 days before the collapse.

If the house had been put into good structural order and weatherproofed before the rest of the site was cleared, it is unlikely that the collapse would have occurred. This was a prime example of a flawed strategy, which no amount of clever tactics could rectify.

Case History 2

A redundant church was being converted into apartments. This involved building a completely new framed structure inside the retained external walls of the church. The new scheme was to have a basement car park, and a ground floor at street level. The church did have an undercroft but, as the main church floor was several feet above ground level, the undercroft floor was less than a full basement depth below street level. In order to create the car park, the whole site had to be excavated by about 1.5 m, and this meant excavating below the level of the foundations of the retained façade. The solution to this was to underpin, in mass concrete, the walls to be retained, and thus create new foundations below the new basement level.

The strategy adopted was to procure the work as a series of package contracts, the first of which was 'demolition' and the second 'underpinning and groundworks'. The demolition package involved the removal of *everything*, except the three walls of the church, that were to be retained. The building dated from the mid-nineteenth century and had been built as a non-conformist chapel, with a multiplicity of rooms around the central worship area with various galleries. The external walls were stone on the outside and brick on the inside and were relatively thin for their height and length. *These walls were not self-stable, but relied on the cross walls, roof and galleries for stability* (unlike in older churches, whose walls tend to be much thicker). The demolition package therefore included the provision of an external scaffold structure to provide temporary stability, until the walls could be tied to the new internal structure. This scaffold was tied to the walls through the window openings and restrained from overturning by large amounts of kentledge.

The underpinning commenced, but within a week or so, one of the three walls that were to be retained had collapsed. Fortunately it fell into the site and the scaffolding and kentledge served to prevent the collapse from being precipitate, so no one was injured. It emerged that an excavation for underpinning had been made under a section of wall that was not capable of arching. This had caused the end of the wall to collapse, but as it was connected to the scaffolding which, in turn, was connected to the rest of the wall, the whole wall collapsed over a period of hours.

It was evident that the bay layout for the underpinning had been worked out on a plan drawing of the wall, but the plan had been drawn at gallery level, and therefore did not show the openings in the wall at lower levels. The excavation that caused the collapse was made under what was effectively an isolated pier between openings.

If the proposed underpinning sequence had been drawn on a true elevation of that wall, the problem would have been immediately obvious. Steps could have been taken to shore up the openings and the underpinning could have been successfully carried out. The risks would have been even less if the underpinning had been done before the demolition, whilst the wall was still stabilized by the cross walls, floors and roof.

Case History 3

A terrace of unlisted houses, in a conservation area, backed onto a re-development site. Planning consent was granted for a scheme in which the front rooms of the houses, including all the construction back to a half-way point on the party walls was to be kept ('party walls' are flank walls, shared by the two adjoining houses). It was then discovered that whilst the façades and the party walls were brick, the fourth walls were a mixture of brick and timber stud that had been much altered and were quite unsuitable as a permanent enclosure to the rooms being retained. However, they supported the floors, so when consent was granted to remove them, the timber floors were also to be removed, leaving just the façade and the front halves of the party walls to be retained.

As the building was to become offices, there was a requirement for office floor loadings, and the replacement floors were to be reinforced concrete, supported on new columns, as the existing brick walls were inadequate for the increased loads. If the new columns had been positioned inside the retained walls, they would have occupied lettable floor space, and they would have made it difficult to reinstate the wood panelling that belonged in the front rooms. It was therefore decided to cut vertical slots in the retained party walls, and locate the columns within the thickness of the walls.

When the first slot had been partly cut, the adjacent section of party wall fell down, on a Sunday afternoon when there were no witnesses.

In this case, an initially rather ambitious proposal became progressively more complex as work proceeded, leading to a situation where something was almost bound to go wrong.

The cause of the collapse in the first case history can be traced back to the developer's and/or the contractor's unsuitable sequencing of the work. The collapse in the second case history, to the failure of the designer to visualize the three-dimensional reality of the wall layout, by relying on an irrelevant plan. The collapse in the third case history seems almost inevitable, unless a radical re-think had been done when the nature of the fourth wall became known. All three case histories show that if the initial strategy is flawed, the outcome is likely to be disastrous.

EPILOGUE: MAINTENANCE

In the preface to the first edition, mention was made of the need for money to finance maintenance. This is because all the care and attention lavished on procuring the best possible conservation of a structure will soon be wasted if that structure is not maintained.

Nothing lasts forever; wind, weather, insects, fungi and bacteria are at work all the time, slowly grinding, dissolving, gnawing and digesting the hardwon fruits of our and our forebears' labour. The preceding chapters should however have indicated how, with a modicum of effort and expenditure, we can make our structures more resistant to these effects.

Whatever protective treatments, that have been applied as part of the conservation work, will however be gradually worn away, degraded or dissolved and they must therefore be renewed from time to time. Similarly, it must be ensured that building components, such as gutters and downpipes, which are designed to keep water away from vulnerable parts, are still doing their job.

Car manufacturers' handbooks recommend owners to have certain checks and adjustments carried out at specific intervals, and many owners do, because they can see everyday the results of not maintaining cars. But many owners and tenants carry on treating their buildings, which represent far greater investment, as if they would last indefinitely without any intervention whatsoever. It is time that building owners were issued with maintenance manuals and this has in fact been done by some architects for new buildings.

For traditional building materials and elements, experience already enables us to draw up a rudimentary schedule of 'checks and adjustments' and the time intervals (corresponding to a car's mileage between services) at which they should be carried out:

Clean out gutters	every autumn
Inspect roof, downpipes, etc	every 3 years approx.
Re-paint outside woodwork	every 3 years approx.
Re-point masonry joints	every 50 years approx.
Re-lay roof slates	every 50 years approx.
Re-lay lead roofing	every 100 years approx.

As shown, this list refers to conditions in Britain; a similar schedule would however apply to buildings in most temperate areas.

The most important items on the schedule above are the maintenance of the roof drainage and the regular inspection. The first prevents the worst of the damage that water will cause during the winter months, the second will discover the odd element that has 'gone out of step', e.g. the brick chimney with mortar joints crumbling prematurely due to sulphate attack from the flue gases.

Neither of these operations is very expensive, nor do they cause any great disruption and yet it is often the failure to carry out these simple tasks that lead to deterioration beyond repair: On a large church in West London, overflowing gutters had led to wholesale rotting of the the roof rafters (which were hidden by the ceiling) to the extent that the entire roof needed re-placing—a task that was completely beyond the means of the dwindling congregation, and an expenditure that could have been avoided by even an occasional clearing of the gutters.

Sadly, in the western world, the pursuit of profit at any price, together with rising rates of pay for those carrying out seemingly menial tasks, will tempt building owners to do away with 'the little maintenance chap', because 'he never seems to find anything serious'. The fact that he quietly deals with the minor problems, and thereby prevents the major ones from arising, is forgotten. The result is that, after he has been made redundant, minor maintenance tasks are left undone, and the minor problems grow in seriousness and are allowed to accumulate until major restoration becomes a matter of urgency.

There is also a growing body of safety legislation, which means that certain maintenance tasks require installation safety devices of one kind or another. It is tempting to defer the (usually modest) expenditure on these and, consequently, postpone maintenance work until a crisis occurs.

Another, related, factor is the advent of the tower crane; this enabled certain building elements to be erected without the need of any scaffold. Where windows had to be cleaned, such as on an office tower block, suspended 'gondolas' were provided, but solid elements were often left to survive as best as they could. This has already had consequences for some buildings.

It should be remembered by everybody responsible for buildings, that for all the efforts of art-historians, conservation officials, conservators, architects and engineers, the continuing conservation of our structural heritage depends in the end on people with ladders, trowels, pen-knives, sealant guns and paint brushes.

REFERENCES AND FURTHER READING

This is not intended as a comprehensive bibliography, but as a pointer to books and papers that may be interesting and/or useful for readers.

GENERAL

Construction Industry Research and Information Association (1986) *Structural Renovation of Traditional Buildings*, Report 111.

Doran, D. (ed.) (1992) *Construction Materials Reference Book*, Butterworth-Heinemann, London.
(Mainly materials' properties with some exposition of historical developments.)

Feilden, Sir Bernard M. (2003) *Conservation of Historic Buildings*, Butterworth-Heinemann, London.
(The acknowledged general reference in English on the subject; it has an extensive bibliography.)

Mainstone, Rowland J. (1999) *Structure in Architecture*, Ashgate, Aldershot.
(A collection of papers on history and development of structural forms with detailed studies of monumental buildings, not much on appraisal or remedial work.)

Rabun, J. Stanley (2000) *Structural Analysis of Historic Buildings*, John Wiley & Sons.
(Largely a detailed survey of past USA design and construction practices, very little on appraisal, or remedial work.)

CHAPTER 1

Brohn, D. (1984) *Understanding Structural Analysis*, Granada Publishing, London.

Gordon, J.E. (1976) *The New Science of Strong Materials—Or Why You Don't Fall Through the Floor*, Penguin Books, Harmondsworth.

Gordon, J.E. (1978) *Structures—Or Why Things Don't Fall Down*, Penguin Books, Harmondsworth.
(These three give good qualitative explanations of structural behaviour and generally avoid mathematics.)

Suenson, E. (1920) *Byggematerialer—Vol. 1 Styrkeprøver, Mettaller*, Jul. Gjellerups Forlag (successors to P.E. Bluhmes Boghandel), Copenhagen.
Timoshenko, S.P. (1953) *History of Strength of Materials*, McGraw-Hill, London.
(Whilst not free of mathematical formulae, gives a good readable account of the development of this science.)

CHAPTER 2

Building Research Establishment (1991) 'Structural appraisal of existing buildings for change of use', *BRE Digest* 366, October.
English Heritage (1994) *Office Floor Loading in Historic Buildings*.
(Distributed with 'The Structural Engineer' of August 1994.)
Fitzpatrick, A. et al. (1992) *An Assessment of the Imposed Loading Needs for Current Commercial Office Buildings in Great Britain*, Ove Arup & Partners, London.
Institution of Structural Engineers (1996) *Appraisal of Existing Structures*—Second Edition, London.
International Standards Organisation (1986) ISO 2394: *General Principles on Reliability for Structures*.
(Explains in some detail the functions of the partial safety factors; the earlier 1973 version is simpler.)
Mitchell, G.R. and Woodgate, R.W. (1971) *Floor Loadings in Office Buildings—The Results of a Survey*, Current Paper 3/71, Building Research Station.

CHAPTER 3

Beckmann, P. (1972) 'Structural analysis and recording of ancient buildings', *The Arup Journal*, Vol. 15, No. 4, December.
Beckmann, P. (1986) 'The step-by-step approach to investigation and remedial work, illustrated by the work at York Minster and Holy Trinity Church, Coventry', in *Stable—Unstable*. R.M. Lemaire and K. van Balen (eds); lectures at the International Updating Course on structural Consolidation, Leuven 19–24 May 1986.
Beckmann, P. and Blanchard, J.C. (1980) 'The spire of Holy Trinity Church, Coventry'. *The Arup Journal*, Vol. 15, No. 4, December.
(Describes investigation of wind stability and calculation of resonance to bell-ringing of a tower and spire.)
Bielby, S.C. (1992) *Site Safety*, Construction Industry Research and Information Association, Special Publication 90.
Building Research Establishment (1989) 'Simple measuring and monitoring of movement in low-rise buildings', BRE Digests 343, April and 344, May.
Construction (Health, Safety and Welfare) Regulations 1996 (Statutory Instruments 1996 No. 1592).
Construction (Design and Management) Regulations 1994 (Statutory Instruments 1994 No. 3140, amended by Statutory Instruments 2000 No. 2380).
Dowrick, D.J. and Beckmann, P. (1971) 'York Minster structural restoration', in *Proceedings of the Institution of Civil Engineers*, Paper 7415S.
Institution of Structural Engineers (1996) *Appraisal of Existing Structures*.
(Contains many useful references on testing.)
Institution of Structural Engineers (1989) *Load Testing of Structures and Structural Components*.
Richardson, C. (1985) *A.J.Guide to structural Surveys*, reprinted from *Architect's Journal*, Architectural Press.

CHAPTER 4

Ashurst, J. and Ashurst, N. (1988) *Practical Building Conservation*, English Heritage Technical Handbook, Vol. 1: *Stone Masonry*, Vol. 2: *Brick, Terracotta and Earth* and Vol. 3: *Mortars, Plasters and Renders*, Gower Technical Press.
(These deal with the conservation of the materials, not so much with structural problems.)

Beckmann, P. and Blanchard, J.C. (1980) 'The spire of Holy Trinity Church, Coventry', *The Arup Journal*, Vol. 15, No. 4, December.
(Describes strengthening against wind loading.)

Berger, F. (1986) *Zur nachträglichen Bestimmung der Tragfestigheit von zentrisch gedrücktem Ziegelmauerwerk*, in *Erhalten historisch bedeutsamer Bauwerke*. Sonderforschungsbereich SFB 315, Universität Karlsruhe, Jahrbuch.
(In German, Development of cylinder-splitting test for brickwork.)

Berger, F. (1989) *Zerstörungsarme Untersuchungen historischen Mauerwerks* in *Internationale Tagung des SFB 315*, Universität Karlsruhe 18–21 October.
(In German, Development work on Ultrasonic Pulse Velocity Testing of Brickwork.)

BS 5628: Part 1 (1992) *Code of Practice for the Use of Masonry*, British Standards Institution.
(Strength data and rules for design of new Masonry.)

Construction Industry Research and Information Association (2003) *Masonry Façade Retention: Best Practice Guide*, CIRIA RP 626.
(Comprehensive advice on design of schemes incorporating retained facades, detail design considerations, temporary works, safety aspects and contract procurement.)

Cook, D.A., Ledbetter, S., Ring, S. and Wenzel, F. (2000) 'Masonry crack damage: its origins, diagnosis, philosophy and a basis for repair', in *Proc. Instn. Civ. Engrs. Structs & Bldgs* **140**, February.

Department of Transport (2001) Design Manual for Roads and Bridges, Departmental Standard BD 21/01, *The assessment of Highway Bridges and Structures*, HMSO.
(Alternative strength data for assessment of masonry.)

Egermann, R. (1997) 'On the load-bearing capacity of natural stone buildings', in *International Colloquium Seriate 1997: Inspection and Monitoring of the Architectural Heritage*, Italian group of IABSE, Publisher: Ferrari Editrice.
(The step-by-step assessment of the strength of sandstone masonry in an existing, war-damaged building, quoting formulae and describing practical use of split-cylinder test; further references.)

Franken, S. et al. (2001) *Historische Mörtel und Reparaturmörtel. Untersuchen, Bewerten, Einsetzen* (Sonderforschungsbereich 315). *Empfehlungen für die Praxis, Vol. 5*. University of Karlsruhe.
(In German, deals with investigations of historic mortars; composition of repair mortars.)

Hendry, A.W. (1990) *Structural Masonry*, Macmillan Education Ltd.

Heyman, J. (1982) *The Masonry Arch*, Ellis Horwood Ltd.
(Clear exposition of principles with references to eighteenth- and nineteenth-century treatises and to earlier papers by the author.)

Heyman, J. (1992) 'Leaning towers', in *Masonry Construction*, C.R. Calladine (ed.), Kluwer Academic Publishers, 1992.

Heyman, J. (1995) *The Stone Skeleton*, Cambridge University Press.

Heyman, J. (1998) *Hooke's Cubico-Parabolic Conoid*. Notes of the Royal Society, London **52**, 39–50.
(Derives the theoretically exact equation, and Hooke's approximation, of the ideal shape of a dome, loaded by self-weight only—some historical notes on St Paul's Cathedral.)

Price, S. (1996) 'Cantilevered staircases', in *Architectural Research Quarterly* **1**, Spring, Emap Construct, 151 Roseberry Avenue, London EC1R 4QX.
(Shows many examples and explains the structural action in some detail.)

Soo, Lydia M. (1998) *Wren's "Tracts" on Architecture and other Writings*, Cambridge University Press.

Sumanov, L. (1999) *Conservation and Seismic Strengthening of Architectural Heritage: Byzantine Churches of the Ninth to the Fourteenth Centuries in Macedonia*, D. Phil. Dissertation at the University of York, Institute of Advanced Architectural Studies.

Warland, E.G. (1953) *Modern, Practical Masonry*, Published by the Stone Federation, London.

Wenzel, F. et al. (2000) *Historisches Mauerwerk. Untersuchen, Bewerten und Instandsetzen* (Sonderforschungsbereich 315) *Empfehlungen für die Praxis, Vol. 2*. University of Karlsruhe.
(In German, gives details of test methods for masonry and case histories of repair/strengthening.)

CHAPTER 5

Bravery, A.F. et al. (1992) *Recognising Wood Rot and Insect Damage in Buildings*, BR Report 232: Building Research Establishment.
(Excellent pictures which greatly help identification.)

BS 5268: Part 2 (1991) *Code of Practice for Structural Use of Timber*, British Standards Institution.
(Design stresses for new construction.)

Brunskill, R.W. (1985) *Timber Building in Britain*, Victor Gollancz.

Charles, F.W.B. and Mary Charles (1990) *Conservation of Timber Buildings*, Stanley Thornes (Publishers) Ltd.

CP 112: Part 2 (1971) *The Structural Use of Timber*, Code of Practice British Standards Institution.
(Superseded for new construction, but useful for appraisal.)

Görlacher, R. et al. (1999) *Historische Holztragwerke Untersuchen, Berechnen und Instandsetzen.* (Sonderforschungsbereich 315). *Empfehlungen für die Praxis, Vol. 1.* University of Karlsruhe.
(In German, gives details of sophisticated test methods, etc.)

Harris, R. (1979) *Discovering Timber-Framed Buildings*, Shire Publications.

Hewett, C.A. (1980) *English Historic Carpentry*, Phillimore.
(Emphasis on geometry of carpentered joints.)

Lavers, G.M. and Moore, G.L. (1983) *The Strength Properties of Timber*, BRS Report 241; Building Research Establishment.

Newlands, J. (1857) *The Carpenter's Assistant*, reprinted by Studio Editions, 1990.

Ross, P. (2002) *Appraisal and Repair of Timber Structures*, Thomas Telford Publishing.
(Detailed procedures for appraisal and remedial measures; case histories.)

Stocker, A.E. and Bridge, L. (1989) 'Church of St Mary the Virgin, Sandwich, Kent—The repair of the seventeenth century timber roof', in *Conference Proceedings for Structural Repair and Maintenance of Historic Buildings*, Florence. The Institute of Computational Mechanics, Southampton.

Suenson, E. (1922) *Byggematerialer—Vol. 2 Træ, Plantestoffer, Varme-og Lydisolering*, P.E. Bluhmes Boghandel, Copenhagen.
(In Danish, Succeeded by J. Gjellerup.)

Timber Research and Development Association (1991) *Hardwoods in Construction*.

CHAPTER 6

Ashurst, J. and Nicola (1988) Ashurst: *Practical Building Conservation*, English Heritage Technical Handbook, Vol. 4: *Metals*, Gower Technical Press.
(Deals mainly with conservation of decorative ironwork.)

Bates, W. (1984) *Historical Structural Steelwork Handbook*, British Constructional Steelwork Association, Publication 11/84.
(Useful tables of section properties of obsolete rolled sections, but somewhat conservative assessment advice.)

Blanchard, J.C., Bussell, M.N. and Marsden, A. (1982/1983) 'Appraisal of existing ferrous metal structures', in *The Arup Journal*, Vol. 1, No. 4, December 1982 and Vol. 18, No. 1, April 1983.

BS 449 (1937) *The Use of Structural Steel in Buildings*, British Standards Institution.

BS 449: Part 2 (1969) *The Use of Structural Steel in Buildings*. British Standards Institution.

Bussell, M.N. (1997) *Appraisal of existing Iron and Steel Structures*, Steel Construction Institute Publication No. SCI—P—183.

CIRIA (1986) *Structural Renovation of Traditional Buildings*. Construction Industry Research and Information Association, Report 111.

Department of Transport, Roads and Local Transport Directorate (2001) 'The assessment of highway bridges and structures', in *Manual for Roads and Bridges*, Departmental Standard BD 21/01. (Strength data and partial safety factors for wrought iron and early steel; working stresses for cast iron—earlier version, practically identical, BD 21/93.)

Derry, T.K. and Williams, T.I. (1960) *A Short History of Technology*, Oxford University Press. (Development of metallurgical and metal forming processes.)

Hamilton, S.B. (1941) 'The Use of Cast Iron in Building', in *Trans. Newcomen Society* **21**.

Käpplein, R. (1991) 'Zur Beurteilung des Tragverhaltens alter gusseiserner Hohlsäulen', in *Berichte der Versuchsanstalt für Stahl*, Holz und Steine der Universität Fridericiana in Karlsruhe, 4. Folge—Heft 23. (In German, Ultrasonic tests for flaws and strength; cylinder-splitting tests compared with tensile tests.)

Käpplein, R. et al. (2001) *Historische Eisen- und Stahlkonstruktionen. Untersuchen, Berechnen und Instandsetzen. (Sonderforschungsbereich 315). Empfehlungen für die Praxis, Vol. 4*. University of Karlsruhe. (In German, gives details of test methods and case histories.)

Moy, S.S. et al. (2001) *F.R.P. Composites: Life Extension and Strengthening of metallic Structures*, ICE design and practice guide, Thomas Telford Publishing.

Skempton, A.W. and Johnson, H.R. (1962) 'The first iron frames', in *Architectural Review* **131**, No. 781. (Early British industrial building Structures.)

Sutherland, R.J.M. (1982) 'The bending strength of cast-iron', in *Colloquium on History of Structures*, International Association for Bridges and Structures.

Sutherland, R.J.M. (1990) 'The impact of structural iron in france', in *The Structural Engineer* **68**, No. 14, July.

Sutherland, R.J.M. (1991) *Appraisal of Cast and Wrought Iron (as Materials)*, Institution of Structural Engineers and Institution of Civil Engineers, National Piers Society Symposium on Seaside Piers 12 November. (Discusses differences between the design stresses in BD 21/93 and the London 1909 Building Act.)

Suenson, E. (1920) *Byggematerialer—Vol. 1 Styrkeprøver, metaller*, Jul. Gjellerups Forlag (successors to P.E. Bluhmes Boghandel), Copenhagen.

CHAPTER 7

Assessment and Repair of Fire-damaged Concrete Structure (1990) Concrete Society Technical Report No. 33.

Bate, S.C.C. (1984) *High Alumina Cement Concrete in Existing Building Superstructures*, Building Research Establishment Report BR 235, HMSO.

Bromley, A. and Pettifer, K. (1997) *Sulphide-related Degradation of Concrete in South-West England (The 'Mundic' Problem)*, Building Research Establishment Laboratory Report.

BS 785 (1938) *Rolled Steel Bars and Hard-Drawn Steel Wire for Concrete Reinforcement*, British Standards Institution, London.

BS 6089 (1981) *Guide to Assessment of Concrete Strength in Existing Structures*, British Standards Institution.

Bungey, J.H. (1992) *Testing Concrete in Structures: A guide to Equipment for Testing Concrete in Structures*, Technical Note 143, Construction Industry Research and Information Association.

Calcium Aluminate Cements in Construction (1997) Concrete Society Technical Report No. 46. ('Rehabilitation' of High-Alumina Cement.)

Concrete Core Testing for Strength (1976) Concrete Society Technical Report No. 11, including addendum 1987.

(Preceded BS 6089, covers similar topics with more practical advice on interpretation of results.)

Concrete in Aggressive Ground, Building Research Establishment Special Digest SD1 (2001) (revised 2003).

deCourcy, J.W. (1987) 'The emergence of reinforced concrete 1750–1910', *The Structural Engineer*, **65A**, No. 9, September.

Design Guidance for Strengthening Concrete Structures, Using Fibre Composite Materials (2000) Concrete Society Technical Report No. 55.

(Sets out advantages and limitations of the various applications, quotes properties of materials, and gives detailed design recommendations.)

Dunster, A. (2003) *HAC Concrete in the UK, Assessment, Durability Management and Refurbishment*, Building Research Establishment special digest SD3.

Dunster, A. and Moss, R. (2002) *BRAC rules: revised 2002*, BRE 451.

Fookes, P.G., Comberbach, C.D. and Cann, J. (1983) 'Field investigation of concrete structures in south-west England' *Concrete Magazine*, April.

Jones, B.E. (ed.) (1920) *Cassell's Reinforced Concrete*, The Waverley Book Company, Ltd, London.

Loov, R.E. (1991) 'Reinforced concrete at the turn of the century', *Concrete International*, December.

Non-structural Cracks in Concrete (1982) Concrete Society Technical Report No. 22.

Protection of Reinforced Concrete by Surface Treatments (1987) Technical Note 130, Construction Industry Research and Information Association.

Repair of Concrete Damaged by Reinforcement Corrosion (1984) Concrete Society Technical Report No. 26.

(Standard initial reference on the subject.)

Report of the Thaumasite Expert Group (1999) Department of the Environment, Transport and the Regions, London.

(Deals in detail with the risks and diagnosis of Thaumasite attack and remedial works; some advice on old buildings.)

Strengthening Concrete Structures, Using Fibre Composite Materials; Acceptance, Inspection and Monitoring (2003) Concrete Society Technical Report No. 57.

Structural Effects of Alkali-Silica Reaction, Technical Guidance on the Appraisal of existing Structures (1992) The Institution of Structural Engineers.

CHAPTER 8

Assessment of Damage in Low-rise Buildings, with Particular Reference to Progressive Foundation Movement (1981) Building Research Establishment: BRE Digest 251, HMSO.

Building Research Establishment (1990) 'Damage to structures from ground-borne vibrations', *BRE Digest* 353, HMSO, London.

Dowrick, D.J. and Beckmann, P. (1971) *York Minster Structural Restoration*. Proceedings of the Institution of Civil Engineers, Paper 7415S.

Goldscheider, M. (2003) *Baugrund und historische Gründungen. Untersuchen, Bewerton und Instandsetzen. (Sonderforschungsbereich 315). Empfehlungen für die Praxis, Vol. ****. University of Karlsruhe.

(In German, deals particularly with difficult ground conditions, such as soft lake deposits.)

Lord, A. (1981) *Underpinning of York Minster*. Proceedings of Xth International Conference on Soil Mechanics and Foundation Engineering, Stockholm, Vol. 3.

Lord, A. (1985) *The Need to Relate Fabric Deformations to both the Soil Conditions and History of the Building*. Proceedings of a 3-day Symposium on Building Appraisal, Maintenance and Preservation, held 10–12 July at the Department of Architecture and Building Engineering of the University of Bath.

Simpson, B. and William J. Grose (1996) *The Effect of Ground Movements on Rigid Masonry Facades*. International Symposium on Geotechnical aspects of underground construction in soft ground, City University.

Stroud, J. and Harrington, R. (1985) *Structure and Fabric: Part 2*. Mitchell's Building Series, Longman.

Structural Renovation of Traditional Buildings (1986) Construction Industry Research and Information Association Report 111.

Subsidence of Low-rise Buildings (2000) The Institution of Structural Engineers: Second Edition. (Advice on causes, investigations, diagnosis, tree management and remedial works, forms of contracts, etc.)

Tomlinson, M.J. (1986) *Foundation Design and Construction*, Longman Scientific & Technical. (Good introduction to soil properties, site investigations and foundation techniques.)

Wiss, J.F. (1981) 'Construction vibrations: state of the art', *ASCE Journal of the Geotechnical Division*, February.

CHAPTER 9

Beckmann, P. (1986) 'The step-by-step approach to investigation and remedial work, illustrated by the work at York Minster and Holy Trinity Church, Coventry', in *Stable—Unstable,*. R.M. Lemaire and K. van Balen (eds), lectures at the International Updating Course on structural consolidation, Leuven 19–24 May.

Construction (Health, Safety and Welfare) Regulations 1996 (Statutory Instruments, 1996, No. 1592).

Construction (Design and Management) Regulations 1994 (Statutory Instruments, 1994, No. 3140, amended by Statutory Instruments, 2000, No. 2380).

GENERAL APPENDIX: WEIGHTS AND MEASURES

SI units have generally been used in this book. These may however not be familiar to all readers. For those who like to work out things for themselves, the three accurate conversions are:

> 1 inch (1 in.) = 25.4 mm
> 1 pound (1 lb) = 0.4536 kg
> 1 kilogramforce (1 kgf or 1 kp) = 9.80665 newton (N)

Accurate conversion factors between derived 'Imperial' and SI units can be found in official tables, but for quick reference (and perhaps even for memorization) approximate conversions are given below. These include some of the oldfashioned metric units, in use in continental Europe until recently.

LENGTH, AREA AND VOLUME

$1\ \text{m} \approx 3\ \text{ft}\ 3\frac{3}{8}\ \text{ins}$	$1\ \text{ft} \approx 0.305\ \text{m}$	$1\ \text{in.} = 25.4\ \text{mm}$
$1\ \text{m}^2 \approx 10\frac{3}{4}\ \text{sq ft}$	$1\ \text{sq ft} \approx 0.09\ \text{m}^2$	$1\ \text{cm} = 10\ \text{mm} \approx 0.39\ \text{in.}$
$1\ \text{m}^3 \approx 35\ \text{cu ft}$	$1\ \text{cu ft} \approx 0.03\ \text{m}^3$	$1\ \text{sq in.} \approx 645\ \text{mm}^2$

MASS

$1\ \text{kg} \approx 2.2\ \text{lbs}$	$1\ \text{metric tonne} = 1000\ \text{kg} \approx 2205\ \text{lbs}$
$1\ \text{lb} \approx 0.45\ \text{kg}$	$1\ \text{cwt (hundredweight, Imp.)} = 112\ \text{lbs} \approx 51\ \text{kg}$
$1\ \text{ton (Imp.)} = 2240\ \text{lbs} \approx 1016\ \text{kg}$	$1\ \text{kip} = 1000\ \text{lbs} \approx 454\ \text{kg}$
$1\ \text{short ton (US)} = 2000\ \text{lbs} \approx 907\ \text{kg}$	

DENSITY

$1000\ \text{kg/m}^3 \approx 62.4\ \text{lbs/cu ft}$	$1\ \text{lb/cu ft} \approx 16\ \text{kg/m}^3$

FORCE

1 kgf ≈ 9.81 N ≈ 2.2 lbsf

1 N ≈ 0.1 kgf ≈ 0.22 lbsf

1 lbf ≈ 4.45 N ≈ 0.454 kgf

1 tonf ≈ 9.96 kN

STRESS, PRESSURE

1 N/mm^2 ≈ 10 kgf/cm^2 ≈ 145 lbsf/sq in. (psi)

1 psi ≈ 0.007 kgf/cm^2 ≈ 0.0069 N/mm^2

1 ton/sq ft ≈ 11000 kgf/m^2 ≈ 107 kN/m^2

1 ton/sq in. ≈ 157 kgf/cm^2 ≈ 15.4 N/mm^2

INDEX

Numbers in **bold** refer to illustrations. Order is letter by letter.